Leukotrienes and Prostanoids in Health and Disease

New Trends in Lipid Mediators Research

Vol. 3

Series Editor
Pierre Braquet, Le Plessis Robinson

 **KARGER**

Basel · München · Paris · London · New York · New Delhi · Bangkok · Singapore · Tokyo · Sydney

Proceedings of the 2nd International Conference on Leukotrienes and Prostanoids in Health and Disease (LPHD), Jerusalem, October 9–14, 1988

Leukotrienes and Prostanoids in Health and Disease

Volume Editors
U. Zor, Rehovot
Z. Naor, Tel-Aviv
A. Danon, Beer-Sheva

89 figures, 2 color plates and 40 tables, 1989

 KARGER

Basel · München · Paris · London · New York · New Delhi · Bangkok · Singapore · Tokyo · Sydney

New Trends in Lipid Mediators Research

Library of Congress Cataloging-in-Publication Data
International Conference on Leukotrienes and Prostanoids in Health and
Disease (2nd: 1988: Jerusalem)
Leukotrienes and prostanoids in health and disease: proceedings
of the 2nd International Conference on Leukotrienes and Prostanoids
in Health and Disease (LPHD), Jerusalem, October 9–14, 1988 / volume
editors, U. Zor, Z. Naor, A. Danon.
p. cm. – (New trends in lipid mediators research; vol. 3)
Includes bibliographies and index.
1. Leukotrienes – Physiological effect – Congresses.
2. Prostanoids – Physiological effect – Congresses. I. Zor, Uriel.
II. Naor, Zvi. III. Danon, A. IV. Title. V. Series.
[DNLM: 1. Leukotrienes – congresses. 2. Prostaglandins –
– congresses. 3. SRS-A – congresses.]
ISBN 3–8055–5011–1

Drug Dosage

The authors and the publisher have exerted every effort to ensure that drug selection and dosage set forth in this text are in accord with current recommendations and practice at the time of publication. However, in view of ongoing research, changes in government regulations, and the constant flow of information relating to drug therapy and drug reactions, the reader is urged to check the package insert for each drug for any change in indications and dosage and for added warnings and precautions. This is particularly important when the recommended agent is a new and/or infrequently employed drug.

Contents

Allergy, Asthma, Immunology and Inflammation

Brain

Metabolism of Leukotrienes and Induction of Anaphylaxis and Shock

Cardiovascular

Contents

Localization of PLA_2: Regulation by Activated Protein Kinase C

Reproductive Biology and Endocrine System

Cancer, Lipoxins

Contents

Preface

This book is based on the plenary and regular lectures presented in Jerusalem during the 2nd International Conference on Leukotrienes and Prostanoids in Health and Disease (LPHD) which was held in Jerusalem, Israel in October, 1988.

In the 'Jerusalem Conference', the International Program Committee emphasized the progress achieved in developing new drugs against allergy, asthma, inflammation, myocardial infarction, stroke and vascular diseases. These diverse drugs mainly belong to 5-lipoxygenase inhibitiors, leukotrienes antagonists, thromboxae antagonists, prostacyclin analogs and other. Approximately 10 articles in this book are devoted to this important subject.

The rapid development in the important and exciting area of molecular biology of 5-lipoxygenase and cycloxygenase, received specific attention. Several articles related to this subject, describing, in addition, the regulation of the synthesis of these enzymes by interleukin 1, glucocorticosteroids and EGF.

Other biological and medical areas that leukotrienes particularly, and eicosanoids generally are involved include allergy, immunology and inflammation, brain, regulation of phospholypase A2 and involvement of protein kinase C in this process. Cardiovascular system, metabolism of leukotrienes during anaphylaxis and shock, reproductive and endocrine system and cancer.

We would like to acknowledge Dr. *P. Braquet*, Institute Henri Beaufour and the Editor of this series, who made a superb effort in helping to publish this conference. Without this effort the book could not emerge.

We share the hope that many scientists, from various disciplines, physicians and students will find interest in the contents of this book. This of course would be the best reward for the tremendous effort put forth by the authors and editors in publishing this book.

The Editors

Molecular Biology, Regulation and Properties of Cycloxygenase and Lipoxygenase

Zor U, Naor Z, Danon A (eds): Leukotrienes and Prostanoids in Health and Disease.
New Trends Lipid Mediators Res. Basel, Karger, 1989, vol 3, pp 1–7

Analysis of Prostaglandin Endoperoxide Synthase Messenger RNA: Altered Levels in the Rabbit Hydronephrotic Kidney

Diana Fagan, Laura McLaughlin, Philip Needleman

Department of Pharmacology, Washington University School of Medicine,
St. Louis. Mo., USA

Introduction

The cyclooxygenase products prostaglandin E_2 (PGE_2) and thromboxane A_2 (TXA_2) play an important role in several models of human renal disease [1]. Ureter occlusion resulting in kidney hydronephrosis caused increased hormone-stimulated PGE_2 and TXA_2 synthesis in the isolated perfused rabbit kidney [2, 3]. TXA_2 induces renal vasoconstriction [4] resulting in decreased perfusion and renal damage. PGE_2 is an important vasodilator that at least partially reduces the effects of TXA_2 on renal blood flow and excretory function [5]. To determine the mechanism of increased arachidonic acid metabolism, Sheng et al. [6] examined the kinetics of kidney microsomal enzymes and the alterations in enzyme activity resulting from hydronephrosis. PGE_2 isomerase activity was not altered, but marked changes were seen in the activity of prostaglandin endoperoxide synthase (cyclooxygenase, COX) and thromboxane synthetase. Thromboxane synthetase activity in the hydronephrotic kidney (HNK) increased approximately 20-fold in both medulla and cortex when compared to the unaltered contralateral kidney (CLK). COX activity increased only slightly in the HNK medulla but increased approximately 20-fold in cortical microsomes. Although thromboxane synthetase activity increased concurrently with COX activity, conversion of arachidonic acid to prostaglandin H_2 by COX was found to be the rate-limiting step in the conversion of arachidonic acid to PGE_2 and TXA_2.

These studies demonstrated an increase in the V_{max} of both COX and thromboxane synthetase with no apparent alteration in the K_m of the enzymes, suggesting the observed alterations in arachidonic acid metabolism resulted from increased enzyme concentrations. This concept is supported by immunohistochemical studies [7] showing increased immunofluorescence in HNK cortical and medullary collecting tubules incubated with antibody directed against COX. In addition, ureter obstruction stimulated synthesis of an immunoreactive protein in the thin limbs of Henle's loop. Increased COX synthesis could result from altered rates of translation from pre-existing mRNA or transcription resulting in new mRNA encoding COX.

Our laboratory and others have recently isolated and sequenced complementary DNA clones encoding sheep vesicular gland COX [8–10]. In the present investigation, this cDNA clone is used as a probe to examine COX mRNA levels in the rabbit HNK and CLK by Northern blot hybridization.

Materials and Methods

RNA Preparation

Ureter obstruction was performed as described [2] in New Zealand rabbits weighing approximately 1–2 kg (M&K Animals, Bentonville, Ark.). The hydronephrotic (HNK) and contralateral (CLK) kidneys were removed after 3 days and the cortex and medulla separated surgically. Tissue fragments were frozen in liquid nitrogen and stored at –70 °C. Sheep tissues were obtained from a local abattoir and stored at –70 °C. RNA was extracted and purified by the guanidine hydrochloride method described by Chirgwin et al. [11]. Poly(A+)RNA was selected by oligo(dt) cellulose chromotography (Boehringer Mannheim Biochemicals, Indianapolis, Ind.) using the protocol described in Maniatis et al. [12].

Detection and Quantitation of COX-Specific Messenger RNA

Total RNA was fractionated by electrophoresis through 1.5% agarose gels containing formaldehyde [13] and blotted onto nylon membranes (GeneScreen, New England Nuclear, Boston, Mass.). Slot blot samples were denatured in 6 × SSC and 7.4% formaldehyde at 60 °C [14] then blotted onto nitrocellulose membrane filters (Scleicher & Schuller, Keene, N.H.). Membranes and filters were hybridized for 2 days with ^{32}P-labelled 1.5 or 0.5 kilobase fragments from our complementary DNA clone encoding COX [8]. The complementary DNA fragment, in the vector M13mp19, was labelled by nick translation (Bethesda Research Laboratories; α^{32}P-dATP, New England Nuclear) to a specific activity of approximately 1.5×10^8 cpm/μg of DNA, added to hybridization solution (30% formamide, 1 × SSC, 2 × Denhardts, 10 mg/ml of yeast transfer RNA, and 0.1% sodium dodecyl sulfate) at approximately 3×10^6 cpm/ml and hybridized for 2 days at 42 °C. Filters were washed at 50 °C in 1 × SSC and 0.1% sodium dodecyl sulfate and exposed to Kodak XAR-5 film with intensifying screens. Intensity of bands on slot blots was determined by laser densitometry (LKB Ultroscan XL, Houston, Tex.).

Results

Distribution of COX Messenger RNA in Sheep Tissues

In Northern blot hybridization experiments [8–10], only one 2.8 kilobase species of COX mRNA was demonstrated in sheep vesicular

Sheep tissue RNA

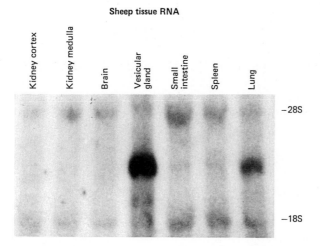

Fig. 1. Hybridization of a sheep complementary DNA clone encoding COX to RNA from sheep tissues. Total RNA was extracted from the indicated tissues and 15 μg (vesicular gland = 4 μg) were fractionated on a 1.5% agarose gel containing formaldehyde. The RNA bands were transferred to a nylon membrane and hybridized to a 0.5 kilobase EcoR1 fragment from a sheep complementary DNA clone encoding COX. The probe was labelled by nick translation to a specific activity of 1.2×10^8 cpm/μg of DNA. Ribosomal RNA bands (28S and 18S) are indicated.

gland. Much of the protein chemistry for COX has examined COX from sheep or bovine vesicular gland as this as a readily available tissue with 2- to 20-fold higher levels of COX than other tissues examined. While only one COX mRNA species appears to exist in vesicular glands, it it not known if this is true for other tissues. In these studies, Northern blot hybridization of various sheep tissues was done to examine the conditions necessary for detection of COX and to identify any tissue variability in mRNA species encoding COX. The tissues examined have been shown to contain COX activity in previous studies measuring PGE_2 synthesis [15] and binding to anti-COX fluorescent antibodies [16]. At concentrations of total RNA as high as 20 μg, we were only able to detect mRNA complementary to our clone in RNA from sheep lung (fig. 1). Lung tissue was also found to have the highest levels of COX immunoreactivity, approximately half that seen in vesicular glands and slightly more than other tissues studied. The mRNA detected in sheep lung comigrated with vesicular gland mRNA with an apparent size of 2.8 kilobases.

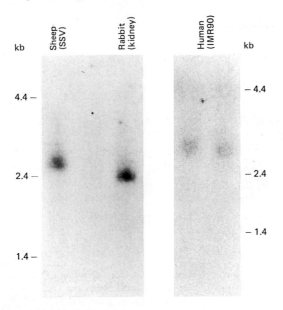

Fig. 2. Northern blot analysis of COX RNA from rabbit and human tissues. Sheep vesicular gland (SSV, 2 μg), rabbit kidney (20 μg) and human fetal lung fibroblast RNA (IMR90, 10 μg) were examined by Northern blot hybridization to the combined 1.5 and 0.5 kilobase EcoR1 fragments from a sheep cDNA clone for COX. Each fragment was labeled by nick translation to a specific activity of 1.5×10^8 cpm per microgram of DNA. An RNA ladder was run on both gels to determine the size of the COX RNA.

Rabbit kidney RNA

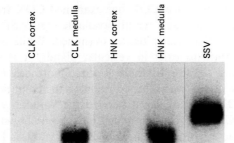

Fig. 3. Effect of unilateral ureter obstruction on COX mRNA levels in rabbit kidneys. Northern blot analysis was performed on medullary and cortical tissue from hydronephrotic (HNK) and contralateral (CLK) rabbit kidneys. A cDNA probe, as described in figure 2, was used to assess COX mRNA levels 3 days after ureter obstruction. RNA was extracted and purified after combining kidney tissue from 3 rabbits.

COX mRNA Synthesis in the Rabbit Hydronephrotic and Contralateral Kidneys

Hybridization of COX cDNA to RNA from rabbit kidneys is shown in figure 2. A Northern blot of RNA from a human fetal lung fibroblast cell line (IMR90) is shown for comparison of size between rabbit, sheep and human RNA. Sheep vesicular gland RNA was previously demonstrated to contain one COX encoding RNA species with a size of approximately 2.8 kilobases [8–10]. In these studies we demonstrate hybridization of our probe to both rabbit and human RNA. Rabbit COX RNA appears to have a size of 2.65 kilobases, and human is slightly larger at approximately 3.1 kilobases.

Northern blot analysis permits an examination of rabbit tissues for alterations in RNA levels corresponding to the increased levels of COX enzyme found in the HNK. As seen in figure 3, high levels of COX RNA are found in both HNK and CLK medullary tissues. COX RNA is seen in much lower amounts in CLK cortical tissues but increases in the HNK. Serial dilutions of poly(A⁺)RNA, followed by slot blot hybridization, allow a semiquantitative measurement of COX RNA levels in rabbit kidney tissues (fig. 4). While some variability is seen between animals, COX

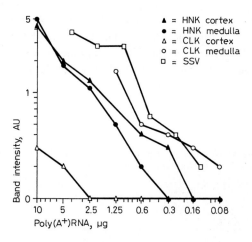

Fig. 4. Comparison of COX mRNA levels in rabbit hydronephrotic and contralateral kidneys by slot blot hybridization. Poly(A⁺)RNA was prepared from the rabbit kidney total RNA used in figure 3. Serial 2-fold dilutions of RNA in 6 × SSC was denatured by incubation for 15 min at 60 °C in the presence of 7.4% formaldehyde. The RNA was blotted onto nitrocellulose filter paper hybridized to the COX DNA probe as described in figure 2. Intensity of the bands after autoradiography was determined by laser densitometry and expressed as arbitrary units (AU).

mRNA increased 5- to 15-fold in the HNK cortex when compared to the CLK cortex. These results do not reflect a nonspecific increase in translation as no corresponding change in RNA expression was seen in slot blots hybridized to rat β-actin DNA (data not shown).

Discussion

These studies examined COX RNA levels in sheep, rabbit and human tissues by Northern blot hybridization to a COX complementary DNA clone prepared from sheep vesicular gland RNA [8]. Although RNA hybridizing to the COX probe varies slightly in size between species, only one RNA band is seen. The interesting possibility of more than one existing form of COX has been suggested from observed differences in enzymatic activity [17]. The results of this study do not support the presence of more than one protein.

Estimated concentrations of COX RNA present in each tissue correlated with COX protein levels determined by immunohistochemistry [7] and enzymatic activity [6]. As much as a 15-fold increase was seen in cortical tissue RNA when CLK were compared to HNK from rabbits 3 days after ureter occlusion. The demonstrated increase in COX mRNA may reflect a specific stimulation of transcription or could result from decreased degradation of pre-existing mRNA [18]. The results of this study suggest that regulation of COX activity in the tissues examined is primarily determined by the amount of COX mRNA available for translation.

Acknowledgements

The authors would like to thank John Merlie for technical advice and Glen Rosen and Mike Holtzman for contributing the IMR90 Northern blot seen in figure 2.

References

1   Stork JE, Rahman MA, Dunn MJ: Eicosanoids in experimental and human renal disease. Am J Med 1986;80(suppl 1A):34–48.
2   Nishikawa K, Morrison A, Needleman P: Exaggerated prostaglandin biosynthesis and its influence on renal resistance in the isolated hydronephrotic rabbit kidney. J Clin Invest 1977;59:1143–1150.
3   Morrison AR, Nishikawa K, Needleman P: Thromboxane A_2 biosynthesis in the ureter obstructed isolated perfused kidney of the rabbit. J Pharmacol Exp Ther 1977;205:2061–2068.

4 Kawasaki A, Needleman P: Contribution of thromboxane to renal resistance changes in the isolated perfused hydronephrotic rabbit kidney. Circ Res 1982;50:486–490.
5 Dibona GF: Prostaglandins and nonsteroidal anti-inflammatory drugs: Effects on renal hemodynamics. Am J Med 1986;80(suppl 1A):12–21.
6 Sheng WY, Lysz TA, Wyche A, et al: Kinetic comparison and regulation of the cascade of microsomal enzymes involved in renal arachidonate and endoperoxide metabolism. J Biol Chem 1983;258:2188–2192.
7 Smith WL, Bell TG, Needleman P: Increased renal tubular synthesis of prostaglandins in the rabbit kidney in response to ureteral obstruction. Prostaglandins 1979;18:269–277.
8 Merlie JP, Fagan D, Mudd J, et al: Isolation and characterization of the complementary DNA for sheep seminal vesicle prostaglandin endoperoxide synthase (cyclooxygenase). J Biol Chem 1988;263:3550–3553.
9 DeWitt DL, Smith WL: Primary structure of prostaglandin G/H synthase from sheep vesicular gland determined from the complementary DNA sequence. Proc Natl Acad Sci USA 1988;85:1412–1416.
10 Yokoyama C, Takai T, Tanabe T: Primary structure of sheep prostaglandin endoperoxide synthase deduced from cDNA sequence. FEBS Lett 1988;231:347–351.
11 Chirgwin JM, Przbyla AE, MacDonald RJ, et al: Isolation of biologically active ribonucleic acid from sources enriched in ribonuclease. Biochemistry 1979;18:5294–5299.
12 Maniatis T, Fritsch EF, Sambrook J: Molecular cloning: A laboratory manual. Cold Spring Harbor, Cold Spring Harbor Laboratory, 1982, p 197.
13 Buonanno A, Merlie JP: Transcriptional regulation of nicotinic acetylcholine receptor genes during muscle development. J Biol Chem 1986;261:11452–11455.
14 Berger SL, Kimmel AR: Guide to molecular cloning techniques; in Methods in Enzymology. San Diego, Academic Press, 1987, vol 152, p 585.
15 Christ EJ, Van Dorp DA: Comparative aspects of prostaglandin biosynthesis in animal tissues. Biochim Biophys Acta 1972;270:537–545.
16 Yoshimoto T, Magata K, Ehara H, et al: Regional distribution of prostaglandin endoperoxide synthase studied by enzyme-linked immunoassay using monoclonal antibodies. Biochim Biophys Acta 1986;877:141–150.
17 Lysz TW, Zweig A, Keeting PE: Examination of mouse and rat tissues for evidence of dual forms of the fatty acid cyclooxygenase. Biochem Pharmacol 1988;37:921–927.
18 Müllner EW, Kühn LC: A stem-loop in the 3′ untranslated region mediates iron-dependent regulation of transferrin receptor mRNA stability in the cytoplasm. Cell 1988;53:815–825.

Philip Needleman, PhD, Department of Pharmacology, Washington University School of Medicine, 660 South Euclid, St. Louis, MO 63110 (USA)

Zor U, Naor Z, Danon A (eds): Leukotrienes and Prostanoids in Health and Disease.
New Trends Lipid Mediators Res. Basel, Karger, 1989, vol 3, pp 8–16

Regulation of Cyclooxygenase Messenger RNA Levels by EGF and Corticosteroids in Cultured Vascular Cells

J. Martyn Bailey, Amar N. Makheja, James Pash, Mukesh Verma

Biochemistry Department, The George Washington University School of Medicine, Washington, D.C., USA

Introduction

The anti-inflammatory activity of the corticosteroids has traditionally been attributed in part to their demonstrated ability to inhibit prostaglandin synthesis. Until recently, the generally accepted explanation for this was that corticosteroids prevented release of the arachidonic acid precursor for prostaglandin synthesis by inhibiting phospholipase A_2 [1]. The proposed mechanism implied that lipocortin, the synthesis of which is induced by corticosteroids, was an inhibitory subunit of the enzyme. Recent reports, however, have questioned this interpretation for it has been shown that inhibition of phospholipase by lipocortin is an artifact caused by binding of phospholipid substrate by lipocortin in the assay systems used [2–4]. Recent reports have identified lipocortin as the primary cellular substrate that is phosphorylated by the epidermal growth factor (EGF) receptor tyrosine kinase activity [5].

Prostanoid synthesis is inactivated by aspirin, which irreversibly inhibits cyclooxygenase by acetylating a serine residue near the active site [6]. Cultures of vascular smooth muscle cells superfused with [^{14}C]arachidonic acid synthesized the antiplatelet substance prostacyclin as the major cyclooxygenase product. Aspirin-inactivated smooth muscle cells recovered their ability to synthesize prostacyclin over a 3-hour period by a process that required either serum-containing medium or serum-free medium supplemented with EGF [7]. EGF, best known for its growth-promoting activity, has also been found to stimulate prostaglandin synthesis in other types of cultured cells [8]. Type β transforming growth factor (TGF-β) has been shown to modulate the effects of EGF on cultured cells.

In this paper we describe the use of cultured vascular smooth muscle cells in which cyclooxygenase has been inactivated by prior treatment with aspirin, as a model to study synthesis of the enzyme. The isolation of a full length cDNA probe for the cyclooxygenase has made it possible to monitor messenger RNA levels for the enzyme. The suppression of cyclooxygenase mRNA by corticosteroids and its enhancement by EGF provides a new molecular mechanism for the inhibition of prostaglandin synthesis by corticosteroids [15].

Materials and Methods

Rat smooth muscle cells were isolated from the thoracic aortas of male Wistar rats aged 3 months and were used between passages 10 and 15. Cyclooxygenase activity was measured by assaying synthesis of [^{14}C]arachidonic acid metabolites from confluent growth-arrested, smooth muscle cell cultures. Synthesis of prostaglandins from endogenous cellular sources was measured by radiolabeling confluent cultures with [^{14}C]arachidonic acid for 24 h. Prostaglandin synthesis from exogenous arachidonic acid was measured by superfusing monolayer cultures with 12.8 μM [^{14}C]arachidonic acid for 10 min and assaying for radioactive products by TLC as previously described [9]. Membrane isolation and phosphorylation experiments were an adaptation of the published procedure of Sheets et al. [10].

A human lung fibroblast (ATCC No. CCL186) cDNA λgt11 expression library was screened with cyclooxygenase antibody and a 27 mer and 42 mer synthetic DNA and a ^{32}P-labelled sheep cyclooxygenase cDNA probe. Using previously described procedures [11] a 2.35 kb cDNA insert was used as probe using hybrid selection technique to isolate a full length cDNA for human cyclooxygenase. Poly(A$^+$)mRNA was isolated from the human lung fibroblast cell line CCL 186 by oligo(dT)cellulose chromatography [12], hybridized to the 2.35 kb cDNA and RNA was eluted from the DNA-RNA hybrid [13]. This cyclooxygenase specific human mRNA was used to make cDNA in λgt11 expression vector using standard protocols [14], giving a cDNA insert of 2.8 kb which was subcloned in PUC18 for restriction analysis.

Northern Hybridization

Total cell mRNA was isolated by the procedure of Chirgwin et al. [14]. RNA was blotted onto nitrocellulose membrane (for dot-blot hybridization) or electrophoresed on formaldehyde-agarose gel followed by transfer on nitrocellulose membrane (for determining the size of mRNA). Hybridization of the blots was done using the ^{32}P-labelled human cDNA (2.8 kb) probe as described by Maniatis et al. [12].

Results

Endogenous cyclooxygenase was inactivated in confluent, 24-hour fed smooth muscle cell cultures by treatment with aspirin (300 μM) for 30 min. There was very little recovery in chemically defined serum-free medium.

Addition of EGF (10 ng/ml) to the serum-free medium resulted in a marked stimulation of enzyme recovery. TGF-β (1 ng/ml) synergized with EGF to give a recovery of activity that fully reproduced that of medium containing 10% fresh fetal bovine serum. Dose-response curves with different concentrations of EGF and TGF-β alone and in combination showed that the mixture of 10 ng/ml EGF and 1 ng/ml TGF-β gave close to optimum stimulation. The time course for recovery of enzyme activity and the influence of cycloheximide and actinomycin D on inhibition of protein and mRNA synthesis was investigated. Cyclooxygenase enzyme levels were measured for 7 h after aspirin inactivation. Rapid induction of cyclooxygenase activity occurred during the first 5 h to levels that were several-fold greater than the baseline levels before aspirin treatment (table 1). Addition of 2 μg/ml cycloheximide completely blocked recovery of activity. Addition of 2 μg/ml actinomycin D to the incubation had essentially no effect on recovery for the first 3 h.

To investigate the effect of EGF and TGF-β on synthesis of prostacyclin from endogenous sources of arachidonic acid, cell lipids were first prelabeled by incubating the cells for 24 h with [^{14}C]arachidonic acid. Prelabeled smooth muscle cell cultures were incubated with 10 ng/ml EGF and 1 ng/ml TGF-β, and labeled prostacyclin release was measured at 30-min intervals over a 3-hour period. Maximal stimulation of prostacyclin release by EGF and TGF-β occurred between 1 and 2 h, and resulted in a cumulative 4-fold greater synthesis of prostacyclin than in controls. Addition of cycloheximide (2 μg/ml) to the recovery medium completely prevented the activation of prostacyclin synthesis by EGF and TGF-β whereas actinomycin D (2 μg/ml) had no effect. Cumulative arachidonate release during 3 h in control cultures averaged 4.7 μg/10^6 cells and was not significantly affected by EGF and TGF-β or by cycloheximide or actinomycin D. Thus, activation of endogenous prostacyclin synthesis by EGF and TGF-β was related to increased cyclooxygenase and not phospholipase activity.

Incubation of cultures with dexamethasone (2 μM) before inactivation by aspirin completely blocked the recovery of cyclooxygenase induced by EGF and TGF-β or fresh serum (table 2).

Total protein synthesis as measured by incorporation of [^{35}S]methionine into TCA precipitable protein was not significantly different in control and EGF/TGF-β-treated cells (91 $\pm$ 1.9 and 89.5 $\pm$ 6.4 μmol/10^6 cells, respectively). Treatment with dexamethasone for 9 h before the experiment and during the 3-hour time course resulted in only a 25 and 22% decrease in total [^{35}S]methionine incorporation in control and growth factor-treated cells, respectively, under conditions in which cyclooxygenase synthesis was completely suppressed. Binding of EGF labeled with ^{125}I to the cells was also unaffected by dexamethasone treatment.

Table 1. Activation of cyclooxygenase synthesis by EGF and TGF-β: influence of cycloheximide and actinomycin D

Treatment	Cyclooxygenase activity U/10^6 cells
None	12.5 ± 1.2
Aspirin 30 min	0.6 ± 0.2
Aspirin followed by 3 h recovery in basal medium with the following additions	
None	6.9 ± 0.4
EGF/TGF-β	22.7 ± 1.6
EGF/TGF-β + actinomycin D	20.2 ± 2.1
EGF/TGF-β + cycloheximide	0.7 ± 0.1

Confluent cultures of vascular smooth muscle cells were treated with aspirin ($300 \mu M$) for
30 min and allowed to recover for 3 h in medium containing the indicated additions. For
further details see text.

Table 2. Inhibition of recovery of prostacyclin synthesis by dexamethasone

Treatment	Cyclooxygenase activity U/10^6 cells
None	11.2 ± 2.1
Dexamethasone	3.5 ± 1.6
Aspirin 30 min followed by 3 h recovery in basal medium plus the following	
None	5.6 ± 0.9
Fresh serum	16.3 ± 1.2
EGF/TGF-β	22.3 ± 4.6
Fresh serum + dexamethasone	0.9 ± 0.2
EGF/TGF-β + dexamethasone	2.1 ± 0.1

Confluent cultures of vascular smooth muscle cells were inactivated by aspirin and allowed to
recover in media containing the indicated additions. For further details see text.

EGF-induced phosphorylation of endogenous cell proteins was studied
by incubating smooth muscle cell membranes at 4 °C for 10 min with either
EGF or EGF plus TGF-β after which [γ-^{32}P]ATP was added for an
additional 5 min [5]. Polyacrylamide gel electrophoresis (PAGE) revealed
an additional, heavily radioactive band in EGF-treated cultures that comigrated with authentic lipocortin and had an apparent molecular mass of

35 kd. Incubation with EGF also increased phosphorylatoin of a 170-kd protein corresponding to the EGF receptor. TGF-β alone did not stimulate protein phosphorylation, but phosphorylation of the 35-kd protein was also observed when EGF was added to the TGF-β incubations.

A full length 2.8 kb cDNA for the human cyclooxygenase was isolated, labelled with ^{32}P and hybridized with the total mRNA extracted from explanted cultures of diploid human lung fibroblasts CCL 186 at daily intervals until the cultures reached confluence (7 days). Two species of mRNA for the human cyclooxygenase were identified in these cultures having approximate sizes of 2.8 and 3.1 kb, with the lung species predominating during the early stages of cell growth and the 2.8 kb species in the confluent cultures (fig. 1).

Cyclooxygenase mRNA content of cultured vascular smooth muscle cells was measured following aspirin inactivation and recovery of cyclooxy-

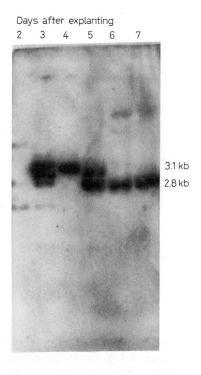

Fig. 1. Identification of two messenger RNA species for human cyclooxygenase. Total mRNA was extracted from cultures of diploid human lung fibroblasts at the indicated times, electrophoresed and analyzed by Northern hybridization using a full length (2.8 kb) ^{32}P-labelled cDNA for the human cyclooxygenase.

Table 3. Suppression of cyclooxygenase messenger RNA levels by the corticosteroid dexamethasone

Treatment of cells	Cyclooxygenase mRNA levels, dpm
Cells incubated in basal medium for 3 h with the following additions	
None	419
Serum	6,309
Serum + dexamethasone	944
EGF/TGF-β	6,374
EGF/TGF-β + dexamethasone	233
Aspirin treatment for 30 min followed by 3 h recovery in basal medium plus the following additions	
None	498
Serum	7,845
Serum + dexamethasone	263
EGF/TGF-β	8,878
EGF/TGF-β + dexamethasone	404

Confluent cultures of vascular smooth muscle cells were washed and incubated with the indicated additions. Northern blots of RNA isolated from the cells were hybridized with ^{32}P-labeled 2.8 kb human cyclooxygenase cDNA as described in the text.

genase activity in the presence of serum and of dexamethasone. Preincubation with dexamethasone suppressed hybridizable cyclooxygenase mRNA to background levels under all conditions of incubation tested. In aspirininactivated cells the failure to recover cyclooxygenase activity in the absence of serum was accompanied by the complete loss of mRNA for the enzyme. Recovery of enzyme activity over a 3-hour period in the presence of fresh serum was associated with an increase in mRNA above the levels in control untreated cultures which was suppressed completely by dexamethasone (table 3).

Discussion

Resynthesis of cyclooxygenase enzyme in cultured vascular smooth muscle cells occurs within 2 h after aspirin inactivation and appears to be regulated by a process that utilizes preexisting mRNA because it is not inhibited by actinomycin D for at least 3 h. This recovery was greatly enhanced in the presence of fresh serum-containing media or with spent media containing EGF. The results described here identify an additional factor, TGF-β. In the presence of optimum concentrations of EGF plus TGF-β, a 40- to 50-fold increase in cyclooxygenase activity was produced in aspirin-inactivated cultures within 7 h. These levels were 3- to 4-fold higher than before aspirin treatment. Cycloheximide totally blocked recovery, but actinomycin D had little effect during the first 3 h.

Previous reports that EGF stimulates prostaglandin synthesis were usually attributed to increased release of the arachidonic acid substrate. This is not the case in the present experiments, because the increased levels of cyclooxygenase are assayed directly using [^{14}C]arachidonic acid as substrate. In addition, the ability of EGF and TGF-β to stimulate prostacyclin synthesis directly without enhancing arachidonic acid release from cellular lipids was also apparent using cultures prelabeled with [^{14}C]arachidonic acid. Cumulative release of [^{14}C]prostacyclin in EGF-/TGF-β-treated cultures was approximately 4-fold greater in the presence of EGF and TGF-β and was not accompanied by any significant activation of arachidonic acid release. Furthermore, cycloheximide reduced prostacyclin synthesis to basal levels, whereas actinomycin D has no effect. Neither cycloheximide nor actinomycin D significantly blocked arachidonic acid release in either control or growth factor-stimulated cultures. These results demonstrate that the enhancement by EGF of prostacyclin synthesis from both exogenous and endogenous substrate is related to an increase in the level of cyclooxygenase enzyme.

Pretreatment of cultures for 12 h with dexamethasone in the 0.1–2 μM range, or cortisone acetate (30 μM), prevented both the serum- and EGF-induced recovery of the enzyme in aspirin-inactivated cultures. Furthermore, both steroids suppressed basal cyclooxygenase activity to only 20–30% of control levels. Since the enzyme activity was measured using ^{14}C-arachidonic acid as substrate, these observations were unrelated to phospholipase and represent direct measurement of cyclooxygenase enzyme. We concluded from these observations that EGF and TGF-β stimulate synthesis of the cyclooxygenase enzyme whereas corticosteroids suppressed it.

Regulation of enzyme synthesis could occur by translational control of mRNA for the enzyme or by regulating levels of mRNA for the enzyme. In order to distinguish these possibilites, a full length cDNA for the human cyclooxygenase was isolated from diploid human lung cells. This cDNA, when used as a hybridization probe against human lung cells, detected two species of mRNA for the human enzyme (fig. 1) of approximately 2.8 and 3.1 kb, respectively. The proportion of the 2 species of mRNA appeared to change in a manner related to the stage of culture growth.

The 2.8 kb cDNA was used to measure mRNA levels in control cultures of vascular cells and in cells inactivated by aspirin and allowed to recover in the presence or absence of serum and corticosteroids. Under the incubation conditions tested, corticosteroids completely suppressed cyclooxygenase mRNA levels, whereas serum or EGF/TGF-β increased the levels of cyclooxygenase mRNA. These changes correlated with the observed level of cyclooxygenase enzyme activity.

Recent work by Pepinsky and Sinclair [5] and by Schlaepfer and Haigler [3] suggests the nature of a possible link between the EGF and the corticosteroid systems. The primary target of the EGF receptor tyrosine kinase in several cell lines was shown to be lipocortin. That a similar mechanism also operates in vascular smooth muscle cells seems probable, in as much as we found that membrane extracts of these cells, when treated with EGF under the conditions used to activate cyclooxygenase synthesis, selectively phosphorylated a 35-kd protein that comigrated with authentic lipocortin.

Conclusions

Aspirin inactivates PGI_2 synthesis in vascular cells. Serum factors are required for recovery. EGF and TGF-β can completely replace these serum factors. Cyclooxygenase recovery is blocked by cycloheximide but not by actinomycin D. Corticosteroids block resynthesis of cyclooxygenase after aspirin treatment and also suppress basal cyclooxygenase level by 70–90%. EGF selectively induces phosphorylation of lipocortin in these cells suggesting a possible mechanistic link between the EGF and corticosteroid systems.

A full length 2.8 kb cDNA for human cyclooxygenase was isolated and used to monitor hybridizable mRNA levels. Human lung cells contain 2 species of mRNA for cyclooxygenase, EGF enhances, and corticosteroids induce complete suppression of the cyclooxygenase mRNA. This indicates that inhibition of prostanoid synthesis by corticosteroids and stimulation by EGF are mediated by regulating cyclooxygenase mRNA levels.

Acknowledgements

We thank Miss Kaewon Kim for assistance in preparation of the manuscript.

References

1   Hirata, F.; Schiffmann, E.; Venkatasubramanian, K.; Solomon, D.; Axelrod, J.: A phospholipase A_2 inhibitory protein in rabbit neutrophils induced by glucocorticoids. Proc. Natl. Acad. Sci. USA 77: 2533–2536 (1980).

2   Davidson, F.F.; Dennis, E.A.; Powell, M.; Glenney, J.R.: Inhibition of phospholipase A_2 by 'lipocortins' and calpactins: an effect of binding to substrate phospholipids. *J. Biol. Chem. 262:* 1698–1705 (1987).

3   Schlaepfer, D.D.; Haigler, H.T.: Characterization of Ca^{2+}-dependent phospholipid binding and phosphorylation of lipocortin I. J. Biol. Chem. *262:* 6931–6937 (1987).

4   Aarsman, A.J.; Munbeek, G.; Van den Bosch, H.; Rothhut, G.; Prieu, B.; Comera, C.; Jordan, L.; Russo-Marie, F.: Lipocortin inhibition of extracellular and intracellular phospholipase A_2 is substrate concentration dependent. FEBS Lett. *219:* 176–180 (1987).

5   Pepinsky, R.B.; Sinclair, L.K.: Epidermal growth factor-dependent phosphorylation of lipocortin. Nature (Lond.) *321:* 81–84 (1986).

6   Roth, G.J.; Stanford, N.; Majerus, P.W.: Acetylation of prostaglandin synthase by aspirin. Proc. Natl. Acad. Sci. USA *72:* 3073–3076 (1975).

7   Bailey, J.M.; Muza, B.; Hla, T.; Salata, K.: Restoration of prostacyclin synthase in vascular smooth muscle cells after aspirin treatment: regulation by epidermal growth factor. J. Lipid Res. *26:* 54–61 (1985).

8   Greene, R.M.; Lloyd, M.R.: Effect of epidermal growth factor on synthesis of prostaglandins and cyclic AMP by embryonic palate mesenchymal cells. Biochem. Biophys. Res. Commun. *130:* 1037–1043 (1985).

9   Bailey, J.M.; Feinmark, S.J.; Whiting, J.D.: Prostacyclin synthesis from isotopically pure [1-^{14}C]arachidonic acid in cultured cells: measurement by radiomass spectrometry. Recent. Dev. Mass Spectrometry Med. *6:* 77–81 (1981).

10   Sheets, E.E.; Guigni, T.D.; Coates, G.G.; Schlaepper, D.D.; Haigler, H.T.: Epidermal growth factor dependent phosphorylation of a 35-kilodalton protein in placental membranes. Biochemistry. *26:* 1164–1172 (1987).

11   Hla, T.; Farrell, M.; Kumar, A.; Bailey, J.M.: Isolation of the cDNA for human prostaglandin H synthase. Prostaglandins *32:* 829–845 (1986).

12   Maniatis, T.; Fritsch, E.F.; Sambrook, J.: Molecular cloning: A laboratory manual, pp. 211–246. Cold Spring Harbor Laboratory Press, Cold Spring Harbor, 1982.

13   Balcarek, J.M.; Theisen, T.W.; Cook, M.N.; Varrichio, A.; Hwang, S.M.; Strohsacker, M.W.; Crooke, S.T.: Isolation and characterization of a cDNA clone encoding rat 5-lipoxygenase. J. Biol. Chem. *263:* 13937–13941 (1988).

14   Chirgwin, J.M.; Przybyla, A.E.; MacDonald, R.J.; Rutter, W.J.: Isolation of biologically active ribonucleic acid from sources enriched in ribonuclease. Biochemistry, *18:* 5294–5299 (1979).

15   Pash, J.M.; Bailey, J.M.: Inhibition by corticosteroids of epidermal growth factor-induced recovery of cyclooxygenase after aspirin inactivation. FASEB J. *2:* 2163–2618 (1988).

J. Martyn Bailey, DSc, Biochemistry Department, The George Washington University School of Medicine, 2300 Eye Street, N.W., Washington, DC 20037 (USA)

Zor U, Naor Z, Danon A (eds): Leukotrienes and Prostanoids in Health and Disease.
New Trends Lipid Mediators Res. Basel, Karger, 1989, vol 3, pp 17–24

Regulation of Cyclooxygenase Synthesis by Interleukin-1, PMA and Glucocorticoids

Amiram Raz, Angela Wyche, Philip Needleman

Department of Pharmacology, Washington University School of Medicine,
St. Louis, Mo., USA

Introduction

Previous studies in our laboratory [1] and others [2] have shown the cytokine interleukin-1 (IL-1) to stimulate formation of PGE_2 in cultured human dermal fibroblasts. The increased PGE_2 production was due to increased V_{max} of fibroblasts prostaglandin endoperoxide synthase (PES) and was blocked by RNA and protein synthesis inhibitors [3]. Recently [4], we have shown that IL-1 stimulates the synthesis of fibroblasts PES enzyme. These results [1–4] suggest an IL-1-dependent transcriptional regulation of PES synthesis. Several studies have shown that the tumor promoter phorbol 12-myristate-13-acetate (PMA) stimulated PGE_2 production in many cell types by a mechanism which involved either increasing PLA_2 activity [5, 6] or inhibiting the reacylation enzymes [7, 8], or increasing the cellular PES enzyme mass [9]. Using specific protein kinase C (PKC) inhibitors, we assessed the role of PKC in the signal transduction mechanism for the IL-1 and PMA induction of PES synthesis.

Glucocorticoids are known to inhibit eicosanoid production when administered both in vivo and in vitro to cells or tissues [10], presumably by inducing the synthesis of lipocortins which inhibit the activity of PLA_2 [10]. Recent reports, however, have raised some doubts regarding the validity of this mechanism, demonstrating that lipocortins do not directly interact with or inhibit the activity of PLA_2 [11, 12], and do not inhibit cellular PLA_2 activity [13] or that glucocorticoids induce synthesis of lipocortin I or its mRNA [14]. Other studies [15, 16] have suggested that glucocorticoids may inhibit PES activity. In support of this, Pash and Bailey [17] recently reported that dexamethasone inhibited the epidermal growth factor stimulation of PES activity in aortic smooth muscle cells. As these studies [17] did not determine the synthesis of PES mass but rather the activity or the enzyme, it was not possible to discriminate whether the steroid-induced protein (lipocortin or otherwise) is inhibiting the PES activity or the synthesis of the enzyme.

The present studies were designed to establish conditions that will afford a temporal separation of the transcriptional and translational phases of PES synthesis. Lacking a PES-cDNA probe for human PES-mRNA, we employed inhibitors of transcription or translation to determine at which step IL-1, PMA and glucocorticoids exert their effects on PES synthesis. Our results suggest that PKC participates in the IL-1-induced transcriptional activation of PES synthesis and that glucocorticoids inhibit the translational phase of PES synthesis.

Methods and Materials

Culture and [^{35}S]Methionine Metabolic Labeling of Cells

Preparation of human dermal fibroblast cultures, labeling of the cells with [^{35}S]methionine, preparation of cell sonicates, immunoprecipitation, SDS-PAGE electrophoresis, and quantitation of [^{35}S]methionine-labeled PES synthesis and assay of PES activity in cell sonicates were done as described by us previously [4]. Actinomycin D (ACD), cycloheximide (CHX), staurosporine and steroids were prepared as stock solutions in dimethylsulfoxide, and 1 μl added per 1 ml of cell media to yield the indicated concentration. H-7 and HA1004 were dissolved to yield a stock solution of 10 mM and 1–2.5 μl added per 1 ml cell media.

Materials

Dexamethasone, triamcinolone acetonide, progesterone, testosterone, ACD, and CHX were obtained from Sigma (St. Louis, Mo.). H-7 and HA1004 were purchased from Seikagaku America (St. Petersburg, Fla.) and staurosporine from Kyowa Hakko (Tokyo, Japan). All other reagents were obtained from suppliers as described [4].

Results and Discussion

Time-Dependent Induction of PES Synthesis by IL-1

We showed previously that in cultured human dermal fibroblasts, IL-1 induces a time- and dose-dependent synthesis of PES enzyme which leads to a 5- to 20-fold higher level of PES mass (and activity) [4]. We now were able to resolve the temporal sequence for IL-1 stimulation of PES synthesis into transcription and translation phases by the use of selective inhibitors. When fibroblasts were incubated for 3–4 h with IL-1, only small increases (50–80%) in PGE$_2$ production and in PES activity were observed. In contrast, fibroblasts incubated with IL-1 for 3–4 h, and then further incubated in the absence of IL-1, exhibited a dramatic 5-fold increase in both PGE$_2$ production and PES activity (fig. 1). Inhibition of transcription with ACD during the initial 4 h incubation period completely abolished

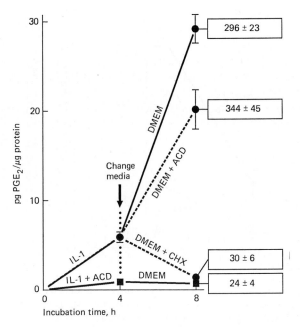

Fig. 1. Time-dependent induction of PES synthesis by IL-1: effect of protein and RNA synthesis inhibitors. Cells were first incubated in Dulbecco's modified Eagle's medium (DMEM) for 4 h with either IL-1 (0.3 U/ml) or IL-1 and ACD (1 μM). The media was then replaced with fresh DMEM, ACD (1 μM), or CHX (10 μM) added to some of the wells and the cells further incubated for 4 h. The amounts of PGE$_2$ synthesized during the first (0–4 h) and second (4–8 h) incubation periods are plotted in the figure. The PES activity (pg PGE$_2$ synthesized/μg protein/10 min) at the end of the second incubation period is given (boxed numbers) next to each treatment. Values are mean $\pm$ SD (n = 4).

subsequent induction of PGE$_2$ production and PES activity whereas ACD addition during the second incubation period (4–8 h) did not significantly affect PES induction or PGE$_2$ synthesis (fig. 1). Addition of the translation inhibitor CHX (0.3 μM) during the initial period did not affect subsequent PES synthesis (data not shown), whereas CHX added during the second incubation period produced total inhibition of PES synthesis and PGE$_2$ production (fig. 1).

PMA Potentiates IL-1-Induced PES Synthesis
PMA was found to produce a significant, albeit modest, increase in PES activity (table 1) and in the synthetic rate of newly formed enzyme (as judged by the synthesis of [^{35}S]methionine-labeled PES; data not shown). This effect of PMA was dose dependent in the range of 1–100 nM and blocked by CHX or ACD when added together with PMA but not when

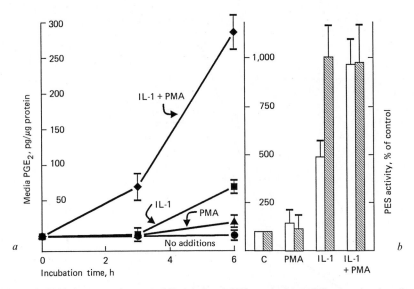

Fig. 2. PMA increases the rate of IL-1-induced PES synthesis. *a* Cells were incubated for 6 h in the absence or presence of either IL-1 (0.3 U/ml), PMA (10^{-7} *M*) or both. Aliquots for media PGE$_2$ were taken at 3 and 6 h after addition of these agents. *b* Cellular PES activity at the end of the 6 h incubation (open bars) and at the end of a similar incubation with the agents for 24 h (hatched). Values are mean $\pm$ SD (n = 4). PES activity of control (C) samples (taken as 100%) was 44 ± 8.

ACD was added subsequent to PMA addition (data not shown). Significantly, addition of PMA together with IL-1 during the transcriptional phase (i.e., 0–4 h) produced a very marked synergistic stimulation of PES induction (table 1). This PMA-potentiating effect is dependent on the time of incubation with the cells, being most pronounced in the initial 3 h of incubation (fig. 2a) but significantly absent after 24 h of incubation (fig. 2b). PMA, thus, appears to accelerate the rate of IL-1-induced PES synthesis, but does not affect the final PES activity induced by IL-1.

PKC Mediates IL-1 Induction of PES Synthesis

PMA exerts its effect on many cells by activation of PKC activity [18]. We employed the PKC inhibitor H-7 [19], and compared its effect on induction of PES synthesis to that of the non-PKC inhibitor HA1004. H-7, but not HA1004, totally inhibited the effects of PMA and of IL-1 (table 1). Similar effects to those of H-7 were also observed with staurosporine, a highly potent inhibitor of PKC, at a concentration of 25 n*M*. H-7 inhibited PES synthesis when added during the initial 4 h transcriptional phase but not if added during the translational phase (fig. 3). We, therefore, conclude

Table 1. IL-1 and PMA induction and manipulation of fibroblasts PES activity

Addition	PES activity, pg PGE$_2$ synthesized/μg protein/10 min
None	58 ± 8
+IL-1 (0.3 U/ml)	355 ± 36
+PMA ($10^{-7} M$)	99 ± 16
+IL-1 + PMA	765 ± 113
IL-1 + H-7 (15 μM)	55 ± 9
IL-1 + HA1004 (15 μM)	327 ± 49
PMA + H-7	54 ± 10
IL-1 + PMA + H-7	59 ± 11

Cells were incubated for 6 h in DMEM in the presence of the various test compounds. The media was removed and the cells further incubated for 4 h in methionine-poor DMEM, containing 0.1 mCi [^{35}S]methionine. Cells were then isolated and the PES activity determined. Results are mean $\pm$ SD (n = 3).

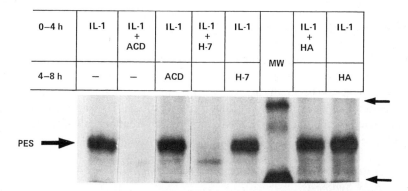

Fig. 3. Effect of protein kinase inhibitors H-7 and HA1004 on IL-1 induction of PES synthesis. Cells were first incubated in DMEM for 4 h with IL-1 (0.3 U/ml). The media was then removed, the cells extensively washed with DMEM, and further incubated in DMEM for 4 h. The media was then replaced with methionine-poor media and the cells labeled with [^{35}S]methionine for 4 h, and then processed for immunoprecipitation and SDS-PAGE electrophoresis. ACD, H-7, or HA1004 were added during either the first incubation (0–4 h) or the second incubation (4–8 h). Small arrows indicate positions of molecular weight (MW) markers phosphorilase B (92 kDa) and BSA (68 kDa).

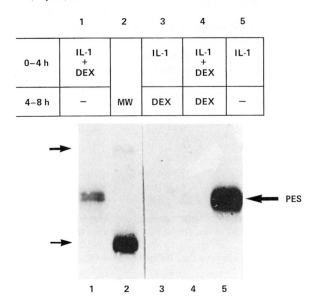

	1	2	3	4	5
0–4 h	IL-1 + DEX		IL-1	IL-1 + DEX	IL-1
4–8 h	–	MW	DEX	DEX	–

Fig. 4. Dexamethasone inhibits the synthesis of PES. Cells were incubated for 4 h in DMEM containing IL-1 (0.3 U/ml). The media was then removed, the cells were washed extensively with DMEM and further incubated in DMEM for 4 h. Dexamethasone (DEX) was added to some samples during either the first incubation, the second incubation, or during both periods. The cells were then labeled with [^{35}S]methionine and processed for immuno-precipitation and SDS-PAGE electrophoresis. Small arrows indicate positions of molecular weight (MW) markers phosphorilase B (92 kDa) and BSA (68 kDa).

that IL-1 signal transduction mechanism to induce PES-mRNA synthesis involves a critical step in which activation of PKC is required. The involvement of PKC in the transcriptional regulation of specific genes expression has been previoulsy documented [20–23]. IL-1 appears to be one of a group of cellular growth mediators/immunoregulators which transduce their specific cellular effects through activation of PKC.

Anti-Inflammatory Steroids Are Potent Inhibitors of PES Synthesis

Addition of the glucocorticoid dexamethasone (2 μM) throughout the entire transcription-translation sequence (i.e., 0–8 h, fig. 1), produced a very marked inhibition (>92%) of IL-1-stimulated PES activity. We found that the full inhibitory effect of the steroid was obtained when it was added only during the presumed translational period (i.e., 4–8 h) (fig. 4). The dexamethasone-induced inhibition (>92% at 20 nM, IC$_{50}$ 1 nM) was completely reversed by ACD, suggesting that it is probably mediated via the synthesis of a new protein(s). Only glucocorticoids inhibited PES synthesis (e.g., also triamcinolone acetonide, 60% at 20 nM, but not testosterone or progesterone). This glucocorticoids effect to suppress PES mRNA translation may be the major route by which they modulate cellular prostaglandin and thromboxane production. The mechanism by which the glucocorticoid-induced protein suppress PES synthesis is unknown. One possibility is that steroids cause an accelerated degradation of PES-mRNA, e.g., via stimulation of an RNAase. Alternatively, if the PES-mRNA is stable, the dexamethasone-induced protein may directly inhibit translation. These and other possibilities need investigation by quantitative analysis of PES-mRNA and by in vitro mRNA translation experiments.

Acknowledgements

We thank Mr. Gary Harrington and Mr. Tom Barthel for valuable technical assistance. This study was supported by NIH grants 5-PO1-DK38111 and 5-RO1-HL20787.

References

1   Albrightson, C.R.; Baenziger, N.L.; Needleman, P.: J. Immunol. *135:* 1872–1877 (1985).
2   Zucali, J.R.; Dinarello, C.A.; Oblon, D.J.; Gross, M.A.; Anderson, L.; Weiner, R.S.: J. Clin. Invest. *77:* 1857–1863 (1986).
3   Jonas-Whitely, P.E.; Needleman, P.: J. Clin. Invest. *74:* 2249–2253 (1984).
4   Raz, A.; Wyche, A.; Siegel, N.; Needleman, P.: J. Biol. Chem. *263:* 3022–3028 (1988).
5   Parker, J.; Daniel, L.W.; Waite, M.: J. Biol. Chem. *262:* 5385–5393 (1987).
6   Burch, R.M.; Axelrod, J.: Proc. Natl. Acad. Sci. USA *84:* 6374–6378 (1987).
7   Goppelt-Strube, M.; Pfaunkuche, H.; Glinsa, D.; Resch, K.: Biochem. J. *247:* 773–777 (1987).
8   Fuse, I.; Tai, H.H.: Adv. Prostaglandin Thromboxane Leuktriene Res. (in press).
9   Goerig, M.; Habenicht, A.J.R.; Heitz, R.; Zeh, W.; Katus, H.; Kommerell, B.; Zeigler, R.; Glomset, J.A.: J. Clin. Invest. *79:* 903–911 (1987).
10   Flower, R.J.: Br. J. Pharmacol. *94:* 987–1015 (1988).
11   Davidson, F.F.; Dennis, E.A.; Powell, M.; Glenney, J.R.: J. Biol. Chem. *262:* 1698–1705 (1987).
12   Aarsman, A.J.; Nunbeek, G.; Van den Bosch, H.; Rothhut, B.; Prieur, B.; Comera, C.; Jordan, L.; Russo-Marie, F.: FEBS Lett. *219:* 176–180 (1987).
13   Russo-Marie, F.; Duval, D.: Biochim. Biophys. Acta *712:* 177–185 (1982).
14   Bronnegard, M.; Andersson, O.; Edwall, D.; Lund, J.; Norstedt, G.; Carlstedt-Duke, J.: Mol. Endocrinol. *2:* 732–739 (1988).
15   Moore, P.K.; Holt, J.R.S.: Nature *288:* 269–270 (1980).
16   Wood, J.N.; Coote, P.R.; Rhodes, J.: FEBS Lett. *174:* 143–146 (1982).
17   Pash, J.H.; Bailey, J.M.: FASEB J. *2:* 2613–2618 (1988).
18   Nishizuka, Y.: Science *233:* 305–312 (1986).

19   Hikada, H.; Inagaki, M.; Kawamoto, S.; Saskai, Y.: Biochemistry *23:* 5036–5039 (1984).
20   Greenberg, M.E.; Ziff, E.B.: Nature (Lond.) *311:* 433–438 (1984).
21   Kelly, K.; Kohran, B.H.; Stile, C.D.; Leder, P.: Cell *35:* 603–610 (1983).
22   Colamonici, O.R.; Trepel, J.B.; Vidal, C.A.; Neckers, L.M.: Mol. Cell. Biol. *6:* 1847–1850 (1986).
23   Fan, X.; Goldberg, M.; Bloom, B.R.: Proc. Natl. Acad. Sci. USA *85:* 5122–5125 (1988).

Amiram Raz, PhD, Department of Pharmacology, Washington University
School of Medicine, St. Louis, MO 63110 (USA)

Zor U, Naor Z, Danon A (eds): Leukotrienes and Prostanoids in Health and Disease.
New Trends Lipid Mediators Res. Basel, Karger, 1989, vol 3, pp 25–29

The Role of Membrane Translocation in the Activation of Human Leukocyte 5-Lipoxygenase

Carol A. Rouzer, Stacia Kargman

Department of Pharmacology, Merck Frosst Canada Inc., Kirkland, Qué., Canada

Introduction

The enzyme 5-lipoxygenase (5-LO) catalyzes the first two steps in the synthesis of leukotrienes (LT) from arachidonic acid [1–4]. These involve the oxidation of arachidonic acid at carbon 5 to form 5-hydroperoxy-6,8,11,14-eicosatetraenoic acid (5-HPETE) and the further conversion of 5-HPETE to 5,6-oxido-7,9,11,14-eicosatetraenoic acid (LTA_4). During the past 5 years, considerable progress has been made in the purification of 5-LO from a variety of sources [1–6], and a cDNA for the human enzyme has now been cloned, sequenced, and expressed [7–9]. Together, these studies have shown that 5-lipoxygenase is a relatively hydrophobic, soluble protein, of approximately 78,000 daltons molecular weight. Its activity is highly dependent on Ca^{2+}, and is stimulated by ATP and phosphatidylcholine micelles. In addition, the human enzyme is stimulated by several impure protein fractions from leukocytes, one of which includes a microsomal membrane preparation (100,000 g pellet) [5, 9–11]. Together, these findings suggest some interesting potential regulatory mechanisms for the enzyme; however, despite this progress, very little is understood about the mechanism of 5-LO activation in the intact leukocyte. Clearly, this question is a critical one if the role of 5-LO in the inflammatory response is to be fully understood. In this article, we address the issue of 5-LO activation in the intact leukocyte, and propose that Ca^{2+}-dependent translocation of the enzyme from cytosol to membrane may be an important step in that activation process.

Methods

One major difficulty encountered with the study of 5-LO activation in intact cells arises from the fact that 5-LO is a suicide enzyme. Utilization of enzyme for LT biosynthesis results in irreversible inactivation in cell-free preparations, and probably also in intact cells, since the

5-LO content of supernatants prepared from leukocytes is reduced in cells that have been stimulated to synthesize LT as compared to control, resting cells [11–14]. This loss of activity is not accompanied by a comparable loss of enzyme protein, indicating that it is not due to proteolytic destruction of 5-LO [14]. These observations indicate that, in order to study the fate of enzyme that has been utilized for LT synthesis, one must employ techniques that do not require the presence of active enzyme. For this purpose, we have developed an immunoblot technique involving gel electrophoresis followed by Western blotting of proteins onto nitrocellulose paper, and detection of the 5-LO by incubation with a rabbit anti-human 5-LO antiserum and ^{125}I-labelled goat anti-rabbit IgG. The labelled 5-LO bands are then visualized by autoradiography. Identity is confirmed by the fact that the labelled band co-migrates with purified 5-LO, and that it is not detected when the anti 5-LO antiserum is replaced by normal rabbit serum. Furthermore, a good linear correlation is obtained between the increases in intensity of the labelling and the quantity of purified 5-LO added to supernatant and pellet samples indicating that the method can provide fairly reliable quantitative results [14]. We therefore have used this method, in conjunction with direct enzyme activity assay, to determine the fate of 5-LO in human leukocytes challenged with ionophore A23187 as an inducer of LT biosynthesis.

Results

Subcellular Localization of 5-LO following Ionophore Challenge

Human peripheral blood leukocytes (PBL), isolated by dextran sedimentation followed by hypotonic lysis of contaminating erythrocytes, were suspended in phosphate-buffered saline (PBS) at a concentration of 4×10^7 cells/ml, and challenged with $2\,\mu M$ ionophore A23187 for 10 min at 37 °C. The PBL synthesized approximately 5.5 nmol of $LTB_4/10^8$ cells under these conditions, along with minor quantities of 5-HPETE, isolated as its reduction product 5-hydroxyeicosatetraenoic acid (5-HETE). Control cells not treated with ionophore produced no detectable 5-LO products. Control and ionophore-treated PBL were then homogenized, and 100,000 g supernatants and pellets were prepared and examined for 5-LO activity and immunoreactive 5-LO protein. The results indicated that ionophore challenge resulted in a loss of 55% of the 5-LO activity and a 35% loss of immunoreactive 5-LO protein from the cytosol (100,000 g supernatant). Concomitantly, there was a 2- to 3-fold increase in immunoreactive 5-LO protein in the 100,000 g pellet that accounted for at least 50% of the lost cytosolic enzyme. However, this membrane-associated enzyme was inactive, as indicated by the fact that the membranes showed a decrease rather than an increase in activity [14].

Quantitative and Temporal Correlation of Translocation with LT Synthesis

The results described above were consistent with the hypothesis that ionophore activation resulted in a translocation of 5-LO from the cytosol to the membrane where the enzyme became activated and began to synthesize LT. Since it was the membrane associated 5-LO that was utilized for LT synthesis, this enzyme pool was irreversibly inactivated so that only inactive enzyme protein could be detected in the membrane of ionophore-challenged cells. The finding that most of the 5-LO protein remaining in the cytosol had retained its activity further suggested that this enzyme pool had not been utilized for LT synthesis, implying that membrane translocation may be a required initial step for 5-LO activation.

Consistent with this hypothesis were our findings that, when the ionophore concentration was increased from 0 to 5 μM, the increase in LT synthesis from 0 to 6 nmol/10^8 cells was accompanied by an increase in the amount of translocating 5-LO from 0 to 45%. Thus, there was a correlation between the quantity of translocated enzyme and the quantity of LT produced over a range of ionophore concentrations. Furthermore, when addition of EDTA in excess of the extracellular Ca^{2+} concentration was used to stop ongoing LT synthesis by ionophore-challenged PBL, we found that translocation of 5-LO to membrane was also stopped. This technique was utilized to study the time course of 5-LO translocation as compared to LT synthesis. Because EDTA did not stop the reaction instanteously, the time courses could not be compared with great accuracy. However, the data did allow the conclusion that the two processes occur with very similar kinetics, and that translocation may precede LT synthesis as would be expected if it is an early enzyme activation step [14].

Inhibition of 5-LO Translocation by L-663,536

The compound L-663,536 is a specific and potent LT biosynthesis inhibitor that has been shown to block LT synthesis by a variety of intact leukocytes in vitro, as well as in a number of animal models in vivo [15]. The compound has no demonstrable effects on phospholipase or any other enzymes in the arachidonic acid metabolic cascade. Furthermore, L-663,536 has no direct effect on 5-LO in any of a variety of cell-free enzyme preparations that have been tested. One explanation for such observations was that L-663,536 might alter the activation of 5-LO. Accordingly the effects of the compound on the translocation of 5-LO were studied. Human PBL (4×10^7 cells/ml) were incubated with varying concentrations of L-663,536 (0–500 μM) and then challenged at 37 °C for 10 min with 2 μM ionophore A23187. LT synthesis by the cells was measured, and then the cells were homogenized for the preparation of 100,000 g supernatants and pellets. 5-LO activity was measured in the supernatants and immunoreactive 5-LO protein was determined in the pellets. The results demonstrated that L-663,536 caused a concentration-dependent inhibition of the membrane translocation of 5-LO, and that this effect correlated with the inhibition of LT synthesis. At concentrations of L-663,536 that totally inhibited LT synthesis, the activity of 5-LO in ionophore-treated cells was restored to 96% of control cells (no ionophore treatment), and the accumulation of enzyme protein in the 100,000 g pellets was nearly totally eliminated.

These initial experiments suggested that the mechanism of action of L-663,536 might be to block the enzyme translocation of 5-LO, and thus prevent its activation. However, other interpretations were possible. For example, L-663,536 could block an activation step such as phosphorylation or selective proteolysis that may precede 5-LO translocation. Alternatively, it is possible that translocation has nothing to do with 5-LO activation, and the membrane is simply serving as a repository for inactive enzyme after LT synthesis has taken place. In this case, L-663,536 could be blocking any step in 5-LO activation, and still appear to inhibit translocation as well. In order to address this question, experiments were performed in which PBL were challenged with ionophore for 10 min at 37 °C and then L-663,536 (500 nM) was added. The cells were homogenized and 100,000 g supernatants and pellets were prepared. The results of immunoblot analysis showed that addition of L-663,536 to the cells after ionophore addition caused 65% of the translocated enzyme to be released back into the supernatant. Thus L-663,536 could not only prevent, but could also reverse the translocation process. This result is more consistent with a direct effect of L-663,536 on the interaction of 5-LO with the membrane rather than an indirect effect of the inhibitor on some prior process.

Conclusions

The data reviewed here are consistent with the hypothesis that activation of leukocytes with ionophore A23187 results in a Ca^{2+}-dependent translocation of 5-LO from cytosol to membrane. The membrane-associated enzyme is then activated, synthesizes LT, and becomes irreversibly inactivated. L-663,536, a potent and specific LT biosynthesis inhibitor, interferes with the translocation of 5-LO to membrane, and thus prevents enzyme activation. Since L-663,536 can also reverse translocation after it occurs, a direct effect of the inhibitor on the translocation process is likely. Although much remains to be learned concerning the role of translocation in 5-LO activation, and the exact mechanism of action of L-663,536, this system offers a unique opportunity to delineate the role of translocation in enzyme function, and may be the first description of a specific translocation inhibitor.

References

1   Rouzer CA, Matsumoto T, Samuelsson B: Single protein from human leukocytes possesses 5-lipoxygenase and leukotriene A_4 synthase activities. Proc Natl Acad Sci USA 1986;83:857–861.

2   Ueda N, Kaneko S, Yoshimoto T, et al: Purification of arachidonate 5-lipoxygenase from porcine leukocytes and its reactivity with hydroperoxyeicosatetraenoic acids. J Biol Chem 1986;261:7982–7988.

3   Hogaboom GK, Cook M, Newton JF, et al: Purification, characterization and structural properties of a single protein from rat basophilic leukemia (RBL-1) cells possessing 5-lipoxygenase and leukotriene A_4 synthase activities. Mol Pharmacol 1986;30:510–519.

4   Shimizu T, Izumi T, Seyama Y, et al: Characterization of leukotriene A_4 synthase from murine mast cells: Evidence for its identity to arachidonate 5-lipoxygenase. Proc Natl Acad Sci USA 1986;83:4175–4179.

5   Rouzer CA, Samuelsson B: On the nature of the 5-lipoxygenase reaction in human leukocytes: Enzyme purification and requirement for multiple stimulatory factors. Proc Natl Acad Sci USA 1985;82:6040–6044.

6   Goetze AM, Fayer L, Bouska J, et al: Purification of a mammalian 5-lipoxygenase from rat basophilic leukemia cells. Prostaglandins 1985;29:689–701.

7   Dixon RAF, Jones RE, Diehl RE, et al: Cloning of the cDNA for human 5-lipoxygenase. Proc Natl Acad Sci USA 1988;85:416–420.

8   Matsumoto T, Funk CD, Rådmark O, et al: Molecular cloning and amino acid sequence of human 5-lipoxygenase. Proc Natl Acad Sci USA 1988;85:26–30, and correction: Proc Natl Acad Sci USA 1988;85:3406.

9   Rouzer CA, Rands E, Kargman S, et al: Characterization of cloned human leukocyte 5-lipoxygenase expressed in mammalian cells. J Biol Chem 1988;263:10135–10140.

10 Rouzer CA, Shimizu T, Samuelsson B: On the nature of the 5-lipoxygenase reaction in human leukocytes: Characterization of a membrane-associated stimulatory factor. Proc Natl Acad Sci USA 1985;82:7505–7509.

11 Rouzer CA, Thornberry NA, Bull HG: Kinetic effects of ATP and two cellular stimulatory components on human leukocyte 5-lipoxygenase. Ann NY Acad Sci 1988;524:1–11.

12 Aharony D, Redkar-Brown DG, Hubbs SJ, et al: Kinetic studies on the inactivation of 5-lipoxygenase by 5(S)-hydroperoxyeicosatetraenoic acid. Prostaglandins 1987;33:85–100.

13 Rouzer CA, Samuelsson B: Reversible calcium-dependent membrane association of human leukocyte 5-lipoxygenase. Proc Natl Acad Sci USA 1987;84:7393–7397.

14 Rouzer CA, Kargman S: Translocation of 5-lipoxygenase to the membrane in human leukocytes challenged with ionophore A23187. J Biol Chem 1988;263:10980–10988.

15 Gillard JW, Girard Y, Morton HE, et al: The discovery and optimization of new classes of thromboxane antagonists and leukotriene biosynthesis inhibitors, in Zor U, Naor Z, Danon A (eds): Leukotrienes and Prostanoids in Health and Disease. New Trends Lipid Mediators Res. Basel, Karger, 1989, vol 3, pp 46–49.

Carol A. Rouzer, MD, Department of Pharmacology, Merck Frosst Canada Inc., PO Box 1005, Dorval, Qué. H9R 4P8 (Canada)

Zor U, Naor Z, Danon A (eds): Leukotrienes and Prostanoids in Health and Disease.
New Trends Lipid Mediators Res. Basel, Karger, 1989, vol 3, pp 30–33

Biochemical and Immunological Studies on Arachidonate 12-Lipoxygenase of Porcine Leukocytes

T. Yoshimoto[a], *N. Ueda*[a], *T. Maruyama*[a], *A. Hiroshima*[a], *F. Shinjo*[a],
K. Natsui[a], *S. Yamamoto*[a], *N. Komatsu*[b], *K. Watanabe*[b],
K. Gerozissis[c], *F. Dray*[c]

Department of Biochemistry, Tokushima University School of Medicine, Tokushima,
Japan, [b]Department of Pathology, Tokai University School of Medicine, Isehara,
Japan, and [c]URIA Institute Pasteur, Paris, France

Arachidonate 12-lipoxygenase introduces a molecular oxygen into C-12 of arachidonic acid to yield 12S-hydroperoxyeicosa-tetraenoic acid. The hydroperoxy acid is further transformed to a variety of hydroxy and epoxy compounds. However, a generalized biological function of 12-lipoxygenase has not yet been clarified.

We found a high activity of 12-lipoxygenase in porcine leukocytes [1], and prepared monoclonal antibodies against this enzyme [2]. These monoclonal anti-12-lipoxygenase antibodies were utilized to develop a peroxidase-linked immunoassay which allows a sensitive quantification of 12-lipoxygenase in various tissues. The highest enzyme content was found in leukocytes, followed by small intestine, thymus, and lymph node. It is of particular interest that 12-lipoxygenase was not detected in porcine platelets [1, 2], since the platelets are well known as the richest source of 12-lipoxygenase in human and other animals.

Recently, we made an attempt to improve the sensitivity of the enzyme immunoassay of 12-lipoxygenase by employing a solid-phase method. One of the monoclonal antibodies was bound to a well of a polystyrene immunoplate. The immobilized antibody was incubated with 12-lipoxygenase. The second antibody recognizing a different epitope of the enzyme was conjugated to biotin. The conjugate was allowed to react with the immobilized 12-lipoxygenase. Next, the biotin-containing complex was allowed to bind to an avidin-peroxidase conjugate. The peroxidase reaction was carried out in the wells, and absorbance of the peroxidase product was measured by an automated microplate reader. The new assay method was about 10 times more sensitive than the previous assay method [2]. This more sensitive assay was applied to the determination of 12-lipoxygenase content in porcine brain tissues which had not been investigated previously. Figure 1 shows the distribution of 12-lipoxygenase in various parts of porcine brain as examined by the enzyme immunoassay. Anterior pituitary showed the highest content of the 12-lipoxygenase, which was about 5% of that of porcine leukocytes, the richest source of this enzyme. The enzyme content in posterior pituitary was about one-tenth of that of anterior pituitary. We have earlier found a prominent 12-lipoxygenase activity in rat pineal gland [3]. It should be noted that only a very small amount of enzyme was present in porcine pineal gland as assessed by the enzyme immunoassay. Thus, the distribution of 12-lipoxygenase in brain may be different from animal to animal as in the case of platelet 12-lipoxygenase.

The presence of 12-lipoxygenase in anterior pituitary was confirmed by the assay of 12-lipoxygenase activity. The cytosol fraction of anterior pituitary was incubated with arachidonic acid, and the products were analyzed by reverse-phase HPLC monitoring absorbance at 235 nm. A peak was observed which coeluted with authentic 12-HETE.

As described above, the enzyme immunoassay demonstrated a wide distribution of 12-lipoxygenase in porcine tissues with the highest content in leukocytes. The next subject to be investigated was to find out what types of cell in individual tissues contained 12-lipoxygenase. In addition, the precise intracellular localization of the enzyme must be elucidated. When we started immunohistochemical studies, we soon recognized that the monoclonal antibodies were not useful due to their lower affinity to the enzyme. Therefore, we prepared polyclonal antibody against the purified 12-lipoxygenase of porcine leukocytes. The polyclonal antibody was raised in rabbit, and its specificity was examined by the immunoblotting analysis. After electrophoresis proteins were transferred onto a nitrocellulose filter, and the filter was incubated with the anti-12-lipoxygenase antibody followed by the peroxidase reaction. When the purified enzyme of porcine leukocytes was analyzed, a major stained band with a molecular weight of about 70,000 was observed. Moreover, when the cytosol fraction of anterior pituitary was examined with the antibody, a similar stained band was observed. The result indicated that the polyclonal antibody specifically recognized the 12-lipoxygenase protein among the cytosolic proteins of anterior pituitary.

When the peripheral blood cells were examined by electron microscopy with the polyclonal anti-12-lipoxygenase antibody, granulocytes and monocytes were positively stained. The enzyme was localized in the cytoplasm of these cells, and was not associated with endoplasmic reticulum, plasma membrane and any other subcellular organelles. In contrast, the enzyme was not detected in lymphocytes. Neither porcine platelets nor red blood cells were stained with anti-12-lipoxygenase antibody.

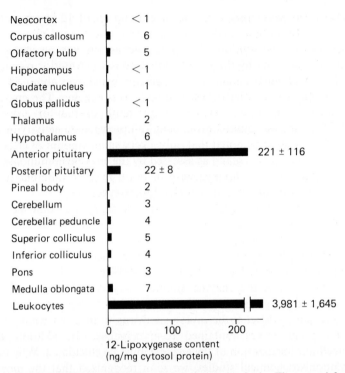

Fig. 1. Distribution of 12-lipoxygenase in porcine brain. The cytosol fraction of each part of brain was subjected to the solid-phase enzyme immunoassay for 12-lipoxygenase with the purified enzyme from porcine leukocytes as a standard.

In addition to leukocytes in peripheral blood, several other porcine tissues were examined immunohistochemically with the anti-12-lipoxyge-nase antibody. When porcine ileum was observed by light microscopy, certain cells in lamina propria mucosae and tunica mucosae were positively stained. These cells were not significantly stained with nonimmune rabbit IgG. In contrast, mucosal surface was totally unstained. Similar results were obtained in duodenum and jejunum. These stained cells were iden-tified as mast cells based upon metachromatic staining with toluidine blue. The positively stained mast cells of ileum were examined by electron microscopy. The enzyme was diffusely localized in the cytoplasm of these cells. Neutrophils and eosinophils were also stained. In contrast, any par-enchymal cell of ileum was not stained under our experimental conditions.

When porcine lung was examined with the anti-12-lipoxygenase anti-body, the enzyme was detected in neutrophils found in capillary vessels and peribronchial interstitium as well as in mast cells in the interstitium. In contrast, there was no stained cell with nonimmune serum. On the other hand, when alveolar macrophages were collected by lavage and examined by electron microscopy, these cells were not stained with the antibody.

When the number of stained cells of various tissues was compared in the same magnified field by light microscopy, a relatively large number of stained cells were found in ileum and other digestive organs. The enzyme was also found in lymph node and thymus. These immunohistochemical findings were in agreement with the distribution of the enzyme determined by the enzyme immunoassay [2]. It should be noted that relatively higher contents of the 12-lipoxygenase in organs of alimentary tract may be explained by the presence of the enzyme in mast cells and infiltrating granulocytes rather than in parenchymal cells of these tissues.

References

1   Yoshimoto T, Miyamoto Y, Ochi K, et al: Arachidonate 12-lipoxygenase of porcine leukocyte with activity for 5-hydroxy-eicosatetraenoic acid. Biochim Biophys Acta 1982;713:638–646.
2   Shinjo F, Yoshimoto T, Yokoyama C, et al: Studies on porcine arachidonate 12-lipoxygenase using its monoclonal antibodies. J Biol Chem 1986;261:3377–3381.
3   Yoshimoto T, Kusaka M, Shinjo F, et al: 12- and 15-lipoxygenases in rat pineal gland. Prostaglandins 1984;28:279–285.

Tanihiro Yoshimoto, MD, Department of Biochemistry, Tokushima University School of Medicine, Kuramoto-cho, Tokushima 770 (Japan)

Zor U, Naor Z, Danon A (eds): Leukotrienes and Prostanoids in Health and Disease.
New Trends Lipid Mediators Res. Basel, Karger, 1989, vol 3, pp 34–41

Mammalian Lipoxygenases: Their Multifunctional Nature

Shozo Yamamoto[a], *Natsuo Ueda*[a], *Yoshitaka Takahashi*[a],
Akira Hattori[b], *Ichiro Fuse*[b]

[a]Department of Biochemistry, Tokushima University, School of Medicine,
Tokushima, and [b]Department of Internal Medicine, Niigata University,
School of Medicine, Niigata, Japan

'In many cases the enzyme acts on one substrate only, and carries out only one reaction' [1]. It is a general belief of most biochemists that an enzyme catalyzes a specific type of reaction. For example, an oxidoreductase catalyzes oxidation or reduction, and does not catalyze hydrolysis. However, it is said that there is no rule without exception.

For a while after arachidonic acid was found to be the precursor of prostaglandins, a microsomal preparation of ovine or bovine vesicular gland was used by many investigators as 'prostaglandin synthetase'. The microsomal fraction catalyzed an overall synthesis of prostaglandin E_2 from arachidonic acid, but the composition and the nature of the enzyme were unclarified. We solubilized the prostaglandin synthetase from bovine vesicular gland microsomes and resolved the enzyme into two components. In 1973, prostaglandin endoperoxides were discovered by the Karolinska [2] and Unilever [3] groups, and we could find out the roles of our two enzyme components in terms of the endoperoxide intermediates [4]. Namely, one of the two enzyme fractions transformed arachidonic acid to PGG_2, which was further converted to PGH_2 by the same fraction. The other enzyme fraction converted PGH_2 to PGE_2. At that time we considered that the reaction from arachidonic acid to PGG_2 and that from PGG_2 to PGH_2 were catalyzed by two separate enzymes and these enzymes would be resolved by further purification of the enzyme. However, when the enzyme was purified to homogeneity, we found that the two enzyme activities were still tightly associated. Several lines of evidence supported that these two enzymatic reactions were catalyzed by a single enzyme [5, 6].

The first step is an oxygenase reaction incorporating two molecules of oxygen and cyclizing the fatty acyl carbon chain. The term of fatty acid cyclooxygenase first proposed by Bengt Samuelsson is properly applicable to this enzyme. The second step is a peroxidase reaction with the hydroperoxide of PGG_2 as a hydrogen acceptor, and we referred to this enzyme as PG hydroperoxidase. The bifunctional nature was confirmed later with the enzyme of ovine vesicular gland [7, 8]. Recently, DeWitt and Smith [9] used a cDNA for ovine cyclooxygenase to express this enzyme protein, and demonstrated that the de novo synthesized enzyme exhibited both the cyclooxygenase and hydroperoxidase activities.

It is now well known that the cyclooxygenase activity is selectively inhibited by nonsteroidal anti-inflammatory drugs. When we found the bifunctional nature of the enzyme, we tested the effect of indomethacin. The cyclooxygenase activity from arachidonic acid to PGG_2 was inhibited by indomethacin, whereas the hydroperoxidase activity from PGG_2 to PGH_2 was not affected [5]. This observation was extended later to other nonsteroidal anti-inflammatory drugs [10]. Cyclooxygenase was inhibited more or less by several anti-inflammatory drugs, but these drugs did not significantly affect PG hydroperoxidase. The effect of these anti-inflammatory drugs was the only experimental observation to distinguish between cyclooxygenase and hydroperoxidase.

Recently, we investigated a clinical case with cyclooxygenase abnormality. The patient is a woman born in 1941. Her platelet aggregation is markedly decreased, and her bleeding time is prolonged [11]. The enzyme solubilized from normal platelets transformed arachidonic acid to PGH_2 in the presence of heme and tryptophan. In contrast, there was no production of PGH_2 by the enzyme from the patient's platelets, and arachidonic acid remained unchanged. However, the thromboxane synthesis from PGH_2 was observed to the same extent in the platelets of both the normal subject and the patient [12]. The PGH_2 synthesis by the solubilized enzyme from the normal subject was not reduced by the presence of the solubilized enzyme from the patient's platelets. The experimental result ruled out a possible presence in the patient's platelets of a compound which inhibited either the catalysis of normal cyclooxygenase or its binding to the antibody.

For further study on the abnormal cyclooxygenase we developed a peroxidase-linked immunoassay of cyclooxygenase [12, 13]. Two species of monoclonal anti-cyclooxygenase antibody which recognized different sites of cyclooxygenase were utilized. The immunoassay allowed the quantitative determination of the amount rather than the activity of platelet cyclooxygenase. Unexpectedly, determinations of the cyclooxygenase by the enzyme immunoassay revealed an almost normal level of the cyclooxygenase protein in the patient's platelets [12]. Since the cyclooxygenase and hydroperoxidase activities were tightly associated as discussed above, we expected that the PG hydroperoxidase activity was not detected in the platelets of the patient with cyclooxygenase abnormality. However, when the solubilized enzyme from the patient's platelets was precipitated by the use of antibody and allowed to react with PGG_2, we detected a normal level of the PG hydroperoxidase activity. The conversion of PGG_2 to PGH_2 was dependent on tryptophan, which was a characteristic hydrogen donor to reduce the peroxide of PGG_2 to a hydroxide [6], and the enzymatic reaction was not stimulated by the addition of glutathione. These findings ruled out the participation of nonspecific peroxidase in the conversion of PGG_2 to PGH_2 by the patient's platelets.

As mentioned above, aspirin or indomethacin inhibited the cyclooxygenase reaction without affecting the PG hydroperoxidase reaction. A serine residue of the enzyme has been presumed to be acetylated by aspirin [14], and the particular serine residue has been located at the amino acid No. 506 by the recent cDNA cloning by three different research groups [15–17]. The platelet cyclooxygenase of the patient may have a structural similarity to the normal cyclooxygenase which is treated with aspirin. The protein chemical nature of this abnormal cyclooxygenase is now under investigations.

Another example of multifunctional enzyme is 5-lipoxygenase. We detected the activity of 5-lipoxygenase in the cytosol of porcine leukocytes, and partially purified the enzyme. This enzyme preparation was used as an antigen to raise monoclonal anti-5-lipoxygenase antibodies [18]. The antibody was then utilized for immunoaffinity chromatography to purify the enzyme to near homogeneity [19]. We found that the purified enzyme catalyzed both the 5-oxygenation of arachidonic acid and the conversion of 5-HPETE to 5,6-epoxyleukotriene by the loss of the elements of water. Our finding together with the concurrent reports from three groups [20–22] confirmed an earlier report by the Karolinska group for the bifunctional nature of potato 5-lipoxygenase [23]. Recently the LTA synthase activity was found in the 5-lipoxygenase protein expressed in osteosarcoma cells transfected with the cDNA for human 5-lipoxygenase [24]. Another catalytic activity of 5-lipoxygenase was found when the enzyme was incubated with 5-HPETE on ice and the products after borohydride reduction were analyzed by HPLC [25]. In addition to two peaks of 5,12-diHETEs which were derived from LTA_4, a big peak appeared, which was identified as 5S,6R-diHETE. This peak disappeared in the absence of air. Without borohydride reduction a less polar product was detected. These observations were interpreted in a way that 5-HPETE was metabolized in two ways by the catalysis of 5-lipoxygenase. A part of 5-HPETE was the substrate for LTA synthase, and another part was oxygenated to 5S,6R-diHPETE. The 6R-oxygenase reaction with 5-HPETE as substrate showed a characteristic requirement of calcium ion and ATP, which was a well-known property of the 5-oxygenase activity, and inhibited by two selective inhibitors of 5-lipoxygenase developed by our group [25].

15-HPETE was a substrate of the 5-oxygenase activity at a considerable reaction rate, and converted to 5S,15S-diHPETE [19]. Considering the 6R-oxygenase activity and the 5,6-LTA synthase activity, we subjected the 5,15-diHPETE to the action of 5-lipoxygenase expecting the production of lipoxins [26]. HPLC analysis of the reaction products demonstrated the synthesis of lipoxin A_4 and other lipoxin isomers. The product profile was essentially unchanged in the absence of air. Therefore, lipoxins were produced predominantly by the anaerobic catalysis of the 5,6-LTA synthase activity. The aerobic lipoxin synthesis contributed by the 6R-oxygenase activity was only a minor pathway.

Another example of the multifunctional enzyme is 12-lipoxygenase. Using a whole cell suspension of porcine leukocytes, the Vanderbilt group demonstrated the transformation of 15-HPETE to various dihydroxy and 14,15-epoxy compounds with conjugated triene. They predicted that 12-lipoxygenase might be involved in these reactions [27]. At that time we had found a high activity of 12-lipoxygenase in the cytosol fraction of porcine leukocytes [28], and purified the enzyme by immunoaffinity chromatography [29] utilizing a monoclonal anti-12-lipoxygenase antibody [30]. When we studied the substrate specificity of the enzyme, we found that 12-HPETE, which was the primary product from arachidonic acid, was not further metabolized by 12-lipoxygenase [29]. This was in contrast to the reactivity of 5-lipoxygenase with 5-HPETE. Instead, the enzyme metabolized 15-HPETE almost as fast as the arachidonate 12-oxygenation [29]. When the reaction products were reduced with borohydride prior to HPLC, several dihydroxy compounds with a conjugated triene were detected, and the product profiles were compared with or without borohydride reduction or in the presence or absence of air. On the one hand, the 12-lipoxygenase catalyzed 8S-oxygenation and 14R-oxygenation of 15-HPETE and produced corresponding dihydroperoxy acids each with a conjugated triene. On the other hand, the bond between the two oxygen atoms of the 15-hydroperoxide was cleaved, and 14,15-epoxy compound with a conjugated triene was produced with the loss of water. The 14,15-LTA$_4$ was hydrolyzed nonenzymatically, and diastereomeric mixtures of 8,15-dihydroxy acids and 14,15-dihydroxy acids were produced. Thus, like 5-lipoxygenase, 12-lipoxygenase exhibited both the oxygenase activity and the LTA synthase activity.

12-Lipoxygenase was also allowed to react with 5,15-diHPETE [31]. HPLC analysis of the reduced products monitoring absorption at 301 nm demonstrated the production of several lipoxin isomers. The highest peak was identified as lipoxin B, and this peak was markedly reduced in the absence of air. These observations suggested that 5,15-diHPETE was converted to 5,14,15-trihydroperoxy derivative of lipoxin B$_4$ by the 14R-oxygenase activity of 12-lipoxygenase. On the other hand, the peaks which were not affected by the absence of air were identified as the degradation products of 5-hydroperoxy-14,15-LTA$_4$. Thus, the lipoxin synthesis by 12-lipoxygenase proceeded in both the aerobic and anaerobic reactions. This was in contrast to the lipoxin synthesis by 5-lipoxygenase which occurred mostly by the anaerobic mechanism.

It is well known that 12-lipoxygenase is present in the platelets of many animal species. However, we could not detect the 12-lipoxygenase activity in porcine platelets [28, 30]. Instead, a high activity of 12-lipoxygenase was found in porcine leukocytes [28, 30]. When we prepared monoclonal antibodies against the 12-lipoxygenase of porcine leukocytes, we found a strict specificity of the antibody [30]. The antibodies did not cross-react with 12-lipoxygenases of bovine and human platelets. This finding led us to consider that there might be two kinds of 12-lipoxygenase in mammalian tissues. The substrate specificity study of porcine leukocyte 12-lipoxygenase revealed a broad specificity of the enzyme in terms of the carbon-chain length of unsaturated fatty acids. The enzyme showed a considerable activity with octadecapolyenoic acids and docosahexaenoic acids in addition to arachidonic acid and other eicosapolyenoic acids [29]. Such a broad substrate specificity of 12-lipoxygenase of porcine leukocytes was also reported by the Unilever group [32], and was in sharp contrast with the property of the eicosapolyenoic acid-specific 5-lipoxygenase [33, 34]. A recent report from Vliegenthart's group [35] described that 12-lipoxygenase of bovine platelets was inactive with linoleic acid whereas the enzyme of bovine leukocytes oxygenated C-13 of linoleic acid. This report suggested the occurrence of two different lipoxygenases in bovine tissues. Two species of our monoclonal anti-12-lipoxygenase antibody were useful tools in the study of this subject.

An antibody, which was raised against 12-lipoxygenase of porcine leukocytes, cross-reacted with the enzyme of bovine leukocytes, but not with the enzyme of bovine platelets [36]. In contrast, an antibody, which was raised against human platelet 12-lipoxygenase, cross-reacted with the enzyme of bovine platelets, but not with the enzyme of bovine leukocytes [36]. These two species of anti-12-lipoxygenase antibody were utilized for immunoaffinity chromatography of the enzymes of bovine platelets and leukocytes [36], and their substrate specificities were investigated with the purified enzymes. The two immunologically different enzymes from the same animal species were distinct in terms of substrate specificity. The leukocyte 12-lipoxygenase has a broad specificity in terms of the carbon-chain length while the platelet enzyme showed a narrow specificity for arachidonic acid and eicosapentaenoic acid [36].

Unlike cyclooxygenase and 5-lipoxygenase, the physiological role of 12-lipoxygenase has not been elucidated to date. Since two distinct 12-

lipoxygenases were found in bovine tissues, we might predict the occurrence of the leukocyte-type 12-lipoxygenase in human tissues and the platelet-type 12-lipoxygenase in porcine tissues. We still do not know the significance that there are two distinct types of 12-lipoxygenase.

Acknowledgements

We are grateful to Dr. Alan R. Brash and his associates of Vanderbilt University for their collaboration to elucidate the reaction mechanisms of 12-lipoxygenase of porcine leukocytes. Thanks are also due to Drs. Joshua Rokach and Brian J. Fitzsimmons of Merck Frosst Canada for their help in identifying various lipoxin isomers.

References

1 Dixon, M.; Webb, E.C.; Thorne, C.J.R.; Tipton, K.E.: Enzymes; 3rd ed., p. 231 (Longman, London 1979).
2 Hamberg, M.; Samuelsson, B.: Detection and isolation of an endoperoxide intermediate in prostaglandin biosynthesis. Proc. Natl. Acad. Sci. USA 70: 899–903 (1973).
3 Nugteren, D.H.; Hazelhof, E.: Isolation and properties of intermediates in prostaglandin biosynthesis. Biochim. Biophys. Acta 320: 448–461 (1973).
4 Miyamoto, T.; Yamamoto, S.; Hayaishi, O.: Prostaglandin synthetase system: Resolution into oxygenase and isomerase components. Proc. Natl. Acad. Sci. USA 71: 3645–3648 (1974).
5 Miyamoto, T.; Ogino, N.; Yamamoto, S.; Hayaishi, O.: Purification of prostaglandin endoperoxide synthetase from bovine vesicular gland microsomes. J. Biol. Chem. 251: 2629–2636 (1976).
6 Ohki, S.; Ogino, N.; Yamamoto, S.; Hayaishi, O.: Prostaglandin hydroperoxidase, an intergral part of prostaglandin endoperoxide synthetase from bovine vesicular gland microsomes. J. Biol. Chem. 254: 829–836 (1979).
7 Hemler, M.; Lands, W.E.M.; Smith, W.L.: Purification of the cyclooxygenase that forms prostaglandins. Demonstration of two forms of iron in the holoenzyme. J. Biol. Chem. 251: 5575–5579 (1976).
8 Van der Ouderaa, F.J.; Buytenhek, M.; Nugteren, D.H.; Van Dorp, D.A.: Purification and characterization of prostaglandin endoperoxide synthetase from sheep vesicular glands. Biochim. Biophys. Acta 487: 315–331 (1977).
9 DeWitt, D.L.; Smith, W.L.: Molecular cloning of prostaglandin G/H synthase. Adv. Prostaglandin Thromboxane Leukotrine Res. 19 (in press).
10 Mizuno, K.; Yamamoto, S.; Lands, W.E.M.: Effect of nonsteroidal anti-inflammatory drugs on fatty acid cylclooxygenase and prostaglandin hydroperoxidase activities. Prostaglandins 23: 743–757 (1982).
11 Nakanishi, K.; Ikeda, K.; Hato, T.; Imai, A.; Kaneko, H.; Murakami, A.; Kawashima, K.; Kaido, H.; Kondo, M.; Hattori, A.; Fujita, S.; Kobayashi, Y.: Platelet cyclo-oxygenase deficiency in a Japanese. Scand. J. Haematol. 32: 167–174 (1984).
12 Ehara, H.; Yoshimoto, T.; Yamamoto; S.; Hattori, A.: Enzymological and immunological studies on a clinical case of platelet cyclooxygenase abnormality. Biochim. Biophys. Acta 960: 35–42 (1988).

13 Yoshimoto, T.; Magata, K.; Ehara, H.; Mizuno, K.; Yamamoto, S.: Regional distribution of prostaglandin endoperoxide synthase studied by enzyme-linked immunoassay using monoclonal antibodies. Biochim. Biophys. Acta *877:* 141–150 (1986).

14 Roth, G.J.; Siok, C.J.; Ozolz, J.: Structural characteristics of prostaglandin synthetase from sheep vesicular gland. J. Biol. Chem. *255:* 1301–1304 (1980).

15 Merlie, J.P.; Fagon, D.; Mudd, J.; Needleman, P.: Isolation and characterization of the complementary DNA for sheep seminal vesicle prostaglandin endoperoxide synthase (cyclooxygenase). J. Biol. Chem. *263:* 3550–3553 (1988).

16 DeWitt, D.L.; Smith, W.L.: Primary structure of prostaglandin G/H synthase from sheep vesicular gland determined from the complementary DNA sequence. Proc. Natl. Acad. Sci. USA *85:* 1412–1416 (1988).

17 Yokoyama, C.; Takai, T.; Tanabe, T.: Primary structure of sheep prostaglandin endoperoxide synthase deduced from cDNA sequence. FEBS Lett. *231:* 347–351 (1988).

18 Kaneko, S.; Ueda, N.; Tonai, T.; Maruyama, T.; Yoshimoto, T.; Yamamoto, S.: Arachidonate 5-lipoxygenase of porcine leukocytes studied by enzyme immunoassay using monoclonal antibodies. J. Biol. Chem. *262:* 6741–6745 (1987).

19 Ueda, N.; Kaneko, S.; Yoshimoto, T.; Yamamoto, S.: Purification of arachidonate 5-lipoxygenase from porcine leukocytes and its reactivity with hydroperoxyeicosatetraenoic acids. J. Biol. Chem. *261:* 7982–7988 (1986).

20 Rouzer, C.A.; Matsumoto, T.; Samuelsson, B.: Single protein from human leukocytes possesses 5-lipoxygenase and leukotriene A_4 synthase activities. Proc. Natl. Acad. Sci. USA *83:* 857–861 (1986).

21 Shimizu, T.; Izumi, T.; Seyama, Y.; Tadokoro, K.; Rådmark, O.; Samuelsson, B.: Characterization of leukotriene A_4 synthase from murine mast cells: Evidence for its identity of arachidonate 5-lipoxygenase. Proc. Natl. Acad. Sci. USA *83:* 4175–4179 (1986).

22 Hogaboom, G.K.; Cook, M.; Newton, J.F.; Varrichio, A.; Shorr, R.G.L.; Sarau, H.M.; Crooke, S.T.: Purification, characterization, and structural properties of a single protein from rat basophilic leukemia (RBL-1) cells possessing 5-lipoxygenase and leukotriene A_4 synthetase activities. Mol. Pharmacol. *30:* 510–519 (1986).

23 Shimizu, T.; Rådmark, O.; Samuelsson, B.: Enzyme with dual lipoxygenase activities catalyzes leukotriene A_4 synthesis from arachidonic acid. Proc. Natl. Acad. Sci. USA *81:* 689–693 (1984).

24 Rouzer, C.A.; Rands, E.; Kargman, S.; Jones, R.E.; Register, R.B.; Dixon, R.A.F.: Characterization of cloned human leukocyte 5-lipoxygenase expressed in mammalian cells. J. Biol. Chem. *263:* 10135–10140 (1988).

25 Ueda, N.; Yamamoto, S.: The 6R-oxygenase activity of arachidonate 5-lipoxygenase purified from porcine leukocytes. J. Biol. Chem. *263:* 1937–1941 (1988).

26 Ueda, N.; Yamamoto, S.; Fitzsimmons, B.J.; Rokach, J.: Lipoxin synthesis by arachidonate 5-lipoxygenase purified from porcine leukocytes. Biochem. Biophys. Res. Commun. *144:* 966–1002 (1987).

27 Mass, R.L.; Brash, A.R.: Evidence for a lipoxygenase mechanism in the biosynthesis of epoxide and dihydroxy leukotrienes from 15(S)-hydroperoxyicosatetraenoic acid by human platelets and porcine leukocytes. Proc. Natl. Acad. Sci. USA *80:* 2884–2888 (1983).

28 Yoshimoto, T.; Miyamoto, Y.; Ochi, K.; Yamamoto, S.: Arachidonate 12-lipoxygenase of porcine leukocyte with activity for 5-hydroxyeicosatetraenoic acid. Biochim. Biophys. Acta *713:* 638–646 (1982).

29 Yokoyama, C.; Shinjo, F.; Yoshimoto, T.; Yamamoto, S.; Oates, J.A.; Brash, A.R.: Arachidonate 12-lipoxygenase purified from porcine leukocytes by immunoaffinity

chromatography and its reactivity with hydroperocyeicosatetraenoic acids. J. Biol. Chem. *261:* 16714–16721 (1986).

30　Shinjo, F.; Yoshimoto, T.; Yokoyama, C.; Yamamoto, S.; Izumi, S.; Komatsu, N.; Watanabe, K.: Studies on porcine arachidonate 12-lipoxygenase using its monoclonal antibodies. J. Biol. Chem. *261:* 3377–3381 (1986).

31　Ueda, N.; Yokoyama, C.; Yamamoto, S.; Fitzsimmons, B.J.; Rokach, J.; Oates, J.A.; Brash, A.R.: Lipoxin synthesis by arachidonate 12-lipoxygenase purified from porcine leukocytes. Biochem. Biophys. Res. Commun. *149:* 1063–1069 (1987).

32　Claeys, M.; Kivits, G.A.A.; Christ-Hazelhof, E.; Nugteren, D.H.: Metabolic profile of linoleic acid in porcine leukocytes through the lipoxygenase pathway. Biochim. Biophys. Acta *837:* 35–51 (1985).

33　Ochi, K.; Yoshimoto, T.; Yamamoto, S.; Taniguichi, K.; Miyamoto, T.: Arachidonate 5-lipoxygenase of guinea pig peritoneal polymorphonuclear leukocytes. J. Biol. Chem. *258:* 5754–5758 (1983).

34　Furukawa, M.; Yoshimoto, T.; Ochi, K.; Yamamoto, S.: Studies on arachidonate 5-lipoxygenase of rat basophilic leukemia cells. Biochim. Biophys. Acta *795:* 458–465 (1984).

35　Walstra, P.; Verhagen, J.; Vermeer, M.A.; Veldink, G.A.; Vliegenthart, J.F.G.: Demonstration of a 12-lipoxygenase activity in bovine polymorphonuclear leukocytes. Biochim. Biophys. Acta *921:* 312–319 (1987).

36　Takahashi, Y.; Ueda, N.; Yamamoto, S.: Two immunologically and catalytically distinct arachidonate 12-lipoxygenases of bovine platelets and leukocytes. Arch. Biochem. Biophys. *266:* 613–621 (1988).

Prof. Shozo Yamamoto, Department of Biochemistry, Tokushima University, School of Medicine, Kuramoto-cho, Tokushima 770 (Japan)

Zor U, Naor Z, Danon A (eds): Leukotrienes and Prostanoids in Health and Disease.
New Trends Lipid Mediators Res. Basel, Karger, 1989, vol 3, pp 42–45

The Role of Lipoxygenase in the Maturation of Red Cells

S.M. Rapoport

Humboldt University, Medical School (Charité), Institute of Biochemistry, Berlin, GDR

The differentiation of erythroid cells involves inactivation and in mammals expulsion of the nucleus. The resulting cell, the reticulocyte, is characterized inter alia by active respiration and the presence of functional mitochondria. During the conversion of the reticulocyte to the erythrocyte, the mitochondria are destroyed. At the same time the ribosomes disappear, receptors and transport systems of the cell membrane are lost and there occurs a selective decay of some cytosolic enzymes [1].

The breakdown of the mitochondria involves an intricate interplay of several proteins, which constitute a cascade of reactions in which the erythroid-specific 15-lipoxygenase plays a key role [2]. The other components are the mitochondrial susceptibility factor (MSF), and an ATP- and ubiquitin-dependent proteolytic system. It is triggered by the action of lipoxygenase.

The delayed synthesis of LOX is due to the existence of the LOX mRNA in a masked state in the form of LOX mRNPs in the nucleated precursors and even in the most immature fraction of reticulocytes. The unmasking may be brought about by deproteinization, proteinase K and high KCl concentrations. Preliminary experiments indicate the presence of a cytosolic system of unmasking which is present in reticulocytes but no longer in erythrocytes [3].

What about the tissue specificity of the erythroid LOX? Immunological tests with a polyclonal antiserum against rabbit erythroid LOX, hybridization tests with the LOX cDNA [4] and as well as the presence of LOX products [5] indicate a specificity for erythroid tissue in a wide range of species including mice, rats, monkeys and man. On the other hand, erythroid LOX was not found expressed in any other tissue of rabbits, nor in cultural fibroblasts or leukoblasts [6]. It was coordinatively expressed in Friend erythroleukemia cells after induction by dimethylsulfoxide together with the globin mRNAs. The erythroid LOX mRNA has been recently cloned. Its translated sequence corresponds to 661 amino acids. Untranslated are 25 bp at the 5' end and 550 bp at the 3' end. This portion is characterized by a block of 175 bp which contains a repeated cytosin-rich motif of about 20 bp. Such a sequence is missing from human 5-LOX and may possibly be of importance for the masking of erythroid LOX mRNA. A comparison of the sequence of erythroid LOX with those from leukocytes shows considerable homology [8–10]. There is 45% identity of the amino acids between rabbit erythroid LOX and human 5-LOX, and even 71% with human eosinophile 15-LOX, as far as the N-terminal amino acids are concerned. The similarity is even greater if conservative exchanges are included.

From these results the following conclusions may be drawn: For one, it is clear that despite similar or identical catalytic mechanisms, their origin from the same stem cell and considerable homology, the lipoxygenases from different types of blood cells have diverged during evolution. Secondly, the apparent conservation of the erythroid enzyme in evolution suggests that the enzyme plays an essential role in the breakdown of mitochondria in all types of erythroid cells. The synchronous expression of mRNAs for lipoxygenase, globin, glutathione peroxidase, and others during differentiation favors the hypothesis of common factors, which activate their genes coordinately.

The structure of the gene of the erythroid LOX is in the process of elucidation. The transcription unit comprises 8.5 kb. It contains more than 10 introns. In the flanking regions, hypersensitive sites to DNase I are present and erythroid specific promotors and enhancer sections have been demonstrated. The multiple regulation by promoters and enhancers may be the explanation for the overexpression of the erythroid LOX caused by stress erythropoiesis.

What are the actions of erythroid LOX and which of them occur in the intact cell and are biologically relevant? The enzyme has the nearly unique property to attack not only free polyenoic acids but also phospholipids and isolated mitochondria [11]. However, it appeared that fresh or ATP-protected preincubated mitochondria are resistant to the enzyme as judged by oxygen consumption whereas the products of LOX action could be demonstrated. Also, the oxygen consumption of intact reticulocytes is nearly an order of magnitude higher than the amount of the products of LOX action. A secondary radical chain reaction, perhaps iron-catalyzed, may be presumed. In keeping with this assumption, the proportion of transisomers and of racemates among the LOX products is considerably higher with mitochondria and intact reticulocytes as substrates as compared with free fatty acids.

The most important physiological action of the erythroid LOX is the triggering of the ATP- and ubiquitin-dependent proteolysis. This effect is demonstrable in intact reticulocytes. Other actions of the enzyme, such as the inhibition or inactivation of the respiratory chain or the destruction of FeS centers of the mitochondrial outer membrane, are probably caused by secondary reactions. The primary products of LOX action are demonstrable even in immature reticulocytes. However, in such cells the ATP-dependent breakdown of mitochondria does not occur. A further protein component (MSF) is necessary, the action of which is permissive for the triggering effect of the LOX. It is formed after the synthesis of the LOX enzyme, i.e. in more mature reticulocytes, provided that iron is present in the incubation medium [12]. Its molecular mass is about 11 kd, the amounts in the cell are about three orders of magnitude less than those of LOX. MSF exhibits four actions: (1) it is permissive for the LOX-associated excessive oxygen consumption; (2) it causes the breakdown of the permeability barrier of the inner mitochondrial membrane; (3) it produces an inhibition of the respiratory chain at the site of antimycin A action; and (4) it is the main factor causing desintegration of the mitochondria observable by electron microscopy. By itself it does not trigger ATP-dependent proteolysis.

What about the fate of LOX formed in reticulocytes? Three options appear to exist: One is the self-inactivation during the enzymatic reaction. Its mechanism is an irreversible inactivation of the enzyme by its products, the hydroperoxy fatty acids. It is accompanied by the conversion of a single methionine, which is presumably located in the active center of the enzymes, to methionine sulfoxide. This reaction is followed by slow secondary alteration of the spatial structure of the protein. A second process is the proteolytic degradation of the LOX by thiol-containing proteases. A small part of the LOX escapes degradation and persists in the red cell as long as the lifetime of the cell population carrying high concentrations of the enzyme following stimulated erythropoiesis [13].

The biological dynamics of the lipoxygenase protein, the system unmasking its mRNA, the MSF protein and of the proteolytic system degrading mitochondria, permits a general conclusion. The occurrence of each of the components listed, appears to be restricted to a defined developmental stage of the red cell. This characteristic feature implies the operation of two separate processes; for one a factor-specific system of the triggering of translation linked with the unmasking of the respective mRNA. Such a mechanism was found both for the LOX mRNA and for the MSF mRNA. On the other hand, inactivating or proteolytic systems must be responsible for the disappearance of the proteins, be it LOX, its mRNA unmasking factor, the MSF or part of the proteolytic system degrading mitochondria.

References

1 Rapoport SM: The Reticulocyte. Boca Raton, CRC Press, 1986.
2 Rapoport SM, Schewe T: The maturational breakdown of mitochondria in reticulocytes. Biochim Biophys Acta 1986;864:471–495.
3 Höhne M, Thiele BJ, Prehn S, et al: Activation of translationally inactive lipoxygenase mRNP particles from rabbit reticulocytes. Biomed Biochem Acta 1988;47:75–78.
4 Thiele BJ, Höhne M, Nack B, et al: Lipoxygenase mRNA during development of red blood cells studied with a cloned probe. Biomed Biochim Acta 1987;46:S124–S125.
5 Kühn H, Ludwig P, Salzmann-Reinhardt U, et al: Metabolism of polyenic fatty acids by red blood cells. Biomed Biochim Acta 1987;46:S156–S159.
6 Thiele BJ, Black E, Fleming J, et al: Cloning of reticulocyte lipoxygenase mRNA. Biomed Biochim Acta 1987;46:S120–S123.
7 Affara N, Fleming J, Goldfarb PS, et al: Analyses of chromatin changes associated with the expression of globin and non-globin genes in cell hybrid between erythroid and other cells. Nucleic Acids Res 1985;13:5629–5644.
8 Matsumoto T, Funk CD, Radmark O, et al: Molecular cloning and amino acid sequence of human 5-lipoxygenase. Proc Natl Acad Sci USA 1988;85:26–30.
9 Dixon RAF, Jones RE, Diehl RE, et al: Cloning of the cDNA for human 5-lipoxygenase. Proc Natl Acad Sci USA 1988;85:416–420.
10 Sigal E, Grunberger D, Croik CS, et al: Arachidonate 15-lipoxygenase (ω-6 lipoxygenase) from human leukocytes. J Biol Chem 1988;263:5328–5332.
11 Schewe T, Rapoport SM, Kühn H: Enzymology and physiology of reticulocyte lipoxygenase: Comparison with other lipoxygenases; in Meister A (ed); Adv Enzym. New York, Wiley, 1986, vol 58, pp 191–272.
12 Rapoport S, Schmidt J, Prehn S: Fe-dependent formation of a protein that makes mitochondria lipoxygenase-susceptible during maturation of reticulocytes. FEBS Lett 1986;198:109–112.
13 Ludwig P, Höhne M, Kühn H, et al: The biological dynamics of lipoxygenase in rabbit red cells in the course of an experimental bleeding anemia. Biomed Biochim Acta 1988;47:593–608.

Prof. Dr. Dr. S.M. Rapoport, Institut für Biochemie der Humboldt-Universität, Hessische Strasse 3–4, DDR–1040 Berlin (GDR)

Zor U, Naor Z, Danon A (eds): Leukotrienes and Prostanoids in Health and Disease.
New Trends Lipid Mediators Res. Basel, Karger, 1989, vol 3, pp 46–49

The Discovery and Optimization of New Classes of Thromboxane Antagonists and Leukotriene Biosynthesis Inhibitors

John W. Gillard, Yves Girard, Howard E. Morton, Christiane Yoakim,
Rejean Fortin, Yvan Guindon, Thomas R. Jones, Anne Lord,
Diane Ethier, Gordon E. Letts, Euan MacIntyre,
Anthony Ford-Hutchinson, Joshua Rokach

Merck Frosst Canada Inc., Kirkland, Qué., Canada

We have recently described the prototypes of a new class of thromboxane antagonists, the indole-2-propionic acids, as exemplified by L-655,240 [1], which possess submicromolar inhibition of platelet aggregation and are active in blocking the effects of TXA_2 on vascular and smooth muscle in vitro and in vivo. This class of compounds was also observed to possess an additional property: potent inhibition of leukotriene biosynthesis in cellular preparations of polymorphonuclear leukocytes, eosinophils, macrophages and mast cells. Through an analysis of conformational preferences and through rational synthesis, we have achieved the complete separation of the above mentioned biochemical properties into discrete chemical entities [2]. Optimization of the potency of the respective series has led to a new class of enantiomerically pure thromboxane antagonists with nanomolar potency and excellent in vivo properties. L-670,596, a tetrahydrocarbazole-4-acetic acid derivative (IC_{50} 2.5 nM; pA_2 9.0 vs. U-44069 induced platelet aggregation and guinea pig trachea contraction) is an agent in this series whose in vitro and in vivo properties justify further development as a clinical candidate [3]. As an important contribution to the theoretical study of the thromboxane receptor, the active and inactive enantiomeric forms of this compound have been used to identify and specific labelling of the thromboxane receptor by a radioligand [4].

Using structure/activity considerations, from the original lead compound, a second chemical series of leukotriene biosynthesis inhibitors has been defined and optimized. Highly specific, and totallly devoid of TXA_2 activity, nanomolar leukotriene biosynthesis inhibitors have been found which are systemically active in in vivo studies. L-663,536 represents the

Table 1. Inhibition of leukotriene synthesis in inflammatory cells by L-663,536 (see fig. 1 for structure)

Cell type	Stimulus, IC_{50} nM					
	Ionoph. ±PMA	Anti-IgE	Zymo.	Staph	FMLP cyto.	PAF
Human PMN (LTB_4)	2.7				≪20	
Rat PMN (LTB_4)	3.5				5–20	
Human mast (LTC_4)[1]	3–5	3–5				
Rat RBL-1 (LTC_4)[2]	3–5					
Human eosinophil[3]						50
Macrophage (LTC_4)						
Mouse peritoneal	4.3		10			
Human peritoneal[4]	40			2.4		

[1] E. Schulman, unpubl results.
[2] E. Schulman, unpubl. results.
  U. Zor, unpubl. results.
[3] P. Borgeat, unpubl. results.
[4] J. Williams, unpubl. results.

compound selected for further development as a leukotriene biosynthesis inhibitor [5]. Its intrinsic potency on a number of cellular systems is given in table 1.

In all cases, there is inhibition of LT synthesis at the low nanomolar level, independent of the method of stimulus. The mechanism of action of L-663,536 is dependent upon inhibition of one of the requisite steps in 5-lipoxygenase activation, namely, translocation of the activated enzyme to the cellular membrane [6]. In the absence of membrane association, the enzyme is incapable of synthesis of 5-LO products. That this mechanism is acting in vivo is seen by the potent inhibition of LT synthesis by L-663,536 in animal models of pleural inflammation induced by neutrophil accumulation in the pleural cavity. In this model, A-23187 stimulated leukotriene synthesis is blocked for up to 10 h following a single oral dose (ED_{50} 0.20 mg/kg).

The compound is extremely potent topically in a skin model of leukotriene synthesis in the guinea pig ear, with a topical ED_{50} of 0.001 mg/ear. Acute systemic administration of L-663,536 ED_{50} (2.5 mg/kg p.o.) 2 h prior to ionophore administration afforded identical protection. Finally, in a model of plantar inflammation of the rat paw, induced by thioglycollate, the compound was shown to be capable of total inhibition of leukotriene

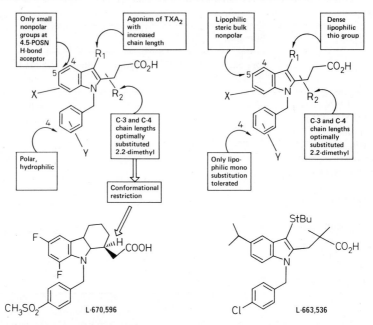

Fig. 1. Structure activity relationships for optimization of thromboxane antagonism and leukotriene biosynthesis inhibition.

biosynthesis induced by subsequent stimulation with A-23187 when dosed orally at a dose of 3 mg/kg. The ED_{50} in this model was found to be 0.8 mg/kg.

Finally, in a rat model of allergic bronchoconstriction the compound is extremely potent ($ED_{50} = 0.015$ mg/kg) with concomitant inhibition of antigen-induced biliary leukotriene secretion ($ED_{50} = 0.05$ mg/kg p.o.). In a primate model of antigen-induced bronchoconstriction, the compound is capable of completely blocking changes in respiratory resistance and dynamic compliance at a 1.0 mg/kg oral dose.

The specificity of the compound for inhibition of leukotriene synthesis has been shown by its inability to inhibit in vitro or in vivo the concomitant synthesis of prostaglandins induced by ionophore in PMN or of thromboxane, or 12-HETE in human platelets when stimulated by ionophore or thrombin.

The compound is extensively protein bound (99.5%) and has a biological half-life of 8.5 h. Oral bioavailability has been demonstrated in two species (rat and dog) to be of the order of 30%. Finally in a model of antigen-induced bronchoconstriction in a sensitized primate, the compound was capable of complete blockade of dynamic compliance changes and bronchial resistance changes induced by the aerosol administration of ascaris antigen.

References

1   Hall, R.A.; Gillard, J.W.; Guindon, Y., et al.: Pharmacology of L-655,240 (3-[1-(4-chlorobenzyl)-5-fluoro-3-methyl-indol-2-yl]2,2-dimethylpropanoic acid); a potent, selective thromboxane/prostaglandin endoperoxide antagonist. Eur. J. Pharmacol. *135:* 193 (1987).
2   Gillard, J.W.; Guindon, Y.; Fortin, R., et al.: Indole-2-alkanoic acids: a new class of biologically active molecules 1. Thromboxane antagonists. J. Med. Chem. (submitted).
3   Gillard, J.W.; Morton, H.E.; Fortin, R., et al.: The use of synthetic thromboxane A$_2$ (TXA$_2$-S) and a new class of thromboxane antagonists, indole-2-propanoic acids to characterize the thromboxane receptor. Proceedings of 4th International Symposium on Prostaglandins (Liss, New York 1989).
4   Gillard, J.W.; Evans, J., et al: Synthetic thromboxanes and the development of potent thromboxane antagonists: their role in identifying and characterizing the thromboxane receptor. Prostaglandins (submitted).
5   Gillard, J.W.; Ford-Hutchinson, A.W.; Chan, C., et al: L-663,535 (3-[3-(4-chlorobenzyl)-3-t-butyl-thio-5-isopropylindol-2-yl]-2,2-dimethylpropanoic acid), a novel, orally active leukotriene biosynthesis inhibitor. Can. J. Physiol. Pharmacol. (in press 1989)
6   Rouzer, C.; Kargman, S.: The role of membrane translocation in the activation of human leukocyte 5-lipoxygenase; in Zor, U., Naor, Z., Danon, A. (eds): New Trends Lipid Mediators Res., vol. 3, pp. 25–29 (Karger, Basel 1989).

John W. Gillard, PhD, Merck Frosst Canada Inc., Kirkland, Qué. H9H 3L1 (Canada)

Zor U, Naor Z, Danon A (eds): Leukotrienes and Prostanoids in Health and Disease.
New Trends Lipid Mediators Res. Basel, Karger, 1989, vol 3, pp 50–55

A-64077, a New Potent Orally Active 5-Lipoxygenase Inhibitor

George W. Carter[a], *Patrick R. Young*[a], *Daniel H. Albert*[a],
Jennifer B. Bouska[a], *Richard D. Dyer*[a], *Randy L. Bell*[a],
James B. Summers[a], *Dee W. Brooks*[a], *Bruce P. Gunn*[b], *Paul Rubin*[c],
James Kesterson[c]

[a]Immunoscience Research, [b]Chemical and Agricultural Products Division, and
[c]Immunoscience Venture, Abbott Laboratories, Abbott Park, Ill., USA

Products resulting from the oxidative metabolism of arachidonic acid by 5-lipoxygenase, most notably the 6-sulfidopeptide leukotrienes (LTC_4, LTD_4 and LTE_4) and LTB_4, have been implicated in the pathophysiology of allergic and inflammatory diseases [1, 2]. Leukotrienes C_4, D_4, and E_4 comprise slow-reacting substance of anaphylaxis, and are highly potent airway constrictor substances both in vitro [3] and in vivo [4–6]. They also have been shown to enhance mucus secretion [7], cause extravasation of plasma proteins [8, 9] and contract vascular smooth muscle [10]. LTB_4 is one of the most powerful chemoattractants for polymorphonuclear leukocytes (PMN) known [11]. Like other chemoattractants, LTB_4 also triggers PMN to adhere to surfaces [12], generate superoxide anion [13] and release lysosomal enzymes [14]. Injection of LTB_4 into skin causes rapid erythema, vascular permeability and cellular accumulation [15].

In addition to their powerful biological actions, leukotrienes have been recovered from tissue sites undergoing pathological reactions. For example, LTB_4 has been recovered from synovial fluids of arthritics [16], colonic mucosa of inflammatory bowel disease patients [17] and psoriatic lesions [18]. Sulfidopeptide leukotrienes have been found in sputum from asthmatics [19], and elevated blood levels of sulfidopeptide leukotrienes have been observed in children during acute asthmatic attack [20]. Regulation of leukotriene biosynthesis by inhibiting 5-lipoxygenase is therefore an attractive approach toward limiting the involvement of leukotrienes in disease.

Based on the assumption that 5-lipoxygenase contains a catalytically important iron, we and others have prepared compounds containing a hydroxamate moiety (hydroxamic acids are known to bind strongly to Fe^{+3}) and have demonstrated them to be potent inhibitors of the enzyme

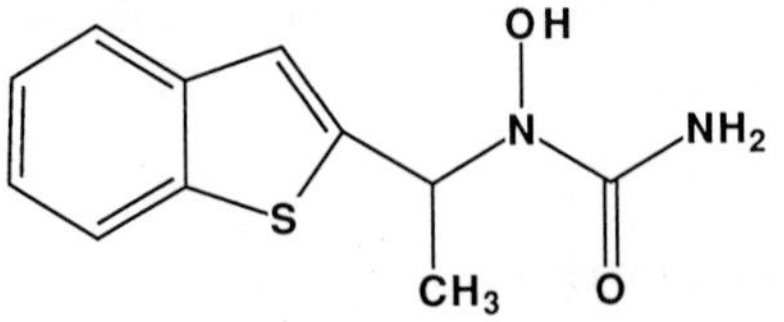

Fig. 1. A-64077, N-(1-benzo[b]thien-2-ylethyl)-N-hydroxyurea.

[21–27]. Initial hydroxamate inhibitors exhibited poor bioavailability. Identification of structural features which led to improved pharmacokinetics properties, ultimately culminated in the discovery of A-64077 (fig. 1). This communication will profile the in vitro and in vivo effects of this new 5-lipoxygenase inhibitor.

Lipoxygenase Inhibition, in vitro

The 20,000 g supernatant from sonicated rat basophilic leukemia (RBL-1) cells is a rich source of 5-lipoxygenase activity and was utilized to evaluate the 5-lipoxygenase inhibitory properties of A-64077 and reference compounds using a radiometric TLC assay [23]. A-64077 produced a concentration-dependent inhibition of 5-HETE formation with an IC_{50} of 0.53 μM. Compared to several commonly cited reference agents, A-64077 was about 4 times more potent than BW755c ($IC_{50} = 2.0\ \mu M$) and approximately half as active as nordihydroguaiaretic acid (NDGA, $IC_{50} = 0.27\ \mu M$). Since the inhibitory activity of A-64077 was reduced upon dilution, the compound appeared to be a reversible inhibitor of the RBL-1 enzyme (data not shown). In contrast to the previous reference agents, A-64077 was a selective inhibitor of 5-lipoxygenase; at concentrations up to 100 μM, the compound produced less than 20% inhibition of platelet 12-lipoxygenase and no significant inhibition of either soybean 15-lipoxygenase or sheep seminal vesicle cyclooxygenase.

A-64077 was also evaluated against calcium ionophore (A-23187)-stimulated eicosanoid biosynthesis using rat leukocytes and human whole blood. In addition to further characterizing the selectivity of A-64077, these assays afforded the opportunity to evaluate the compound's ability to inhibit 5-lipoxygenase activity expressed by intact cells, and to function in a complex biological milieu, such as blood. Stimulated LTB_4 biosynthesis in rat leukocytes and human blood was inhibited in a concentration-dependent manner by A-64077 with IC_{50}s of 0.16 and 0.60 μM, respectively. Inhibition of cyclooxygenase products (prostaglandin E_2, PGE_2 from rat leukocytes and thromboxane A_2, measured as TXB_2 from blood) required much higher concentrations of A-64077 (approximately 30-fold higher) than needed to prevent LTB_4 biosynthesis.

Inhibition of Blood 5-Lipoxygenase ex vivo

To assess the onset and duration of LTB_4 inhibition by A-64077, dogs were orally dosed and blood samples collected at various times were stimulated ex vivo with calcium ionophore and LTB_4 biosynthesis measured by RIA. As shown in figure 2, the synthesis of LTB_4 was markedly inhibited soon after the administration of A-64077 at oral doses of 0.5, 1.0 and 5.0 mg/kg. At the 5 mg/kg dose, LTB_4 biosynthesis was essentially abolished (>95% reduced) for more than 24 h postdosing. LTB_4 synthesis was reduced by $\geq 50\%$ for approximately 2.5 and 7 h following administration of A-64077 at 0.5 and 1 mg/kg, respectively.

Inhibition of Leukotriene Formation in vivo

The ability of A-64077 and reference compounds to inhibit leukotriene formation in vivo following oral administration was evaluated using a rat peritoneal anaphylaxis reaction [24]. In this model, rabbit antiserum to bovine serum albumin (BSA) was injected into the rat peritoneal cavity and 3 h later, BSA was injected intraperitoneally. Fifteen minutes following antigen challenge, the peritoneal fluids were collected and the levels of sulfidopeptide leukotrienes measured by radioimmunoassay. A-64077 and reference compounds were dosed 1 h prior to antigen administration.

At oral doses from 1 to 30 mg/kg, administered 1 h prior to antigen challenge, A-64077 produced a dose-related inhibition of leukotriene formation with an ED_{50} of 3.1 mg/kg. Compared to reference compounds, A-64077 was approximately 5-fold more potent than phenidone, and more than 10-fold more active than BW755c in blocking leukotriene bisoynthesis (table 1). NDGA produced no significant oral activity. The duration of leukotriene inhibition caused by A-64077 at 10 mg/kg p.o in this rat model was in excess of 4 h (data not shown).

Anti-inflammatory Activity

The Arthus reaction is an acute inflammatory response which has been proposed as a particularly relevant model to search for novel agents to treat immune complex diseases, such as rheumatoid arthritis [28]. The reverse

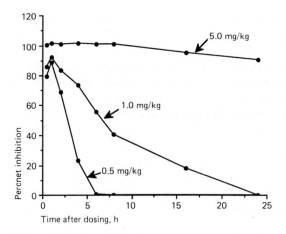

Fig. 2. Inhibition of calcium ionophore-stimulated blood LTB_4 biosynthesis (ex vivo) in the dog following oral administration of A-64077. Each point is the mean inhibition from 3 dogs.

Table 1. Comparison of the inhibitory effects of A-64077 and reference agents in the rat peritoneal anaphylaxis model

Compound	Oral ED_{50}[a] mg/kg
A-64077	3.1 (2.2–4.1)
BW755c	35.7 (14.2–52.2)
Phenidone	16.9 (10.4–27.5)
Indomethacin	NA at 20 mg/kg
NDGA	NA at 100 mg/kg

[a] Dose calculated by linear regression to produce 50% inhibition of LT generation with 95% confidence limits in parenthesis.
NA = Not active

passive Arthus reaction was elicited in the rat pleural cavity and the ability of A-64077 to reduce cellular influx and edema formation was assessed simultaneously at 3 h. Significant inhibition of cellular accumulation (ranging from 35 to 55%), but not of fluid formation, was produced by A-64077 at doses from 10 to 100 mg/kg. These results with A-64077 as well as our studies with other 5-lipoxygenase inhibitors have led us to conclude that 5-lipoxygenase products play an important although not exclusive role in leukocyte accumulation in the Arthus reaction.

A-64077 has been shown to be a potent, selective, orally active inhibitor of 5-lipoxygenase. In view of the accumulating evidence suggesting the involvement of 5-lipoxygenase products in a number of clinical conditions, compounds such as A-64077 hold great promise for the treatment of several important diseases. Early clinical studies with A-64077 have already demonstrated the compound to be effective in reducing blood leukotriene biosynthesis, ex vivo, following oral administration.

References

1   Samuelsson B: Leukotrienes: mediators of immediate hypersensitivity reactions and inflammation. Science 1983;220:568–575.
2   Piper PJ: Formation and actions of leukotrienes. Physiol Rev 1984;64:744–761.
3   Drazen JM, Austen KF, Lewis DA, et al: Comparative airway and vascular activities of leukotrienes C-1 and D in vivo and in vitro. Proc Natl Acad Sci USA 1980;77:4354–4358.
4   Weiss JW, Drazen JM, Coles N, et al: Bronchoconstrictor effects of leukotriene C in humans. Science 1982;216:196–198.
5   Griffin M, Weiss JW, Leitch AG, et al: Effects of leukotriene D on airways in asthma. N Engl J Med 1983;308:436–439.
6   Holroyde MC, Althounyan REC, Cole M, et al: Bronchoconstriction produced in man by leukotrienes C and D. Lancet 1981;ii:17–18.
7   Marom Z, Shelhamer JH, Bach MK, et al: Slow-reacting substances, leukotrienes C_4 and D_4, increase of mucus from human airways in vitro. Am Rev Respir Dis 1982;126:449–451.
8   Dahlen SE, Bjork J, Hedqvist P, et al: Leukotrienes promote plasma leakage and leukocyte adhesion in postcapillary venules: in vivo effects with relevance to the acute inflammatory response. Proc Natl Acad Sci USA 1981;78:3387–3891.
9   Soter NA, Lewis RA, Corey EJ, et al: Local effects of synthetic leukotrienes (LTC_4, LTD_4, LTE_4 and LTB_4) in human skin. J Invest Dermatol (Tokyo) 1983;80:115–118.
10 Ford-Hutchinson AW, Letts G: Biological actions of leukotrienes. Hypertension 1986; 8(suppl 2):44–49.
11 Ford-Hutchison AW, Bray MA, Doig MV, et al: Leukotriene B, a potent chemokinetic and aggregating substance released from polymorphonuclear leukocytes. Nature 1980;286:264–265.
12 Palmblad JC, Malmsten CL, Iden AM, et al: Leukotriene B_4 is a potent stereoselective stimulator of neutrophil chemotaxis and adherence. Blood 1984;58:658–661.
13 Palmblad JC, Gyllenhammer JA, Lindggren JA, et al: Effects of leukotrienes and f-met-leu-phe on oxidative metabolism of neutrophils and eosinophils. J Immunol 1984;132:3041–3045.
14 Showell HJ, Naccache PH, Borgeat P, et al: Characterization of the secretory activity of LTB_4 toward rabbit neutrophils. J Immunol 1982;128:811–816.
15 Movat HZ, Rettl C, Burrowes CE, et al: The in vivo effect of leukotriene B_4 on polymorphonuclear leukocytes and microcirculation. Am J Pathol 1984;115:233–244.
16 Klickstein LB, Shapleigh C, Goetzl EJ: Lipoxygenation of arachidonic acid as a source of polymorphonuclear leukocyte chemotactic factors in synovial fluid and tissue in rheumatoid arthritis and spondyloarthritis. J Clin Invest 1980;66:1166–1170.

17   Sharon P, Stenson WF: Enhanced synthesis of leukotriene B_4 by colonic mucosa in inflammatory bowel disease. Gastroenterology 1984;86:453–460.

18   Brain SD, Camp RDR, Doud PM, et al: The release of leukotriene B_4-like material in biologically active amounts from the lesional skin of patients with psoriasis. J Invest Dermatol 1984;83:70–73.

19   Zakrzewski JT, Barnes NC, Piper PJ, et al: Quantitation of leukotrienes in asthmatic sputum. Br J Pharmacol 1985;19:574P.

20   Schwartzberg SB, Shelov SP, Van Praag D: Blood leukotriene levels during acute asthmatic attack in children. Prostaglandins Leukotrienes Med 1987;26:143–155.

21   Corey EJ, Cashman JR, Kantner SS, et al: Rationally designed, potent competitive inhibitors of leukotriene biosynthesis. J Am Chem Soc 1984;106:1503–1504.

22   Kerdesky FAJ, Holms JH, Schmidt SP, et al: Eicosatetraenehydroxamates: inhibitors of 5-lipoxygenase. Tetrahedron Lett 1985;26:2143–2146.

23   Summers JB, Mazdiyasni H, Holms JH, et al: Hydroxamic acid inhibitors of 5-lipoxygenase. J Med Chem 1987;30:574–580.

24   Summers JB, Gunn BP, Mazidiyasni H, et al: In vivo characterization of hydroxamic acid inhibitors of 5-lipoxygenase. J Med Chem 1987;30:2121–2126.

25   Summers JB, Gunn BP, Martin JG, et al: Orally active hydroxamic acid inhibitors of leukotriene biosynthesis. J Med Chem 1988;31:3–5.

26   Summers JB, Gunn BP, Martin JG, et al: Structure-activity analysis of a class of orally active hydroxamic acid inhibitors of leukotriene biosynthesis. J Med Chem 1988;31:1960–1964.

27   Tateson JE, Randall RW, Reynolds CH, et al: Selective inhibition of arachidonate 5-lipoxygenase by novel acetohydroxamic acids: biochemical assessment in vitro and ex vitro. Br J Pharmacol 1988;94:528–539.

28   Carter GW, Martin MK, Krause RA, et al: The effects of anti-inflammatory drugs and other pharmacological agents on the rat dermal Arthus reaction. Res Commun Pathol Pharmacol 1982;35:189–207.

George W. Carter, PhD, Area Head, Immunoscience Research, Abbott Laboratories, D–464 AP9, Abbott Park, IL 60064 (USA)

Zor U, Naor Z, Danon A (eds): Leukotrienes and Prostanoids in Health and Disease.
New Trends Lipid Mediators Res. Basel, Karger, 1989, vol 3, pp 56–61

Development of an Orally Active Antiallergic Drug: TMK-688

S. Murota[a], *H. Tomioka*[b], *S. Ozawa*[c], *T. Suzuki*[c], *T. Wakabayashi*[c]

[a]Tokyo Medical and Dental University, Yushima, Bunkyo-ku, Tokyo;
[b]School of Medicine, Chiba University, Chiba City, and [c]Technical Research and
Development Division, Terumo Corp., Shibuya-ku, Tokyo, Japan

Introduction

Leukotrienes are known to be deeply involved in the processes of allergic asthma and inflammation. Because all the leukotrienes are formed from arachidonic acid through 5-lipoxygenase, there is the possibility that the discovery of new specific inhibitors of 5-lipoxygenase may lead to development of new antiasthmatic and anti-inflammatory drugs. We established a unique in vitro screening system for 5-lipoxygenase inhibitors [1]. Using this screening system, we examined several natural compounds isolated from certain Chinese herbs and found such compounds as caffeic acid, eupatilin, 4′-demethyleupatilin, and esculetin to have very strong inhibitory activities for 5-lipoxygenase [2, 3]. We focused our attention on caffeic acid because caffeic acid was the easiest compound to work with in terms of chemical synthesis and modification of its molecular structure. Because we found that caffeic acid methylester was approximately 10 times as potent as caffeic acid [4], we synthesized further derivatives of caffeic acid methylester. We have synthesized nearly 250 derivatives of caffeic acid methylester (TMK numbering) and obtained some more active inhibitors of 5-lipoxygenase [5, 6]. Our strategy was to design such drugs as having both anti-5-lipoxygenase activity and antihistamine activity in the same molecule. In addition, our purpose was to develop orally active drugs. Finally, we selected TMK-688, 1-(3-(5-(3-methoxy-4-ethoxycarbonyloxy)-1-oxo-2,4-pentadienyl)-aminoethyl)-4-benzhydryloxypiperidine, as the most suitable, orally active antiallergic drug [7].

Results and Discussion

The ID_{50} value of TMK-688 was 320 nM in the in vitro assay system [1], while that of TMK-777, a major metabolite of TMK-688, was 17 nM.

TMK-777 was about 300 times as potent as caffeic acid. When TMK-688 was incubated with rat plasma, TMK-777 was detected as early as 10 min, and increased linearly with time just like the mirror image of the decrease in TMK-688. From these results we estimate TMK-688 as a kind of pro-drug. Owing to the better intestinal absorption, crystalline conditions, etc., TMK-688 is more potent than TMK-777 after oral administration, though the order of the potency is the other way round in the in vitro assay system. TMK-688 showed significant inhibition in various models of allergic reaction as described in our previous paper [7], i.e., (1) inhibition of anaphylactic leukotriene release and bronchoconstriction in guinea pigs (30 mg/kg, p.o.); (2) inhibition of IgE-mediated passive cutaneous anaphylaxis in rats for as long as over 6 h (30 mg/kg, p.o.) and (3) inhibition of direct passive Arthus reaction in rats (100 mg/kg, p.o.), etc.

We examined whether TMK compounds were active on human tissue. Human chopped lung fragments were sensitized with IgE myeloma protein overnight, and then challenged with anti-IgE serum. Drugs were given 10 min before the challenge. The amount of LTC_4 and LTD_4 released was measured by radioimmunoassay. The results are shown in figure 1. Both TMK-688 and TMK-777 inhibited leukotriene release from the human lung in a dose-dependent manner. TMK-777 was almost equipotent to AA-861.

Since TMK compounds were designed to have an antihistamine activity in their moiety, we next examined the antihistamine activity of these compounds. Both TMK-688 and TMK-777 were able to block histamine-induced tracheal smooth muscle contraction in guinea pigs. The potency of the antihistamine activity in TMK-688 ($pD'_2 = 7.5$) was between that in ketotifen ($pD'_2 = 7.9$) and oxatomide ($pD'_2 = 6.7$).

TMK-688 was also found to inhibit PAF-induced bronchoconstriction in guinea pigs. Intravenous injection of PAF ($0.5\,\mu g/kg$) caused transient increase in airway resistance. However, this increase was inhibited by the pretreatment of the animals with TMK-688 (30 mg/kg, p.o.) 1 h before the PAF injection.

We next examined the in vivo metabolism of TMK-688 by using ^{14}C-labeled TMK-688. The radioactive TMK-688 was given orally to guinea pigs, and the radioactivity was chased. The blood concentration of the radioactivity changed with time as shown in figure 2. It was found that TMK-688 can remain for a fairly long time in the blood. This may be the reason why the duration of the antiallergic effect of TMK-688 is over 6 h after oral administration.

TMK-688 was given orally to dogs and the form of its metabolites was analyzed. As shown in figure 3, the major metabolite was TMK-777, but mostly conjugated with glucuronic acid.

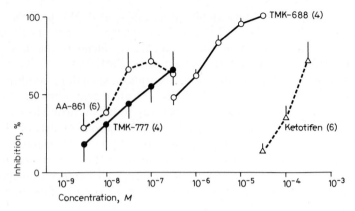

Fig. 1. Inhibitory effect of various compounds on IgE-mediated LTC$_4$ and D$_4$ release from human chopped lung. Numbers in parentheses indicate the number of independent experiments (see text for details).

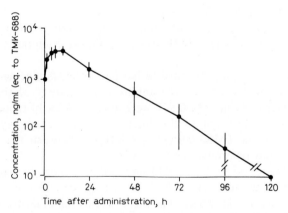

Fig. 2. Time course of blood concentration of radioactivity after oral administration of [^{14}C]-TMK-688 (10 mg/kg) to guinea pigs (mean ± SD, n = 3).

Since the majority of TMK-688 given orally is converted to TMK-777-glucuronide, we next examined whether the glucuronide form had antiallergic activity. TMK-777-glucuronide was injected intravenously and we examined its effect on passive anaphylactic bronchoconstriction. As shown in figure 4, TMK-777-glucuronide showed an inhibitory effect almost similar to the TMK-777-free form.

We next examined the reason why the glucuronized form showed a similar effect to the free form of TMK-777. Figure 5 shows the effect of

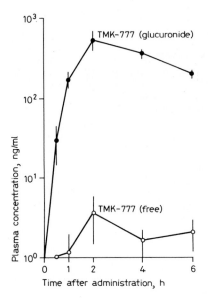

Fig. 3. Time course of plasma concentration of the metabolites of TMK-688 (20 mg/kg, p.o.) in dogs (mean ± SE, n = 4)

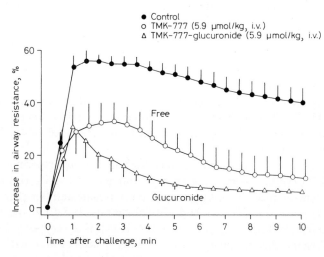

Fig. 4. Inhibitory effect of TMK-777 in free and glucuronide form on passive anaphylactic bronchoconstriction in guinea pigs (mean ± SE, n = 8–12).

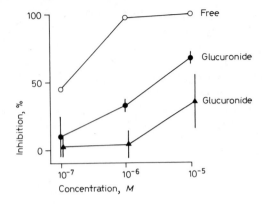

Treatment	n	Preincubation	Incubation
TMK-777 (o)	2	10 min	15 min
TMK-777g (●)	3	60 min	15 min
TMK-777g (▲)	3	10 min	60 min

Fig. 5. Inhibitory effect of TMK-777 in free and glucuronide form on Ca^{2+}-ionophore induced LTC_4 release from human leukocytes (mean $\pm$ SE).

TMK-777 in both free form and glucuronized form on Ca^{2+}-induced LTC_4 release from human leukocytes. Without preincubation, the glucuronized form was essentially inactive; however, after 60 min preincubation, it could inhibit the release of LTC_4. These results suggest that during the preincubation period, the conjugate form of TMK-777 was hydrolyzed to the free form of TMK-777, and the free form showed the inhibitory effect.

From the data mentioned above, the in vivo metabolism of TMK-688 can be speculated as follows: TMK-688 given orally converts to TMK-777 while it is passing through the intestinal membrane, and then the TMK-777 is rapidly conjugated with glucuronic acid in the portal system and the glucuronized TMK-777 circulates in the blood for a fairly long time. Leukocytes can hydrolyze the conjugated form and produce TMK-777 again at the site of action.

The LD_{50} value of TMK-688 was very high in various experimental animals, i.e., LD_{50} values were > 5 g/kg (s.c. and p.o.), 1.7 g/kg (i.p.) in both male and female mice, > 5 g/kg (s.c. and p.o.) in male and female rats, 2.5 g/kg (i.p.) in male rats, 2.2 g/kg (i.p.) in female rats, > 5 g/kg (p.o.) in both male and female dogs, suggesting that TMK-688 is a very safe drug. One and a half years' chronic toxicity test has already been cleared without any trouble. The phase 1 clinical study is almost finished without any special problems.

In conclusion, we tried to develop antiallergic drugs, based on the study on 5-lipoxygenase inhibitors, and we selected TMK-688 as a final one. TMK-688 showed strong inhibitory activities in various allergic experimental models. It is noteworthy that in all these in vivo assay systems, TMK-688 was very active in oral administration. TMK-688 has quite a different antiallergic profile from other drugs in clinical use at present. TMK-688 has already been proved to be quite safe in use after one and a half years safety tests. The phase 1 clinical study is almost finished now without any trouble. We hope TMK-688 will become a unique orally active antiallergic drug for clinical use in the near future.

References

1   Koshihara, Y.; Murota, S.; Petasis, N.A.; Nicolaou, K.C.: Selective inhibition of 5-lipoxygenase by 5,6-methanoleukotriene A_4, a stable analogue of leukotriene A_4. FEBS Lett. *143:* 13–16 (1982).

2   Koshihara, Y.; Neichi, T.; Murota, S.; Lao, A.N.; Fujimoto, Y.; Tatsuno, T.: Selective inhibition of 5-lipoxygenase by natural compounds isolated from Chinese plants, *Artemisia rubripes Nakai.* FEBS Lett. *158:* 41–44 (1983).

3   Neichi, T.; Koshihara, Y.; Murota, S.: Inhibitory effect of esculetin on 5-lipoxygenase and leukotriene biosynthesis. Biochim. Biophys. Acta *753:* 130–132 (1983).

4   Koshihara, Y.; Neichi, T.; Murota, S.; Lao, A.N.; Fujimoto, Y.; Tatsuno, T.: Caffeic acid is a selective inhibitor for leukotriene biosynthesis. Biochim. Biophys. Acta *792:* 92–97 (1984).

5   Murota, S.; Koshihara, Y.: New lipoxygenase inhibitors isolated from Chinese plants: Development of new anti-allergic drugs. Drugs Exp. Clin. Res. *11:* 641–644 (1985).

6   Murota, S.; Koshihara, Y.; Wakabayahshi, T.; Arai, J.: Anti-allergic action of inhibitors for leukotriene biosynthesis. Adv. Prostaglandin Thromboxane Leukotriene Res. *15:* 221–223 (1985).

7   Wakabayashi, T.; Ozawa, S.; Arai, J.; Takai, M.; Koshihara, Y.; Murota, S.: Anti-allergic action of TMK-777, a leukotriene biosynthesis inhibitor. Adv. Prostaglandin Thromboxane Leukotriene Res. *17:* 186–188 (1987).

S. Murota, PhD, Tokyo Medical and Dental University, 5–45, Yushima 1-chome, Bunkyo-ku, Tokyo 113 (Japan)

Zor U, Naor Z, Danon A (eds): Leukotrienes and Prostanoids in Health and Disease.
New Trends Lipid Mediators Res. Basel, Karger, 1989, vol 3, pp 62–66

L-660,711, a Potent Selective and Orally Active Antagonist of Leukotriene D_4

R.N. Young, R. Zamboni, M. Belley, E. Champion, L. Charette,
R. Dehaven, R. Frenette, A.W. Ford-Hutchinson, J.Y. Gauthier,
T.R. Jones, S. Leger, C.S. McFarlane, P. Masson, H. Piechuta,
S.S. Pong, J. Rokach, H. Williams

Merck Frosst Centre for Therapeutic Research, Pointe Claire-Dorval, Que., Canada

The peptide leukotrienes LTC_4, D_4 and E_4 collectively account for the biological activity known as slow-reacting substance of anaphylaxis (SRS-A). These metabolites of arachidonic acid exhibit potent actions on respiratory smooth muscle, effect mucocillary clearance and vascular permeability and inflammatory processes in general [1]. The biological activity of these leukotrienes and their demonstrated production following antigen challenge in airways of allergic patients have led to the hypothesis that leukotrienes may be mediators of human asthma. The biological activities of the leukotrienes are mediated by specific receptors and in human lung the actions of the peptide leukotrienes are mediated by a common receptor (or receptors) on which LTD_4 exhibits the most potent activity [2]. The hypothesis that a specific receptor antagonist for the LTD_4 receptor might serve as a novel and effective therapy of asthma has prompted a number of research groups to attempt to develop potent and safe LTD_4 receptor antagonists. Early clinical trials on two orally active LTD_4 antagonists (L-649,923 [3] and LY171883 [4]) and an aerosol-active antagonist (L-648,051 [5]) have been generally disappointing. However, these drugs may not be potent enough to fully block leukotriene action at maximum tolerated or deliverable doses. In this report the development and pharmacological profile of a novel specific and very potent LTD_4 antagonist, L-660,711, is described.

Discovery of L-660,711

Screening of compounds from the Merck sample collection identified a series of styrylquinoline analogs as weak LTD_4 antagonists. Evolution of

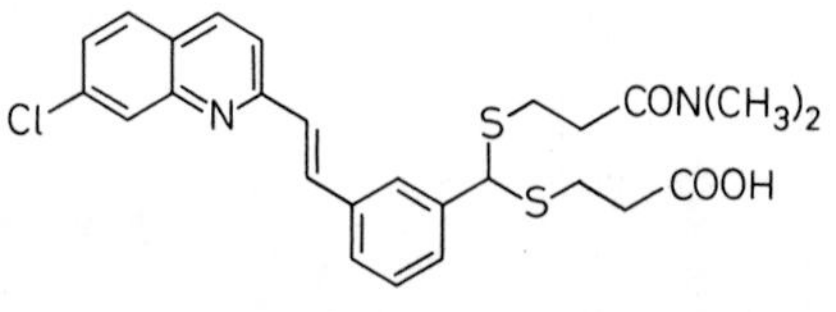

Fig. 1. Structure of L-660,711.

the initial lead, 3-(2-(2-quinolinyl)ethenyl)pyridine, to optimize lipophilic, polar and ionic binding in keeping with a model we have developed for the LTD_4 receptor [6], led to the identification of L-660,711 (fig. 1).

Pharmacological Profile of L-660,711 [7]

In vitro. L-660,711 was shown to interact in a competitive manner on guinea pig and human lung membranes with K_i values of 0.22 ± 0.15 nM (n = 35) and 2.1 ± 1.8 nM (n = 29) versus [^{3}H]-LTD$_4$, respectively. The compound was essentially inactive versus [^{3}H]-LTC$_4$, (IC_{50} values of 23 ± 11 μM (n = 16) and 32 μM (n = 1) in guinea pig and human lung, respectively). Scatchard analysis in guinea pig lung for [^{3}H]-LTD$_4$ binding showed that L-660,711 significantly increased the apparent K_D without affecting the B_{max} and Schild analysis gave a slope of 0.97 and a K_B for L-660,711 of 0.23 nM. The compound was a competitive antagonist of LTD$_4$-induced contraction of guinea pig ileum ($pA_2 = 10.5$, slope 0.91) and of guinea pig trachea ($pA_2 = 9.4$, slope 0.81). Minimal or nonsignificant shifts to the right of the dose response curves to PGF$_{2\alpha}$, U44069, PGD$_2$, serotonin, histamine or acetylcholine were observed at 1.9×10^{-5} M. The compound was also a competitive antagonist of LTD$_4$-induced contraction of isolated human trachea ($pA_2 = 8.45$, slope 1.09, in the presence of atropine, 10^{-7} M; mepyramine, 7×10^{-7} M).

Pretreatment of guinea pig trachea (previously sensitized with ovalbumin) with L-660,711 (1.9×10^{-7} to 1.9×10^{-5} M) blocked only a small component of the response to antigen. However, L-660,711 (1.9×10^{-5} M) produced complete inhibition of anti-IgE-induced contraction of human trachea pretreated with atropine, mepyramine and indomethacin. With lower concentrations of L-660,711, partial but significant inhibitions of the response was observed (fig. 2).

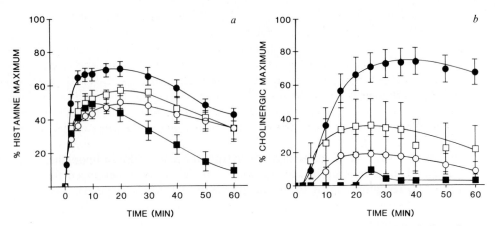

Fig. 2. Effect of L-660,711 on contractile responses of ovalbumin (0.1 µg/ml) challenged isolated trachea from sensitized guinea pigs *(a)* and anti-human IgE (179 µg/ml protein) challenged isolated human trachea *(b)*. Responses were determined in the absence (control ●; n = 7 in *a*; n = 6 in *b*) and in the presence of 19 µM (■; n = 3 in *a*; n = 2 in *b*), 1.9 µM (○; n = 3 in *a*; n = 4 in *b*) and 0.19 µM (□; n = 3 in *a*; n = 6 in *b*) L-660711 (30 min pretreatment). Symbols represent mean responses expressed as percent of maximum histamine *(a)* or methacholine *(b)* contraction; vertical lines show SE mean. All studies were carried out in the presence of 0.1 µM atropine, 7 µM mepyramine and 1.4 µM indomethacin.

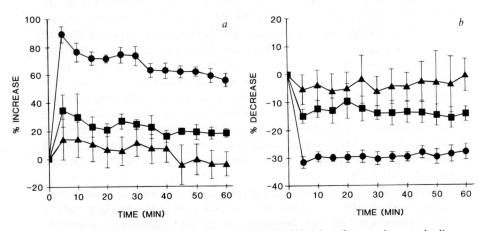

Fig. 3. Effect of a 2-hour pretreatment with L-660,711, 1.0 mg/kg, p.o., in normal saline (▲; n = 5) and a 4-hour pretreatment with 0.1 mg/kg p.o. in 1% methocel (■; n = 5) on control (●; n = 5) increases in resistance *(a)* and control (●; n = 5) decreases in compliance *(b)* induced in conscious squirrel monkeys by an aerosol of LTD$_4$ (50 µ/ml; 5 min aerosol). Symbols represent mean ± SE mean responses from (n) monkeys.

In vivo. Intravenously administered L-660,711 effectively antagonized bronchoconstriction to intravenous LTC$_4$, LTD$_4$ and LTE$_4$ in anesthetized, artifically ventilated guinea pigs with mean ED$_{50}$ values of 1.6 ± 0.7, 1.1 ± 0.2, and 1.0 ± 0.8 μg/kg, respectively. It was highly selective for this blockade relative to other contractile agonists. L-660,711 up to 3 mg/kg produced no consistent or significant inhibition of the responses to histamine, acetylcholine, serotonin, U44069 or arachidonic acid. When administered by the intraduodenal route in the same model at 0.1 mg/kg, L-660,711 effectively inhibited bronchoconstriction induced by 0.2 μg/kg of LTD$_4$ (i.v.) at 10 min (59%) to 70 min (80%) after drug administration.

L-660,711 was orally active for inhibition of the duration of antigen-induced dyspnea in hyperreactive rats [8] with an ED$_{50}$ value of 0.068 (95% CI; 0.03–0.14) mg/kg (dosed 4 h before antigen challenge). L-660,711 was also orally effective for the inhibition of bronchoconstriction induced by LTD$_4$ in conscious squirrel monkey and caused significant inhibition ($p < 0.05$) of the increases in resistance (R$_L$) and decreases in dynamic compliance (C$_{DYN}$) at doses as low as 0.1 mg/kg p.o. (4 h pretreatment) (fig. 2). The compound also significantly inhibited ($p < 0.05$) bronchonconstriction induced by an aerosol of ascaris antigen when administered at 0.5 mg/kg p.o. (2 h pretreatment).

Conclusions

L-660,711 has been shown to be an extremely potent and highly selective leukotriene D$_4$ receptor antagonist both in vitro and in vivo in a variety of species. It is also a potent inhibitor of anti-IgE-induced contraction of human trachea in vitro and of antigen-induced dyspnea and bronchoconstriction in the hyperreactive rat and squirrel monkey, respectively. L-660,711 shows excellent activity when dosed by the oral route (or i.d.) in the rat, guinea pig or monkey. Unpublished studies have shown that the compound is highly bioavailable in the rat, dog and squirrel monkey. The compound can be formulated for delivery by the intravenous, oral or aerosol route of administration (the compound has also been shown to be active following aerosol administration in the monkey and guinea pig). The excellent pharmacological profile of L-660,711 and its compatibility for administration by these three routes suggest that it may be an ideal compound to clearly define the role of LTD$_4$ in human disease. Ongoing clinical studies in normal and asthmatic patients will serve to define the potential role of L-660,711 as a novel therapy for human asthma.

References

1   Piper, P.J.: Formation and actions of leukotrienes. Physiol. Rev. *64:* 744–761 (1984).
2   Buckner, C.K.; Krell, R.D.; Laravusa, R.B.; Coursin, D.B.; Bernstein, P.R.; Will, J.N.: Pharmacological evidence that human intralobar airways do not contain different receptors that mediate contractions to leukotriene C_4 and leukotriene D_4. J. Pharmacol. Exp. Ther. *237:* 558–562 (1986).
3   Britton, J.R.; Hanley, S.P.; Tattersfield, A.E.: The effect of an oral leukotriene D_4 antagonist L-649,923 on the response to inhaled antigen in asthma. J. Allergy Clin. Immunol. *79:* 811–816 (1987).
4   Cloud, M.; Eras, G.; Kemp. J.; Platts-Mills, T.; Altnan, L.; Townley, R.; Tinkelman, D.; King, T.; Middleton, E.; Sheffer, A.; McFadden, E.; Efficacy and safety of LY17883 in patients with mild chronic asthma. J. Allergy Clin. Invest. *79:* 525 (1987).
5   Evena, J.M.; Barnes, N.C.; Piper, P.J.; Costello, J.K.: The effect of a single dose of inhaled L-648,051, a leukotriene D_4 antagonist in mild asthma. Br. J. Pharmacol. *25:* 112–113 (1988).
6   Young, R.N.: The development of new anti-leukotriene drugs: L-648,051 and L-649,923, specific leukotriene D_4 antagonists. Drugs Future *13:* 745–759 (1988).
7   Jones, T.R.; Zamboni, R.; Belley, M.; Champion, E.; Charette, L.; Ford-Hutchinson, A.W.; Frenette, R.; Gauthier, J.Y.; Leger, S.; Masson, P.; McFarlane, C.S.; Piechuta, H.; Rokach, J.; Williams, H.; Young, R.N.: Pharmacology of L-660,711: A novel potent and selective leukotriene D_4 antagonist. Can. J. Pharmacol. Physiol. (in press).
8   Ford-Hutchinson, A.W.; Brunet, G.; Hamel, R.; Piechuta, H.; Holme, G.: Respiratory responses to leukotriene and biogenic amines in normal and hyperactive rats. J. Immunol. *131:* 434–438 (1983).

R.N. Young, PhD, Merck Frosst Centre for Therapeutic Research, PO Box 1005, Pointe Claire-Dorval, Que. H9R 4P8 (Canada)

Zor U, Naor Z, Danon A (eds): Leukotrienes and Prostanoids in Health and Disease.
New Trends Lipid Mediators Res. Basel, Karger, 1989, vol 3, pp 67–71

Binding of ^{3}H-LTD$_4$ and the Peptide Leukotriene Antagonist ^{3}H-ICI 198,615 to Receptors on Human Lung Membranes

D. Aharony, R.C. Falcone

ICI Pharmaceuticals Group, Department of Pharmacology, Section of Pulmonary
Pharmacology, Wilmington, Del., USA

We have recently characterized ICI 198,615 as a highly potent and selective leukotriene (LT) antagonist [1, 2]. Utilizing guinea pig lung (GPL) as a rich source of LT receptors, we demonstrated by both functional receptor assay (i.e., LT-induced contraction) and radioligand binding studies that ICI 198,615 is a competitive antagonist of LTD$_4$/LTE$_4$ ($K_i = 0.3$ nM) receptors. The inhibition by ICI 198,615 was highly selective as demonstrated in experiments in which it failed to antagonize the response to a variety of non-LT agonists. Recently, the binding of ^{3}H-ICI 198,615 to LTD$_4$ receptors on GPL membranes was demonstrated [3] and shown to be of high-affinity and saturable nature. Moreover, only LTs and selective LT antagonists [4] inhibited binding of ^{3}H-ICI 198,615 (with good correlation with their ability to inhibit ^{3}H-LTD$_4$ binding), confirming the high selectivity of this antagonist for LTD$_4$/LTE$_4$ receptors. Further studies [5] have also demonstrated that ICI 198,615 potently and competitively antagonizes LT-induced contractions in isolated human intralobar airways. The existence of high-affinity, saturable and distinct LT receptors on membranes prepared from human lung, was directly demonstrated in radioligand binding experiments with ^{3}H-LTD$_4$ [6, 7] and ^{3}H-LTC$_4$ [8].

In this communication we report experiments on binding of ^{3}H-LTD$_4$ and ^{3}H-ICI 198,615 to human lung membranes (HLM) and the inhibition of both ligands by ICI 198,615 and other selective and structurally diverse LT antagonists.

Methods

Human lung tissue was obtained from carcinoma patients after lung resection and removal of abnormal portions by the Departments of Pathology at the University of Wisconsin Hospitals and Clinics (Madison, Wisc.). The tissues were immersed in Krebs buffer and kept on ice. Large airways and blood vessels were dissected and the lung parenchyma was cut into small (20 cm^2) sections, rinsed with PBS, immediately frozen in liquid nitrogen and stored at -70 °C prior to use. Crude membrane fractions were prepared as described elsewhere [6, 7]. Briefly, 50 g of a single human lung were thawed, minced, chopped with a McIlwain tissue chopper and washed several times with 10 mM potassium-phosphate buffer, pH 7.4, until visibly clear of blood. Tissue was suspended in 250 ml of Tris-HCl (20 mM, pH 7.5, Tris)/sucrose (0.25 M) buffer, containing several protease inhibitors [7] and homogenized with a Brinkman PT-20 Polytron (3 × 20 s bursts at setting 6). The homogenate was centrifuged at 1,500 g (10 min at 4 °C) and the supernatant was pooled and further centrifuged at 100,000 g (40 min at 4 °C). The crude membrane pellets were resuspended in Tris buffer, containing 10% sucrose and layered onto a 10-ml cushion of 40% sucrose. The membranes were centrifuged at 100,000 g (60 min at 4 °C) and the fraction concentrated at the interphase was collected and washed (100,000 g 40 min at 4 °C). The HLM pellet was resuspended in Tris buffer and immediately assayed for binding. Protein concentration was adjusted to > 250 µg/ml assay. For determination of ^{3}H-LTD$_4$ (2 nM) binding, HLM were incubated in Tris (containing 3 mM MgCl$_2$ 10 mM cysteine and 10 mM glycine) at 30 °C for 30 min in total volume of 0.310 ml. Total and nonspecific binding were determined as the amount of ^{3}H-LTD$_4$ bound to HLM in the absence or presence of 1 µM LTD$_4$, respectively. The ^{3}H-ICI 198,615 (1 nM) binding assay was conducted in Pipes buffer (10 mM, pH 6.8) under similar conditions as described above. Nonspecific binding was determined in the presence of 1 µM unlabeled ICI 198,615. Separation of receptor bound ^{3}H-ligand was achieved by dilution into ice-cold buffer followed by immediate filtration and washing. Data analysis and statistical calculations were as described elsewhere [1, 3]. ^{3}H-LTD$_4$ (39 Ci/mmol) and ^{3}H-ICI 198,615 (60 Ci/mmol) were from Du Pont-NEN. SKF104,353 was kindly supplied by SK&F. LTs, ICI 198,615 and LY171,883 were made in the Medicinal Chemistry Department at ICI Pharmaceuticals Group.

Results and Discussion

Preliminary binding experiments with crude membranes were difficult and not always reproducible. Saturable binding sites for ^{3}H-LTD$_4$ or ^{3}H-ICI 198,615 could not be consistently demonstrated and poor correlation was observed between K_i values obtained in competition studies and K_d values for either ligand that were determined in Scatchard experiments. This inconsistency prompted us to use the sucrose gradient method in order to further purify the membranes.

Utilizing HLM we found that a protein concentration of >0.25 mg/ml assay is required to accurately measure specific binding (SB). This was due to the low density of LTD$_4$ receptors in HLM, similar to a published report [6]. The assay also had to be performed on the same day of preparation to avoid freezing of the membranes, since in our hands, freezing and thawing had a detrimental effect on ligand binding (data not shown). In kinetic experiments, ^{3}H-LTD$_4$ rapidly ($t^1/_2 = 1.5$ min) bound to HLM and could be slowly dissociated by 1 µM LTD$_4$ or ICI 198,615. In addition, we confirmed an earlier report by Lewis et al. [6] that binding of ^{3}H-LTD$_4$ to HLM is stimulated by magnesium ions and inhibited by Gpp(NH)p (IC$_{50}$: 100 µM). In contrast, binding of ^{3}H-ICI 198,615 was not influenced by either agent, similar to results reported in GPL membranes [3]. Scatchard analysis of ^{3}H-LTD binding to HLM (fig. 1a) illustrates that it binds to a single class of receptors in a saturable manner ($B_{max} = 84 \pm 10$ fmol/mg protein) and with high affinity ($K_d = 1.5 \pm 0.4$ nM, mean $\pm$ SEM, n = 8). These data agree well with a published report [6] which also demonstrated high affinity but very low density ^{3}H-LTD$_4$ receptors (68 fmol/mg protein) in both adult and fetal human lung. Figure 1b illustrates that ^{3}H-ICI 198,615 also bound to a single class of high-affinity receptors on HLM in a saturable manner ($B_{max} = 186 \pm 35$ fmol/mg protein and $K_d = 0.72 \pm 0.14$ nM, mean $\pm$ SEM, n = 6). The presence of a single binding site for ^{3}H-ICI 198,615 in HLM is different from our finding in GPL membranes where we observed two binding sites and may suggest a species difference. In addition, as observed in GPL membranes, a higher density of antagonist binding compared with agonist binding is evident. Having demonstrated specific binding sites for ^{3}H-LTD$_4$ and ^{3}H-ICI 198,615 on HLM, we proceeded to characterize their pharmacology in drug competition assays. Figure 2a and b illustrates the inhibition of ^{3}H-LTD$_4$ (fig. 2a) or ^{3}H-ICI 198,615 (fig. 2b) binding to HLM by LTD$_4$ and LTD$_4$ antagonists.

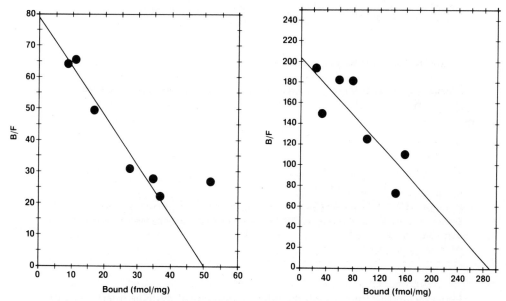

Fig. 1. Scatchard analysis of ^{3}H-ligands binding to HLM. *a* ^{3}H-LTD$_4$ (0.15–1.93 nM) was incubated in Tris-HCl (20 mM, pH 7.5) with HLM (0.34 mg/ml) for 30 min at 30 °C in the presence of 10 mM cysteine, 10 mM glycine and 3 mM MgCl$_3$. *b* ^{3}H-ICI 198,615 (0.12–2.1 nM) was incubated with HLM (0.38 mg/ml) for 30 min at 30 °C in Pipes buffer, pH 6.8. Results with either ligand are mean of triplicate determination.

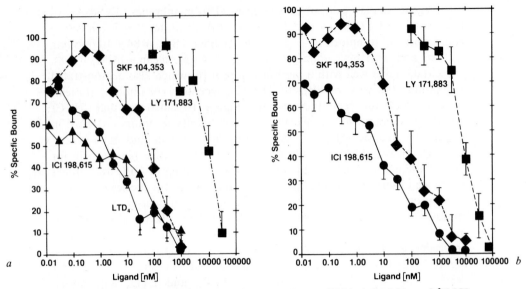

Fig. 2. a Inhibition of ^{3}H-LTD$_4$ (2 n*M*) binding to HLM. *b* Inhibition of ^{3}H-ICI 198,615 (0.5 n*M*) binding to HLM. Assay conditions are as in figure 1 for each ligand respectively. Each data point is a mean (± SEM) of N experiments in duplicates.

Table 1. Inhibition of ^{3}H-ligand binding to HLM

Inhibitor	K$_i$ (n*M*) vs. ^{3}H-LTD$_4$	Relative potency	n	K$_i$ (n*M*) vs. ^{3}H-ICI 198,615	Relative potency	n
LTD$_4$	0.94 ± 0.31	1.5	4	ND	ND	–
ICI 198,615	0.63 ± 0.24	1	3	1.13 ± 0.4	1	3
SKF104, 353	43 ± 17	68	4	51 ± 14	39	3
LY171, 883	2,246 ± 450	3,565	3	2,973 ± 1,173	2,287	3

Results are mean ± SEM of duplicate determinations in individual lungs.
K$_i$ values calculated from the Cheng–Prusoff equation: K$_i$ = IC$_{50}$/(1 + L/K$_d$), where L is the ligand concentration in the assay and K$_d$ is the equilibrium binding constant, determined by Scatchard analysis.
ND = Not determined.

However, in the studies with the labeled antagonist, cysteine interfered with binding and thus, inhibition by LTD$_4$ could not be determined accurately. Table 1 summarizes the apparent K$_i$ values obtained in several batches and illustrates that generally these antagonists possess lower (2- to 5-fold) affinity towards LTD$_4$ receptors on HLM compared with LTD$_4$ receptors on GPL. The relative potency (table 1) for these antagonists against either ligand agrees well and is generally similar to their relative potency in GPL. Moreover, affinity constants (i.e., K_i or K_d) for LTD_4 and ICI 198,615, derived either from competition assays or direct Scatchard analysis of the two ligands, also agree well and suggest that the majority of the binding sites are indeed LTD_4 receptors.

In conclusion, these experiments directly demonstrate the existence of high affinity but low density receptors for agonists and antagonists in human lung membranes which are consistent with specific LTD_4 receptors.

Acknowledgement

We wish to thank Dr. S. Mong (SK&F) for helpful advice on preparation of HLM.

References

1   Aharony, D.; Falcone, R.C.; Krell, R.D.: Inhibition of ^3H-LTD_4 binding to guinea-pig lung receptors by the novel leukotriene antagonist ICI 198,615. J. Pharmacol. Exp. Ther. *243*: 921–926 (1987).

2   Snyder, D.W.; Giles, R.E.; Keith, R.A.; Yee, Y.K.; Krell, R.D.: The in vitro pharmacology of ICI 198,615: A novel, potent and selective peptide leukotriene antagonist. J. Pharmacol Exp. Ther. *243:* 548–556 (1987).

3   Aharony, D.; Falcone, R.C.; Yee, Y.K.; Hesp, B.; Giles, R.E.; Krell, R.D.: Biochemical and pharmacologic characterization of the binding of the selective peptide-leukotriene antagonist ^3H-ICI 198,615 to LTD_4 receptors in guinea-pig lung membranes. Ann. N.Y. Acad. Sci. *524:* 162–180 (1988).

4   Aharony, D.; Falcone, R.C.; Yee, Y.K.; Krell, R.D.: ^3H-ICI 198,615: a novel, potent, selective and stable peptide-leukotriene antagonist. Biotechnol. Update *3:* 1–2 (1988).

5   Buckner, C.K.; Saban, R.; Castleman, W.L.; Will, J.A.: Analysis of leukotriene receptor antagonists on isolated human intralobar airways. Ann. N.Y. Acad. Sci. *524:* 181–186 (1988).

6   Lewis, M.A.; Mong, S.; Vessella, R.L.; Crooke, S.T.: Identification and characterization of leukotriene D_4 receptors in adult and fetal human lung. Biochem. Pharmacol. *34:* 4311–4317 (1985).

7   Mong, S.; Wu, H.L.; Miller, J.; Hall, R.F.; Gleason, J.G.; Crooke, S.T.: SKF 104,353, a high affinity antagonist for human and guinea pig lung leukotiene D_4 receptor, blocked phosphatidylinositol metabolism and thromboxane synthesis induced by leukotriene D_4. Mol. Pharmacol. *32:* 223–229 (1987).

8   Civelli, M.; Oliva, D.; Mezzetti, M.; Nicosia, S.J.: Characteristics and distribution of specific binding sites for leukotriene C_4 in human bronchi. Pharmacol. Exp. Ther. *242:* 1019–1024 (1987).

David Aharony, PhD, ICI Pharmaceuticals Group, Department of Pharmacology, Section of Pulmonary Pharmacology, Wilmington, DE 19897 (USA)

Zor U, Naor Z, Danon A (eds): Leukotrienes and Prostanoids in Health and Disease.
New Trends Lipid Mediators Res. Basel, Karger, 1989, vol 3, pp 72–76

LY171883 and Newer Leukotriene Receptor Antagonists: A Quest for Novel Therapeutic Agents

Jerome H. Fleisch, Anthony R. Dowell, Winston S. Marshall

Lilly Research Laboratories, Eli Lilly & Co., Indianapolis, Ind., USA

Six years ago, only one leukotriene receptor antagonist, FPL 55712 [1], was known and little was appreciated about the various subtypes of leukotriene receptors. Since then, numerous chemical entities have been discovered that antagonize the pharmacologic effects of LTD_4 and LTE_4 in tissues from experimental animals [2–4]. Recent evidence suggests that at least some of these agents might be LTC_4 receptor antagonists in human airways [5]. In addition, potent and specific receptor antagonists for LTB_4, a noncysteinyl-containing leukotriene, have now been developed [6–8].

LY171883 [9, 10] followed FPL 55712 and was the first orally bioavailable leukotriene receptor antagonist with a long metabolic half-life, 30 min in rats to 15 h in sheep. LY171883 was tested in a variety of clinical settings prior to its discontinuation due to long-term toxicity in female mice. Information gathered from these investigations is proving invaluable to the eventual development of a leukotriene receptor antagonist as a therapeutic modality in the treatment of asthma.

In early studies, single doses of LY171883 up to 700 mg and multiple doses up to 1,200 mg/day were well tolerated in normal volunteers [11]. Clinically significant alterations in vital signs, electrocardiograms, or clinical laboratory tests were not observed. This suggested that LTD_4, LTE_4, and perhaps even LTC_4 are not involved in major physiological processes since leukotriene receptor antagonism probably would have resulted in an unacceptable side effect profile had this been the case.

Phillips et al. [12] examined the inhibitory activity of LY171883 on LTD_4-induced bronchospasm in 12 normal subjects. A single oral 400-mg dose of LY171883 resulted in a dose-related rightward shift of the inhaled LTD_4 bronchoconstrictor dose-response curve in 10 of 12 normal human volunteers. FEV_1 and $\dot{V}_{p30}$, indices of pulmonary function, were shifted 4.6- and 6.1-fold respectively. This was in line with a 4-fold shift obtained by

Barnes et al. [13] who gave 1 g L-649,923, another LTD_4/LTE_4 receptor antagonist, to a similar group of individuals. These relatively modest reductions in leukotriene-induced bronchospasm may relate to insufficient potency of LY171883 and L-649,923, to significant differences between cysteinyl leukotriene receptors in humans and experimental animals, to the different routes of administration of agonist and antagonists, or perhaps to the lack of a pharmacologic steady state due to the single dose regimen. Future studies in humans with more potent leukotriene receptor antagonists should differentiate between these possibilities.

Israel et al. [14] reported the activity of LY171883 in 19 asthmatic patients with cold, dry air-induced bronchospasm. The geometric mean provocation dose (PD_{20}) for respiratory heat exchange in drug-treated vs. placebo-treated patients indicated that LY171883 administration had a modest protective effect against cold air challenge in mild asthmatics. Shaker et al. [15] administered LY171883 for 14 days to 10 subjects with exercise-induced bronchospasm. Half of these individuals with the mildest form of asthma responded to LY171883 administration with a reduction in the degree of exercise-induced fall in FEV_1. LY171883 was also compared to placebo in 138 mild asthmatic patients for 6 weeks [16]. At the end of the treatment period, FEV_1 increased in treated patients concomitant with decreases in diurnal wheezing and breathlessness. In those individuals most dependent on inhaled metaproterenol, there was a marked decrease in the need for this bronchodilator beta-receptor stimulant.

In addition to its leukotriene receptor antagonist properties, LY171883 is a phosphodiesterase inhibitor [9]. Certain inhibitors of this enzyme potentiate beta-receptor-mediated responses [17], possibly explaining the apparent potentiation of metaproterenol in the latter clinical study. Further in vitro studies with guinea pig trachea have now clearly dissociated the ability of LY171883 to antagonize LTE_4 and to potentiate isoproterenol, the prototypic beta-receptor agonist [18]. Similar results were obtained in anesthetized guinea pigs using changes in total pulmonary impedance as an index of LTE_4 receptor and beta-receptor activation. Two well-characterized phosphodiesterase inhibitors, isobutylmethylxanthine and theophylline were employed in accompanying experiments and proved unable to antagonize LTE_4 in doses that potentiated isoproterenol-induced bronchodilation. When taken in its entirety, the preclinical and clinical studies with LY171883 support further development of more potent chemicals with a similar pharmacologic profile as potential antiasthma drugs.

The pace of revelation of new molecules capable of acting as leukotriene receptor antagonists has accelerated during the past few years. The earliest antagonists were derived from the acetophenone series from which FPL 55712 evolved [1]. Later, following structural elucidation of the

Acetophenone

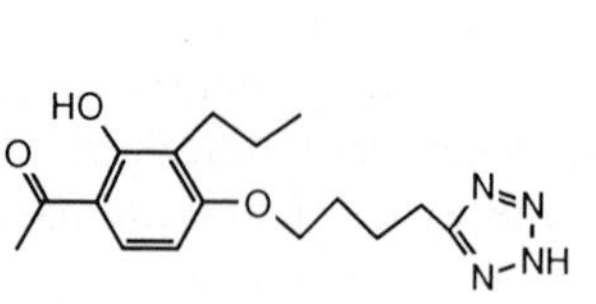

LY171883

Structural Analog

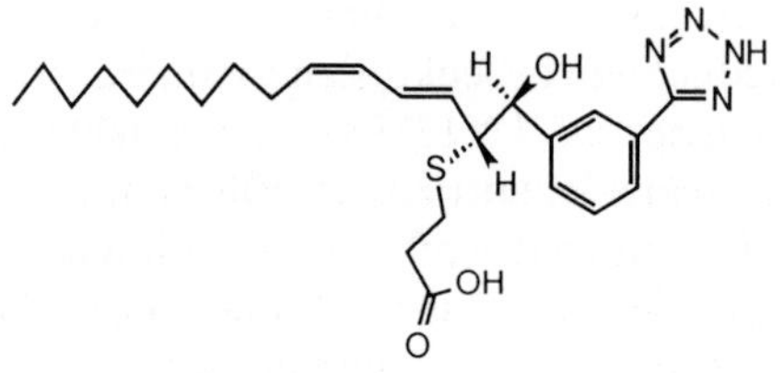

LY170680

Other Structural Types

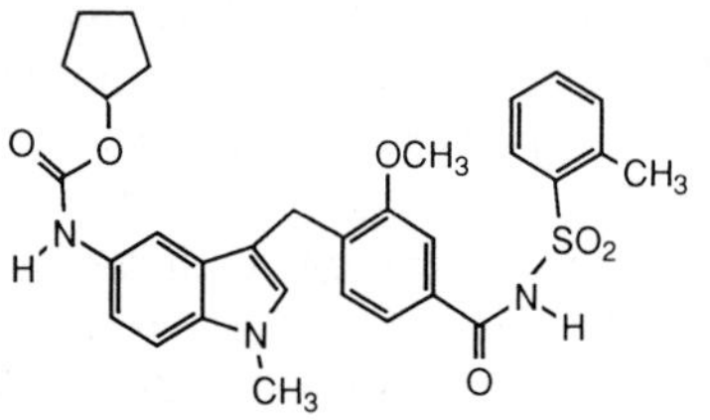

ICI 204,219

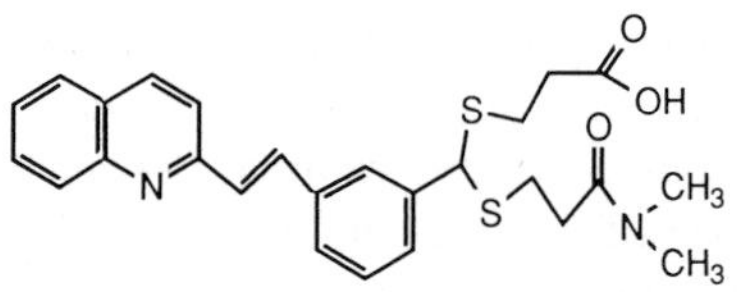

L-660,711

Fig 1. Cysteinyl leukotriene receptor antagonists.

leukotrienes, antagonists were developed by modification of the agonist structure [19]. More recently, chemically dissimilar compounds not obviously related to each other have shown leukotriene antagonist properties (fig. 1). They may differ from one another in potency, absorption, metabolism and elimination from the body, but not in selectivity for blocking LTD_4/LTE_4 receptors. Three representatives of this chemically diverse group are: ICI 204,219 [20], an acylsulfonamidoindole; L-660,771, a styrylquinoline [21], and LY170680, a structural analog of the cysteinyl leukotrienes [22]. All three compounds are considerably more potent than their predecessors. This attribute might help overcome some of the earlier difficulties experienced with FPL 55712, LY7171883, and L-649,923. This next generation of leukotriene receptor antagonists will provide additional insight into the contribution of the cysteinyl leukotrienes to human pathophysiology. The challenge is to identify the critical pharmacophores in these very different molecules which seek out active sites on leukotriene receptors, competing with the agonists but not activating the receptors. Furthermore, we must test the hypothesis of whether an SRS-A antagonist, that is a compound with near equal potency against LTC_4, LTD_4, and LTE_4, will have therapeutic advantages over existing LTD_4/LTE_4 receptor antagonists.

Considering the rapid conversion of LTC_4 to the other two eicosanoids [23], and the possibility of unique receptor subtypes for these substances [5, 24], the likelihood of therapeutic benefits from a drug with a high degree of specificity for only one of the cysteinyl leukotrienes receptors is remote.

Just as drugs that adequately antagonize the actions of LTC_4 are not currently available, LTB_4 receptor antagonists had, until very recently, eluded development. Effects of LTB_4 are mediated through stereospecific receptor sites distinct from those activated by the cysteinyl leukotrienes [25]. SM-9064 [26] and U 75302 [27], two analogs of LTB_4, were reported to be LTB_4 receptor antagonists. More recently, LY223982, a benzophenone, and LY255283, an acetophenone, were described as potent inhibitors of ^{3}H-LTB_4 binding to human neutrophils [6, 7, 8]. These compounds and their successors will prove useful in delineating the role of LTB_4 in psoriasis, inflammatory bowel disease, and perhaps asthma. Experimentally, as was the case with the cysteinyl leukotriene receptor antagonists, they will help gather information on the nature of LTB_4 receptor subtypes.

Thus, a large number of diverse chemicals have been shown to antagonize pharmacologic effects of LTB_4, LTC_4, LTD_4, and LTE_4. Many more agents with subtle differences in their pharmacologic activities are on the horizon. By judicious use of these drugs, the intricacies of leukotriene receptor pharmacology will unfold and the therapeutic utility of some of these agents may become a reality.

References

1   Augstein J, Farmer JB, Lee TB, et al: Selective inhibitor of slow reacting substance of anaphylaxis. Nature New Biol 1973;245:215–217.
2   Gillard JW, Guindon Y: The leukotrienes: prospects for therapy against a unique family of pathophysiological mediators. Drugs Future 1987;12:453–474.
3   Perchonock CD, Torphy TJ, Mong S: Peptidoleukotrienes: pathophysiology, receptor biology and receptor antagonists. Drugs Future 1987;12:871–889.
4   Fleisch JH, Rinkema LE, Whitesitt CA, et al: Development of cysteinyl leukotriene receptor antagonists; in Lewis A, Ackerman N, Otterness I (eds): Advances in Inflammation Research. New York, Raven Press, 1988, vol 12, pp 173–189.
5   Buckner CK, Saban R, Castleman WL, et al: Analysis of leukotriene receptor antagonists on isolated human intralobar airways. Ann NY Acad Sci 1988;524:181–186.
6   Jackson WT, Boyd RJ, Froelich LL, et al: Inhibition of LTB_4 binding and aggregation of neutrophils by LY255283 and LY223982. FASEB J 1988;2:A1110.
7   Herron DK, Bollinger, NG, Swanson-Bean D, et al: LY255283: a new leukotriene B_4 antagonist. FASEB J 1988;2:A1110.
8   Gapinski DM, Mallett BE, Froelich LL, et al: LY223982: a potent and selective antagonist of leukotriene B_4. Structure activity relationships for the inhibition of LTB_4 binding to human neutrophils. FASEB J 1988;2:A1110.

9   Fleisch, JH, Rinkema LE, Haisch KD, et al: LY171883, 1-[2-hydroxy-3-propyl-4-[4-(1H-tetrazol-5-yl)butoxy]phenyl]ethanone, an orally active leukotriene D_4 antagonist. J Pharmacol Exp Ther 1985;233:148–157.

10 Marshall WS, Goodson T, Cullinan GJ, et al: Leukotriene receptor antagonists. 1. Synthesis and structure activity relationships of alkoxyacetophenone derivatives. J Med Chem 1987;30:682–689.

11 Callaghan JT, Farid NA, Bergstrom RF, et al: Clinical observations and the single dose and steady state pharmacokinetics of LY171883, a new leukotriene D_4 antagonist, in man. Ann Allergy 1985;55:279.

12 Phillips GD, Rafferty P, Robinson C, et al: Dose-related antagonism of leukotriene D_4-induced bronchoconstriction by p.o. administration of LY171883 in nonasthmatic subjects. J. Pharmacol Exp Ther 1988;246:732–738.

13 Barnes N, Piper PJ, Costello J: The effect of an oral leukotriene antagonist L-649,923 on histamine and leukotriene D_4-induced bronchoconstriction in normal man. J Allergy Clin Immunol 1987;79:816–821.

14 Israel E, Juniper EF, Morris MM, et al: A leukotriene D_4 (LTD_4) receptor antagonist, LY171883, reduces the bronchoconstriction induced by cold air challenge in asthmatics: a randomized, double-blind, placebo controlled trial. Am Rev Respir Dis 1988;137S:77.

15 Shaker G, Glovsky MM, Kebo D, et al: Reversal of exercise induced asthma by LTD_4, LTE_4 antagonists (LY171883). J Allergy Clin Immunol 1988;81:315.

16 Cloud M, Enas G, Kemp J, et al: Efficacy and safety of LY171883 in patients with mild chronic asthma. J Allergy Clin Immunol 1987;79:256.

17 Lorenz KL, Wells JN: Potentiation of the effects of sodium nitroprusside and of isoproterenol by selective phosphodiesterase inhibitors. Mol Pharmacol 1983;23:424–430.

18 Rinkema LE, Roman CR, Bemis KG, et al: Leukotriene (LT) E_4 receptor antagonism and phosphodiesterase (PDE) inhibition of LY171883, isobutylmethylxanthine (IBMX), and theophylline (THEO) in vitro and in vivo. Physiologist 1988;31:A90.

19 Hay DWP, Muccitelli RM, Tucker SS, et al: Pharmacologic profile of SK&F 104353: a novel, potent and selective peptidoeukotriene receptor antagonist in guinea pig and human airways. J Pharmacol Exp Ther 1987;243:474–481.

20 Krell RD, Buckner CK, Keith RA, et al: ICI 204,219: a potent, selective peptide leukotriene receptor antagonist. J Allergy Clin Immunol 1988;81:276.

21 Charette L, Jones TR, Champion E, et al: In vitro pharmacology of L-660,771, a new LTD_4 receptor antagonist. FASEB J 1988;2:A1264.

22 Baker SR, Boot JR, Jamieson WB, et al: The development and synthesis of LY170680, a new potent leukotriene antagonist. Proc 2nd Conf on Leukotrienes and Prostanoids in Health and Disease, Jerusalem 1988, Abstract book, p 117.

23 Keppler D, Huber M, Hagmann W, et al: Metabolism and analysis of endogenous cysteinyl leukotrienes. Ann NY Acad Sci 1988;524:68–74.

24 Fleisch JH, Cloud ML, Marshall WS: A brief review of preclinical and clinical studies with LY171883 and some comments on newer cysteinyl leukotriene receptor antagonists. Ann NY Acad Sci 1988;524:356–368.

25 Goldman DW: Regulation of the receptor system for leukotriene B_4 on human neutrophils. Ann NY Acad Sci 1988; 524:187–195.

26 Namiki M, Yukinobu I, Sakamoto K, et al: Profiles of a potential LTB_4-antagonist, SM-9064. Biochem Biophys Res Commun 1986;138:540–546.

27 Lin AH, Morris J, Wishka DG, et al: Novel molecules that antagonize leukotriene B_4 binding to neutrophils. Ann NY Acad Sci 1988;524:196–200.

Jerome H. Fleisch, PhD, Lilly Research Laboratories, Eli Lilly & Co.,
Indianapolis, IN 46285 (USA)

Zor U, Naor Z, Danon A (eds): Leukotrienes and Prostanoids in Health and Disease.
New Trends Lipid Mediators Res. Basel, Karger, 1989, vol 3, pp 77–83

Pharmacological Properties of YM-17551, an Orally Active Leukotriene Antagonist

Kenichi Tomioka[a], *Toshimitsu Yamada*[a], *Toshiyasu Mase*[b],
Ryuji Tsuzuki[b], *Hiromu Hara*[b], *Kiyoshi Murase*[b]

Departments of [a]Pharmacology and [b]Chemistry, Central Research Laboratories,
Yamanouchi Pharmaceutical Co., Ltd., Itabashi-ku, Tokyo, Japan

Introduction

It has been suggested that leukotrienes (LTs) have pathological roles in asthma, other types of immediate hypersensitivity reactions, inflammation, ischemic heart and brain diseases, psoriasis and peptic ulcer [1–5]. We have recently reported that YM-16638, a 1,3,4-thiadiazol-substituted acetophenone, shows potent anti-LTs effect in isolated guinea pig tissues and human bronchi [6]. Furthermore, YM-16638 administered orally inhibits antigen-induced bronchoconstriction in guinea pigs [6], sheep [7] and man [unpubl. data]. Further studies on YM-16638-related compounds revealed that 5-alkylthio-1,3,4-thiadiazolyl-thiomethyl acetophenones had more potent anti-LTs effect than that of YM-16638 [unpubl. data].

In the present study, we describe the pharmacological properties of *p*-[[[5-[(4-acetyl-3-hydroxy-2-propylbenzyl)thio]-1,3,4-thiadiazol-2-yl]thio]-methyl]benzoic acid (YM-17551, fig. 1).

Materials and Methods

Anti-LT Effect in Isolated Tissues

The preparations used were: guinea pig ileum, guinea pig trachea prepared according to Constantine [8], and human bronchi obtained from patients with lung carcinoma after surgery and prepared according to Constantine [8]. Preparations were suspended with 1.0 g tension in an organ bath containing 10 ml Tyrode solution equilibrated with a mixture of 95% O_2 and 5% CO_2 at 37 °C. The tissues were equilibrated for 90 min during which the Tyrode solution was replaced every 30 min and the loading tension adjusted to 1 g. The developed tension of the tissue was measured isometrically with a strain gauge transducer (SB-1T, Nihon Kohden), and recorded on a Recticorder (RJG-4008, Nihon Kohden) through a carrier amplifier (RP-5, Nihon Kohden). The contractile response to agonist was measured in the absence and then the presence of various concentrations of the test compound. The incubation time of the test

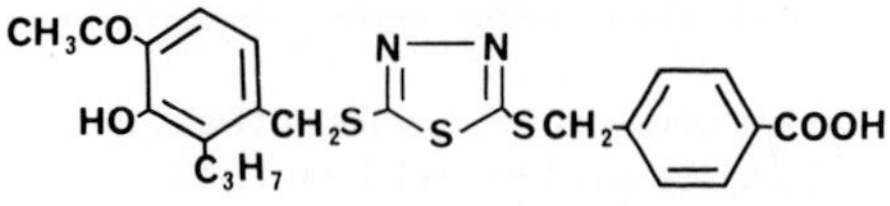

Fig. 1. Chemical structure of YM-17551.

compound was 20 min. The percentage inhibition was calculated by comparing the responses before and after addition of the test compound. The IC_{50} values were calculated by the Probit method.

LTD_4-Induced Skin Reaction in Guinea Pigs

Male Hartley guinea pigs weighing 270–310 g, starved for 16 h, were injected intradermally with 5 ng LTD_4 in a volume of 0.01 ml into two sites on the shaved back. In addition, 0.1 ml of vehicle was injected intradermally in each animal to see nonspecific irritation. One milliliter of saline containing 1% Evans blue was injected intravenously 2 min before LTD_4 injection. Thirty minutes later the animals were sacrificed by decapitation. The dye leaked at the site of LTD_4 or vehicle injection was extracted according to Harada et al. [9] and measured photometrically at 620 nm. The LTD_4-induced skin reaction was expressed as a difference in the amount of dye that leaked at the injection site of LTD_4 and vehicle. The test compound, emulsified in 0.5% methylcellulose (MC), was administered orally 15 min to 24 h before LTD_4 injection.

Antigen-Induced and LTs-Mediated Asthma in Guinea Pigs

Male Hartley guinea pigs weighing 370–420 g were passively sensitized by intravenously injecting 1 ml/kg of rabbit antibovine serum albumin (BSA) serum (PHA titer: 20480) [10]. Twenty-four hours after the sensitization the animals were pretreated with indomethacin (2 mg/kg), mepyramine (2 mg/kg) and propranolol (0.3 mg/kg), 20, 5 and 5 min, respectively, prior to antigen challenge by injecting these drugs into the saphenous vein. The animals were then placed in an 11-liter chamber connected to a glass nebulizer (KG-20, Kinoshita), and 1% solution of BSA was sprayed into the chamber for 30 s. The animals were exposed to the antigen aerosol for 2 min and observed for 20 min after challenge. The time from the start of the inhalation to cause cough and the mortality of animals were recorded. The test compound, emulsified in 0.5% MC, was administered orally 15 min to 24 h before antigen challenge.

Results

Isolated Guinea Pig Ileum. As shown in figure 2, YM-17551 dose dependently inhibited the LTC_4 (10^{-9} M)-, LTD_4 (10^{-9} M)- and LTE_4 (3×10^{-8} M)-induced ileal contractions and its IC_{50} values were 1.3×10^{-7}, 1.1×10^{-8} and 5.9×10^{-8} M, respectively. However, the compound at 10^{-5} M showed only 12–46% inhibition of ileal contractions induced by histamine (10^{-6} M), acetylcholine (5×10^{-7} M), 5-hydroxytryptamine (3×10^{-6} M), PGE_2 (3×10^{-6} M) and $PGF_{2\alpha}$ (10^{-5} M) (fig. 2).

Isolated Guinea Pig Trachea. YM-17551 inhibited the tracheal contractions induced by LTC_4 (10^{-8} M), LTD_4 (10^{-8} M) and LTE_4 (10^{-7} M)

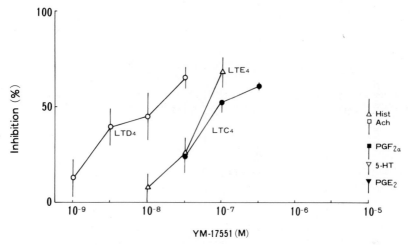

Fig. 2. Effect of YM-17551 on the contractions induced by LTC_4, LTD_4, LTE_4, histamine (Hist), acetylcholine (Ach), 5-hydroxytryptamine (5-HT), PGE_2 and $PGF_{2\alpha}$ in isolated guinea pig ileum. Each point represents the mean $\pm$ SE of 3–4 experiments.

with IC_{50} values of 1.1×10^{-8}, 8.4×10^{-9} and 4.8×10^{-8} M, respectively (fig. 3).

Isolated Human Bronchi. YM-17551 at doses of 3×10^{-9} to 10^{-7} M dose dependently inhibited LTD_4 (10^{-9} M)-induced contraction of human bronchi and its IC_{50} value was 1.7×10^{-8} M (data not shown).

LTD_4-Induced Skin Reaction in Conscious Guinea Pigs. YM-17551 at doses between 0.3 and 10 mg/kg p.o. showed a dose-dependent inhibition of LTD_4 (5 ng/site)-induced skin reaction (fig. 4). When the anti-LTD_4 effect of YM-17551 was plotted as a function of time, the drug action was evidenced within 15 min of treatment, and reached the peak effect at 1 h and ED_{50} value was 1.9 mg/kg p.o. Duration of anti-LTD_4 effect of YM-17551 was prolonged as the dose increased from 0.3 to 10 mg/kg p.o. Namely, the effect of YM-17551 at 0.3 mg/kg returned to the control level at 2 h, but the compound at the doses of 1, 3 and 10 mg/kg p.o. showed 28, 38 and 55% inhibition, respectively, even 16 h after treatment.

Antigen-Induced and LTs-Mediated Asthma in Conscious Guinea Pigs. Guinea pigs in the control group exhibited cough about 250 s following inhalation of antigen and 4/8 to 7/8 of animals died within 20 min (table 1). YM-17551 at doses above 0.3 mg/kg p.o. significantly prolonged the time to cause cough and tended to reduce the mortality of animals (table 1). In

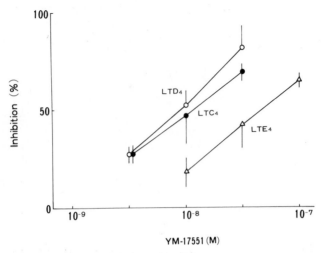

Fig. 3. Anti-LTC$_4$, LTD$_4$ and LTE$_4$ effect of YM-17551 in isolated guinea pig trachea. Each point represents the mean ± SE of 3 experiments.

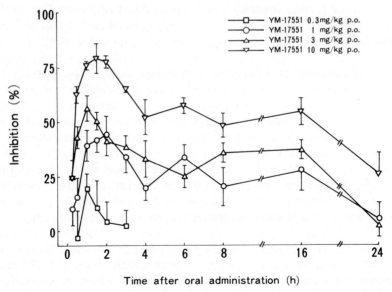

Fig. 4. Time course of the inhibitory effect of YM-17551 on LTD$_4$-induced skin reaction in conscious guinea pigs. Each point represents the mean ± SE of 5–10 experiments.

Table 1. Effect of YM-17551 on antigen-induced and LTs-mediated asthma in conscious guinea pigs

Compound[1]	Dose mg/kg p.o.	n	Time to onset of cough, s[2]	Mortality
Control	–	8	256 ± 11	4/8
YM-17551	0.1	8	274 ± 5	4/8
Control	–	8	267 ± 11	7/8
YM-17551	0.3	8	$324 \pm 19*$	5/8
Control	–	8	256 ± 10	4/8
YM-17551	1.0	8	$368 \pm 41*$	2/8
Control	–	8	231 ± 12	5/8
YM-17551	3.0	8	$396 \pm 48*$	2/8
Control	–	8	249 ± 10	5/8
YM-17551	10.0	7	$528 \pm 40***$	2/7

* $p < 0.05$, *** $p < 0.001$; significantly differed from the value of the control group.
[1] YM-17551 was administered orally 1 h before antigen challenge.
[2] When animals did not cough over 10 min after antigen challenge, time was calculated as 600 s.

order to see the duration of antianaphylactic effect of YM-17551, the compound (10 mg/kg p.o.) was administered 15 min to 24 h before antigen challenge. As shown in figure 5, YM-17551 significantly inhibited the antigen-induced and LTs-mediated asthma in guinea pigs from 30 min to 16 h after administration.

Discussion

The results of this study indicate that YM-17551 possesses the profile of an LT antagonist in several in vitro and in vivo models. YM-17551 produced a dose-dependent inhibition of contractions induced by LTC_4, LTD_4 and LTE_4 in guinea pig ileum (fig. 2), guinea pig trachea (fig. 3) and human bronchi. IC_{50} values of anti-LTC_4, LTD_4 and LTE_4 effects of YM-17551 were 8.4×10^{-9} to 1.3×10^{-7} M. Furthermore, YM-17551 produced a dose-dependent inhibition of $[^3H]LTD_4$ binding to the guinea pig lung membranes determined according to Tomioka et al. [11] and its pKi value was 6.81 [unpubl. data]. The potency of inhibitory effect of

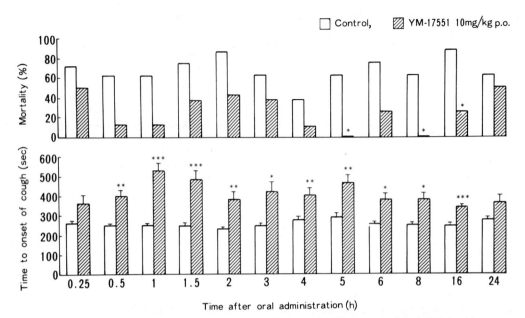

Fig. 5. Time course of the inhibitory effect of YM-17551 on antigen-induced and LTs-mediated asthma in conscious guinea pigs. Each data point represents the mean of 8 animals. *p < 0.05, **p < 0.01, ***p < 0.001; significantly different from the value of the control group by Student's t test or Fisher's exact test.

YM-17551 on LTs-induced contractions in isolated tissues and [³H]LTD₄ binding to the guinea pig lung membranes was about 10 times larger than those of YM-16638 and LY171883 [unpubl. data]. The selectivity of YM-17551 toward LTs was indicated by its inability to antagonize the action of other contractile agents such as histamine, acetylcholine, 5-hydroxytryptamine, PGE_2 and $PGF_{2\alpha}$ (fig. 2). Thus, YM-17551 is a selective and potent sulfidopeptide LT antagonist. However, it must be kept in mind that LTC_4 receptors are distinct from those of LTD_4 and LTE_4 in guinea pig trachea [12], but the receptors mediating LTC_4- and LTD_4-induced contractions of human airways are similar [13].

YM-17551 at doses of 0.3–10 mg/kg p.o. showed a dose-dependent inhibition of LTD_4-induced skin reaction (fig. 4) and antigen-induced and LTs-mediated asthma (table 1) in guinea pigs. Inhibitory effect of YM-17551 after 10 mg/kg p.o. persisted more than 16 h in these guinea pig models (fig. 4, 5), indicating that the action of YM-17551 is long-lasting. In fact, the plasma half-life of YM-17551 after 10 mg/kg p.o. was 8.6 h in guinea pigs [unpubl. data].

In summary, YM-17551 is a selective, potent, long-lasting and orally active LT antagonist. Therefore, YM-17551 could be used in a variety of pathological conditions associated with excessive production of LTs.

References

1 Samuelsson, B.: Leukotrienes: mediators of immediate hypersensitivity reactions and inflammation. Science 220: 568–575 (1983).

2 Piper, P.J.: Formation and actions of leukotrienes. Physiol. Rev. 64: 744–761 (1984).

3 Goetzl, E.J.; Payan, D.G.; Goldman, D.W.: Immunopathogenetic roles of leukotrienes in human diseases. J. Clin. Immunol. 4: 79–84 (1984).

4 Ford-Hutchinson, A.W.: Leukotrienes: their formation and role as inflammatory mediators. Fed. Proc. 44: 25–29 (1985).

5 Whittle, B.J.R.; Oren-Walman, N.; Guth, P.H.: Gastric vasoconstrictor actions of leukotriene C_4, $PGF_{2\alpha}$, and thromboxane mimetic U-46619 on rat submucosal mircrocirculation in vitro. Am. J. Physiol. 248: G580–G586 (1985).

6 Tomioka, K.; Yamada, T.; Mase, T.; Hara, H.; Murase, K.: Pharmacological properties of the orally active leukotriene antagonist [[5-[[3-(4-acetyl-3-hydroxy-2-propylphenoxy)propyl]thio]-1,3,4-thiadiazol-2-yl]thiol]acetic acid. Arzneimittelforschung 38: 682–685 (1988).

7 Tomioka, K.; Garrido, R.; Stevenson, J.S.; Abraham, W.M.: The effect of an orally active leukotriene (LT) antagonist YM-16638 on antigen-induced early and late airway responses in allergic sheep. Prostaglandins Leukotrienes Essent. Fatty Acid (in press).

8 Constantine, J.W.: The spirally cut tracheal strip preparation. J. Pharm. Pharmacol. 17: 384–385 (1965).

9 Harada, M.; Takeuchi, M.; Fukao, T.; Katagiri, K.: A simple method for the quantitative extraction of dye extravasated into the skin. J. Pharm. Pharmacol. 23: 218–219 (1971).

10 Tomioka, K.; Yamada, T.; Ida, H.: Anti-allergic activities of the β-adrenoceptor stimulant formoterol (BD40A). Arch. Int. Pharmacodyn. Ther. 250: 279–292 (1981).

11 Tomioka, K.; Yamada, T.; Teramura, K.; Terai, M.; Hidaka, K.; Mase, T.; Hara, H.; Murase, K.: Isolated tissue and binding studies of YM-17690, a novel and non-analogous leukotriene agonist. J. Pharm. Pharmacol. 39: 819–824 (1987).

12 Synder, D.W.; Krell, R.D.: Pharmacological evidence for a distinct leukotriene C_4 receptor in guinea-pig trachea. J. Pharmacol. Exp. Ther. 231: 616–622 (1984).

13 Buckner, C.K.; Krell, R.D.; Laravuso, R.B.; Coursin, D.B.; Bernstein, P.R.; Will, J.A.: Pharmacological evidence that human intralobar airways do not contain different receptors that mediate contractions to leukotriene C_4 and leukotriene D_4. J. Pharmacol. Exp. Ther. 237: 558–562 (1986).

Kenichi Tomioka, PhD, Department of Pharmacology, Central Research Laboratories, Yamanouchi Pharmaceutical Co., Ltd., 1-1-8 Azusawa, Itabashi-ku, Tokyo 174 (Japan)

Zor U, Naor Z, Danon A (eds): Leukotrienes and Prostanoids in Health and Disease.
New Trends Lipid Mediators Res. Basel, Karger, 1989, vol 3, pp 84–88

Effects of the Leukotriene B$_4$ Antagonist U-75,302 on Guinea Pig Eosinophil Chemotaxis and Activation in vitro and in vivo

Frank F. Sun, Bruce M. Taylor, N.J. Crittenden, C.I. Czuk,
J.A. Oostveen, I.M. Richards

Department of Hypersensitivity Diseases Research, The Upjohn Company,
Kalamazoo, Mich., USA

Introduction

Leukotriene B$_4$ (LTB$_4$) has multiple effects on polymorphonuclear leukocyte functions. It stimulates leukocytes to aggregate, to migrate, to adhere onto natural or artificial substrates, to release granular enzymes and to generate superoxide anion [1]. LTB$_4$ has been detected in the exudate of several sites of inflammation at concentrations sufficient to alter leukocyte functions suggesting this compound may serve as a mediator in the cellular phase of the inflammatory response. Therefore, inhibitors of LTB$_4$ action may be useful in treating symptoms associated with inflammatory cell infiltration such as that encountered in asthma, inflammatory bowel diseases or skin diseases.

Lin et al. [2] reported that a pair of pyridine analogues of LTB$_4$, U-75,302 and U-75,485, inhibited specific binding of radiolabelled LTB$_4$ to receptors on human neutrophils as well as LTB$_4$-induced human neutrophil aggregation. These preliminary results suggested that the two compounds may be LTB$_4$ antagonists at the receptor level. In this study, we examined the effects of U-75,302 and U-75,485 on several neutrophil and eosinophil function in vitro and in antigen-induced eosinophilia in sensitized guinea pigs in vivo.

Materials and Methods

Guinea pig eosinophils were isolated from horse serum-treated guinea pigs by peritoneal lavage. The cells were purified by centrifugation over a discontinuous gradient of Percoll with density from 1.050 to 1.100. Cells sedimenting out at the 1.100 and 1.090 density layers were essentially pure eosinophils. Viability was greater than 95% by trypan blue exclusion.

Guinea pig eosinophil chemotaxis was assessed using a modified Boyden chamber technique [3] and cells labelled with ^{51}Cr. The chemoattractants were added to the bottom compartment and the inhibitors were added to both the top and the bottom compartments. The assembly was incubated at 37 °C for 2 h and the radioactive cells migrating into the lower membrane were quantitated by counting.

Generation of superoxide anion by guinea pig eosinophils was measured by the reduction of cytochrome c. The cells were incubated with 5 μg/ml cytochalasin b in the presence or absence of the inhibitor for 5 min before stimulated with appropriate agonist. The baseline absorbance was determined with the addition of superoxide dismutase.

Human neutrophils were isolated from citrated venous blood of normal donors according to standard techniques. Neutrophil aggregation was assessed in a Payton Dual Channel Aggregometer according to the method of Hammerschmidt et al. [4]. The cells were treated with cytochalasin B (5 μg/ml) for 3 min. The inhibitor or vehicle was subsequently added to obtain a baseline response before addition of the agonists to start the reaction.

In vivo experiments were performed with male Hartley guinea pigs sensitized to ovalbumin according to previously reported methods [5]. Groups of 6 animals were pretreated with pyrilamine maleate, 2 mg/ml, and challenged with 1.0% w/v ovalbumin aerosol administered with a face mask for 1 min. After 24 h following antigen challenge, bronchoalveolar lavage was performed on these animals with 5 ml of warm phosphate-buffered saline twice. The lavage fluids were combined and total and differential cell counts were performed on the cells recovered from the fluid. The percentage of eosinophils in lavage samples was used as the index of bronchoalveolar eosinophilia in vivo. U-75,302 was administered orally to animals at doses of 1.0, 10.0 and 30.0 mg/kg at 1 h before and 7 h after antigen challenge.

Results and Discussion

Five active agonists, LTB$_4$, fMLP, PAF, human recombinant C5a and PMA, were tested for their activity in inducing human neutrophil aggregation and guinea pig eosinophil superoxide generation and chemotaxis. For human neutrophils, fMLP, LTB$_4$ and PAF dose dependently induced aggregation responses. LTB$_4$, PAF, C5a and PMA dose dependently and consistently induced superoxide generation by guinea pig eosinophils. fMLP was effective only in some animals. LTB$_4$ and C5a were shown to be active chemoattractants for guinea pig eosinophils. fMLP activity was inconsistent, and PAF demonstrated no activity in all experiments. Therefore, only the agonists which actively induced responses were used to test the effects of the antagonists U-75,302 and U-75,485.

When cytochalasin B-treated human neutrophils were stimulated with fMLP (0.1 μM) or LTB$_4$ (1 μM), a rapid onset of aggregation occurred and the response was completed within 30 s. When the cells were treated with U-75,302 or U-75,485 at 1–100 μM, the drugs induced a slow aggregation response which lasted several minutes. Further addition of fMLP still induced full responses quantitatively similar to those elicited by the control cells without the inhibitors. However, the addition of LTB$_4$ to the drug-treated cells induced much smaller responses than the control and the

Table 1. EC_{50} (μM) of U-75,302 and U-75,485 against leukocyte functions in vitro

	U-75,302	U-75,485
Human PMN aggregation[1]	0.75 ± 0.40 (5)	0.17 ± 0.12 (3)
Guinea pig eosinophils Superoxide generation[1]	0.17 ± 0.07 (5)	0.13 ± 0.1 (2)
Guinea pig eosinophils Chemotaxis[2]	11.5 ± 5.5 (5)	5.4 ± 2.3 (3)

[1] Against 1 μM of LTB$_4$-induced response.
[2] Against 0.02 μM of LTB$_4$-induced response.

decreases in LTB$_4$ response was proportional to the concentrations of the drug added. The results clearly showed that both U-75,302 and U-75,485 are weak agonists that stimulate neutrophil aggregation be themselves. However, the two drugs possess significant inhibitory activity specifically against LTB$_4$-induced response. The EC_{50} are summarized in table 1.

In a similar manner, LTB$_4$ (1 μM), PAF (1 μM), C5a (60 nM) or PMA (1 nM) stimulated cytochalasin B-treated guinea pig eosinophils to generate maximal or near maximal levels of superoxide anion. When the cells were pretreated with either U-75,302 or U-75,485 (0.1–10 μM), the responses induced by PAF, C5a, or PMA were either unaffected or slightly elevated. The overall response induced by LTB$_4$ was suppressed by 25–30%. However, both inhibitors were active agonists when added to the cells, generating superoxide anion dose dependently by themselves. If the agonist activities were subtracted from the overall response, both compounds now appeared to be potent inhibitors of LTB$_4$-induced effects in a dose-dependent and highly correlated manner (table 1). Even with the correction for agonist activity, the drugs did not significantly affect the responses induced by the other stimuli.

In order to determine if the in vivo inhibition of guinea pig eosinophil migration into the airway by U-75,302 could be correlated with in vitro effect of the drug on chemotaxis, we investigated the effects of U-75,302 and U-75,485 on guinea pig eosinophil chemotaxis using a modified Boyden chamber technique. Eosinophils were induced to migrate using 20 nM of LTB$_4$ or a 1:10 dilution of zymosan-activated guinea pig plasma in the lower compartment of the apparatus. When U-75,302 or U-75,485 were present in both the upper and lower compartments, the LTB$_4$-induced chemotaxis of guinea pig eosinophils was dose dependently inhibited. The EC_{50} for U-75,485 (5.4 ± 2.3 μM) indicates that it was slightly more effective than U-75,302 (EC_{50} = 11.5 ± 5.5 μM). The specificity of U-75,302 and U-75,485 was demonstrated by their inability to block zymosan-activated plasma-induced eosinophil chemotaxis at 10 μM. However, at 100 μM, the compounds appeared to be toxic to the cells as both LTB$_4$ and zymosan-activated plasma-induced responses were inhibited almost completely.

It should be noted that both U-75,302 and U-75,485 are chemoattractants for guinea pig eosinophils in vitro if they were added into the lower compartment by themselves. However, when they were presented to the cells in the upper compartment first, the chemotactic effect was negated and the compounds behaved as typical antagonists.

Dunn et al. [5] have previously demonstrated that inhalation of antigen in sensitized guinea pig induced a massive accumulation of eosinophils into the lungs and airways. The eosinophilia occurs at 18–24 h after antigen challenge and may be similar to the late-phase airway eosinophilia observed in some asthmatic patients. In control, unsensitized guinea pigs challenged with antigen, bronchoalveolar lavage revealed only 8.2 ± 1.9% of the total cells recovered were eosinophils. If the animals were sensitized with ovalbumin for 5 weeks, the eosinophil count in the post-challenged lavage fluid was elevated 3.3-fold to 27.8 ± 4.6%. In sensitized guinea pigs pretreated with U-75,302 orally at the doses of 1.0, 10.0 and 30.0 mg/kg 1 h before and 7 h after antigen inhalation, the percentage of eosinophils in the lavaged cells was markedly and dose dependently inhibited, approaching the nonsensitized, control level at the 30.0 mg/kg dose. The inhibition was statistically significant (p < 0.05) at the 10.0 and 30.0 mg/kg dose levels. If the animals were treated only once at 1 h after antigen inhalation, the inhibition of the increased eosinophil counts in lavage fluid reached a significant level only at the 30.0 mg/kg dose. U-75,485, administered orally at the dose of 30 mg/kg to sensitized guinea pigs, also effectively inhibited antigen-induced eosinophil accumulation in the bronchoalveolar lavage. The neutrophil counts in lavage fluid were not affected by treatment with either U-75,302 or U-75,485.

Our data demonstrated that the two pyridine analogs of LTB$_4$, U-75,302 and U-75,485 possess both partial agonist and antagonist effects against LTB$_4$-induced leukocyte functions including aggregation, superoxide anion generation and chemotaxis. Although the overall effect against LTB$_4$-induced human neutrophil aggregation and guinea pig eosinophil superoxide anion generation was partially obscured by the agonist effect of these compounds, the effect of these compounds against in vitro and in vivo guinea pig eosinophil chemotaxis was highly significant. Due to the presence of this agonist activity, it is difficult to rule out the possibility that U-75,302 or U-75,485 may be acting, in part, through desensitization. Moreover, the inhibitory activities of the two compounds were dose-dependent and stimulus-specific suggesting that their action is at the receptor level. The effectiveness of these compounds in inhibiting the antigen-induced late-phase bronchopulmonary accumulation of eosinophils in allergic guinea pigs not only indicates that LTB_4 may be an obligatory mediator in late-phase airway inflammation, but also suggests that LTB_4 inhibitors or antagonists may be potentially useful in treating chronic airway inflammation associated with asthma or other hypersensitivity diseases.

References

1   Hansson, G.; Malmsten, C.; Radmark, O.: The leukotrienes and other lipoxygenase products; in Pace-Asciak, C.R.; Granström, E. (eds): Prostaglandins and Related Substances, pp. 127–169 (Eslevier, Amsterdam 1983).

2   Lin, A. H.; Morris, J.; Wishka, D.G.; Gorman, R.R.: Novel molecules that antagonize leukotriene B_4 binding to neutrophils. Ann. N.Y. Acad. Sci. *524:* 196–200 (1988).

3   Gallin, J.I.; Clark, R.A.; Kimball, H.R.: Granulocyte chemotaxis: An improved in vitro assay employing ^{51}Cr-labelled granulocytes. J. Immunol. *110:* 233–240 (1973).

4   Hammerschmidt, P.E.; Bowers, T.K.; Kammi-Kepfe, C.J.; Jacob, H.S.; Craddock, P.R.: Granulocyte aggregometry: A sensitive technique for the detection of C5a and complement activation. Blood *55:* 898–912 (1980).

5   Dunn, C.J.; Elliott, G.A.; Oostveen, J.A.; Richards, I.M.: Development of a prolonged eosinophil-rich inflammatory leukocyte infiltration in the guinea-pig asthmatic response to ovalbumin inhalation. Am. Rev. Respir. Dis. *137:* 541–547 (1988).

Frank F. Sun, PhD, Department of Hypersensitivity Diseases Research,
The Upjohn Company, Kalamazoo, MI 49001 (USA)

Zor U, Naor Z, Danon A (eds): Leukotrienes and Prostanoids in Health and Disease.
New Trends Lipid Mediators Res. Basel, Karger, 1989, vol 3, pp 89–93

Studies with SQ 30,741:
A Thromboxane Receptor Antagonist

Martin L. Ogletree

Pharmacology Department, Squibb Institute for Medical Research,
Princeton, N.J., USA

Evidence continues to accumulate implicating thromboxane (Tx) A_2,
and especially TxA_2 receptors, in circulatory disorders. The potent vasocon-
strictive and platelet aggregating activities of TxA_2 appear to serve modulatory and perhaps nonessential roles in physiological hemostais. However, in
pathophysiological thrombosis and vasospasm, TxA_2 receptor stimulation
may play a pivotal mediating role [1]. Although aspirin-like drugs inhibit
TxA_2 synthesis, and Tx synthase inhibitors may be highly selective in
blocking TxA_2 production, TxA_2 receptor antagonists (TRAs) are receiving
close attention due to advantages in their specificity, potency and quality of
activity. The special quality of TRAs derives from blocking the activation
of TxA_2 receptors, which may be stimulated by substances other than TxA_2,
without redirecting arachidonic acid and prostaglandin metabolism, which
occurs with inhibitors of TxA_2 synthesis [1]. Several TRAs are currently
under investigation in human beings. SQ 30,741 [1S-[1α,2α(5Z),3α,4α]]-7-[3-
[[[[(1-oxoheptyl)amino]acetyl]amino]methyl]-7-oxabicyclo[2.2.1]hept-2yl]-
5-heptenoic acid, is a TRA [2, 3] now entering phase II clinical trials in the
United States. Its structure is shown in figure 1, and its synthesis has been
detailed previously [4]. A preliminary report described its pharmacokinetics
and pharmacodynamics in normal human volunteers [5]. This report will
describe preclinical studies with SQ 30,741 which support its use for
prevention and treatment of cardiovascular disorders.

Platelets have abundant TxA_2 receptors which participate in platelet
activation and aggregation and in release of platelet granule contents. SQ
30,741 displayed potent inhibition of human platelet aggregation responses
to arachidonic acid (800 μM), U-46,619 (10 μM), and collagen (10 $\mu g/ml$)
in platelet-rich plasma, with IC_{50} values of 0.2, 0.2 and 0.2 μM, respectively
[2, 6]. Like other TRAs, it did not inhibit platelet aggregation in response
to ADP (20 μM), nor the primary aggregation response to epinephrine

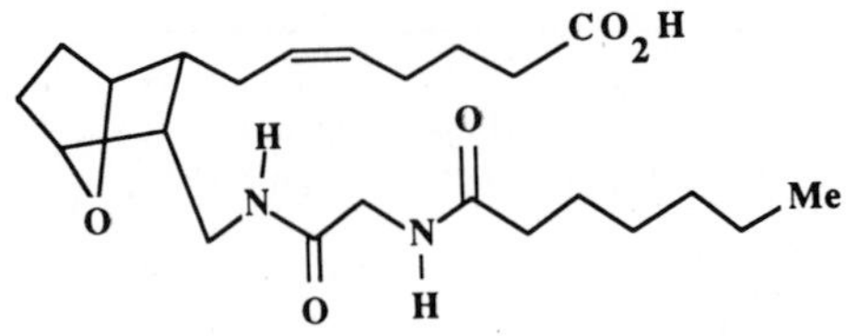

Fig. 1. Structure of SQ 30,741.

(10 μM). However, it inhibited the secondary phase of epinephrine-induced aggregation with an IC_{50} of 0.04 μM. SQ 30,741 competitively antagonized U-46,619-induced platelet aggregation in human platelet-rich plasma with a drug-receptor dissociation constant (K_B) of 54 nM [6]. SQ 30,741 also competitively inhibited specific binding of the TxA_2 receptor radioligand [5,6-^{3}H]SQ 29,548 to human washed platelets and platelet membranes with K_d values of 28 and 50 nM, respectively [3]. K_B values for competitive antagonism by SQ 30,741 of U-46,619-induced contraction of rat aortas and portal veins were 9 and 17 nM, respecitvely, and those of guinea pig aortas and portal veins were 15 and 3 nM, respectively [7]. Thus, SQ 30,741 is a potent, competitive antagonist of TxA_2 receptor-mediated platelet aggregation and vasoconstriction in vitro.

We evaluated the relevance of SQ 30,741's antiplatelet profile in a model of thrombotic renal artery occlusion in anesthetized cynomolgus monkey [6]. Crush injury at a site of renal artery stenosis initiated a pattern of thrombotic renal blood flow reduction which reliably progressed to occlusion unless the thrombus was mechanically disrupted. In 11 (79%) of 14 monkeys, a 1 mg/kg i.v. dose of SQ 30,741 inhibited the thrombotic reductions in blood flow for an average of 224 min. The threshold dose of SQ 30,741 required to interrupt the progression to occlusive thrombosis was 0.2 mg/kg. Additional experiments with this model revealed that the threshold antithrombotic dose of SQ 30,741 antagonized approximately 90% of both platelet (ex vivo) and blood vessel (in vivo) TxA_2 receptors [8]. Since the antithrombotic activity of the minimally effective dose persisted for an average of 109 min, we estimate that antagonism of only 50% of TxA_2 receptors is sufficient to maintain an established antithrombotic state. These studies indicate that SQ 30,741 can reversibly inhibit arterial thrombosis in primates.

The finding from Lefer's group [9] that SQ 29,548, a TRA related to SQ 30,741, exhibited cardioprotective properties in a model of myocardial ischemic injury, led to extensive studies by Grover and Schumacher [10–12] with SQ 30,741 in models of myocardial ischemia and reperfusion injury in dogs. The basic model they employed involved a 90-min occlusion of the left circumflex coronary artery in anesthetized, open-chest dogs, followed by 5 h of reperfusion. Myocardial infarct size and areas at risk were determined by triphenyltetrazolium staining and patent blue violet dye exclusion, respectively. In a preliminary study with this model, infarct size (expressed as percent of the left ventricular area at risk) was reduced 41% by SQ 29,548 compared to vehicle [10]. Areas at risk were the same in the two groups. Studies in this model with SQ 30,741 infusion beginning 10 min after coronary occlusion have consistently shown >40% reductions in myocardial infarct size compared to vehicle treatment [11, 12]. When SQ 30,741 infusion was begun just 2 min before reperfusion, it reduced infarct size by 30% compared to vehicle treatment [12]. The dose of SQ 30,741 in these studies (1 mg/kg + 1 mg/kg/h, i.v.) antagonized 95% of the canine TxA_2 receptors responsible for U-46,619-induced platelet shape change ex vivo [11].

During myocardial ischemia, SQ 30,741 did not change collateral blood flow measured by radioactive microspheres, but after 1 h of reperfusion, SQ 30,741-treated dogs displayed significantly improved ischemic region subendocardial blood flow compared to comparable tissue in vehicle-treated dogs [11, 12]. SQ 29,548 produced similar results [10]. Also after 1 h of reperfusion, SQ 30,741 significantly improved coronary flow reserve in the ischemic region, measured both as reactive hyperemia and as flow in response to maximally dilating doses of adenosine [12]. These findings indicate that SQ 30,741 mitigates the 'no-reflow' phenomenon, but the physiological mechanism for this effect is not known. Since the anti-ischemic activities of both SQ 29,548 and SQ 30,741 are associated with preservation of subendocardial blood flow, TxA_2 receptor activation appears to be responsible for a damaging and preventable compromise of myocardial reflow during reperfusion. Moreover, since both SQ 29,548 and SQ 30,741 reduced myocardial infarct size when infused only during reperfusion [12, 13], 'reperfusion injury' appears to derive in part from TxA_2 receptor stimulation [13]. Interestingly, aspirin treatment did not alter ultimate myocardial infarct size compared to vehicle (63 and 60% of left ventricular area at risk, respectively), nor did aspirin improve subendocardial reflow after 1 h of reperfusion [12].

The anti-ischemic and antithrombotic properties of SQ 30,741 are particularly relevant in the setting of thrombolysis. Since restenosis occurs in a significant fraction of patients after thrombolytic therapy, Schumacher and Heran [14, 15] sought to evaluate whether SQ 30,741 would improve the rate of development, extent and duration of thrombolytic reflow in a primate. They developed a model in the cynomolgus monkey carotid artery, in which an occlusive thrombus was lysed by close arterial infusion of streptokinase (40,000 U over 1 h) in the presence of heparin and either SQ 30,741 (2 mg/kg + 2 mg/kg/h, i.v.) or its vehicle. SQ 30,741 did not accelerate streptokinase-induced reflow, but it enhanced the average rates of flow by 60 and 159%, respectively, during the second and third hours after beginning the streptokinase treatment. In vehicle-treated monkeys, reflow consistently decreased between the second and thrid hours after treatment, but enhanced blood flow was sustained by SQ 30,741. Schumacher and Heran [15] concluded that although TxA$_2$ antagonism may not accelerate thrombolysis in a primate, it 'can increase and sustain reflow once thrombolysis is established'.

Clinical safety studies with SQ 30,741 in healthy human volunteers have shown it to act as a TRA [5]. It inhibited U-46,619-induced platelet aggregation ex vivo in a dose- and time-dependent manner. At the 100-mg dose, platelet aggregation was measurably inhibited for 8 h. As expected, SQ 30,741 also slightly prolonged template bleeding time.

To summarize: SQ 30,741 is a drug candidate TRA. It is a potent inhibitor of TxA$_2$ receptor-dependent platelet function and vasoconstriction. Occlusive thrombosis at a site of arterial injury and stenosis could be blocked by SQ 30,741 in a majority of monkeys. The surprising anti-ischemic activity of SQ 30,741 in dogs has been associated with inhibition of reperfusion injury and of the no-reflow phenomenon. The antithrombotic and anti-ischemic profile of SQ 30,741 is particularly attractive for potential benefit in conjunction with thrombolytic therapy after acute myocardial infarction.

Acknowledgements

The author thanks Drs. William A. Schumacher and Gary J. Gover for reviewing the manuscript before its submission for publication and for maintaining highly productive and innovative scientific programs in support of SQ 30,741.

References

1   Ogletree ML: Overview of the physiological and pathophysiological effects of thromboxane A$_2$. Fed Proc. 1987;46:133–138.
2   Ogletree ML, Harris DN, Hedberg A, et al: SQ 30,741, a selective TxA$_2$ receptor antagonist in vitro. Pharmacologist 1986;28:186.
3   Hedberg A, Hall SE, Ogletree ML, et al: Characterization of [5,6-^{3}H]SQ 29,548 as a high affinity radioligand, binding to thromboxane A$_2$/prostaglandin H$_2$-receptors in human platelets. J Pharmacol Exp Ther 1988;245:786–792.
4   Harris DN, Hall Se, Hedberg A, et al; 7-Oxabicycloheptane analogs: Modulators of the arachidonate cascade. Drugs Future 1988;13:153–169.
5   Friedhoff LT, Fry J, Ivashkiv E, et al: Pharmacokinetics and pharmacodynamics of a thromboxane A$_2$ receptor antagonist. Clin Pharmacol Ther 1988;43:139.

6   Schumacher WA, Goldenberg HJ, Harris DN, et al: Effect of thromboxane receptor antagonists on renal artery thrombosis in the cynomolgus monkey. J Pharmacol Exp Ther 1987;243:460–466.

7   Ogletree ML, Allen GT: Interspecies, not intertissue, differences in thromboxane receptors: Studies with rat and guinea pig smooth muscles. Proc 2nd Conf on Leukotrienes and Prostanoids in Health and Disease, Jerusalem 1988, Abstract book, p 165.

8   Schumacher WA, Heran CL, Goldenberg HJ, et al: Magnitude of thromboxane receptor antagonism necessary for antithrombotic activity in monkeys. Am J Physiol, in press.

9   Hock CE, Brezinski ME, Lefer AM: Anti-ischemic actions of a new thromboxane receptor antagonist, SQ-29,548, in acute myocardial ischemia. Eur J Pharmacol 1986;122:213–219.

10   Grover GJ, Schumacher WA, Simon M, et al: Effect of the thromboxane antagonist SQ 29,548 on myocardial infarct size in dogs. J Cardiovasc Pharmacol 1988;11:29–35.

11   Grover GJ, Schumacher, WA, Simon M, et al: The effect of the thromboxane A_2/prostaglandin endoperoxide receptor antagonist SQ 30,741 on myocardial infarct size and blood flow during myocardial ischemia and reperfusion. J Cardiovasc Pharmacol 1988;12:701–709.

12   Grover GJ, Schumacher WA: Effect of the thromboxane A_2 receptor antagonist SQ 30,741 on ultimate myocardial infarct size, reperfusion injury, and coronary flow reserve. J Pharmacol Exp Ther 1989;248:484–491.

13   Grover GJ, Sleph PG, Parham CS: The role of thromboxane A_2 in reperfusion injury. Proc Soc Exp Biol Med 1988;188:504–508.

14   Schumacher WA, Heran CL: Effect of thromboxane antagonism on recanalization during streptokinase-induced thrombolysis in anesthetized monkeys. J Cardiovasc Pharmacol, in press.

15   Schumacher WA, Heran CL: Effect of thromboxane receptor antagonism on reflow during thrombolysis in monkeys. Circulation 1988;78:II-455.

Martin L. Ogletree, PhD, Section Head, Pharmacology Department,
Squibb Institute for Medical Research, Princeton, NJ 08543-4000 (USA)

Zor U, Naor Z, Danon A (eds): Leukotrienes and Prostanoids in Health and Disease.
New Trends Lipid Mediators Res. Basel, Karger, 1989, vol 3, pp 94–99

New Results in the Development of Prostaglandin Analogues

E. Schillinger

Research Laboratories of Schering AG, Berlin (West) and Bergkamen, FRG

Natural prostaglandins produce a wide variety of physiological responses in a great number of cell types and organs. In comparison with classic hormones, the range of activity is small and restricted to cells within the close vicinity. Cellular stimulation is only short-lasting due to the chemical and metabolic lability of the prostaglandins. Structural modification of PGE_2 and prostacyclin led to a number of new stable compounds with interesting pharmacological profiles (fig. 1).

Sulprostone has been characterized as a specific EP_1 receptor agonist [1] with potent contractile effects on smooth muscle cells in rat, guinea pig and human uterus [2]. The lack of $PGF_{2\alpha}$- and TXA_2-like properties resulted in a high tissue specificity which proved to be useful clinically in life-threatening conditions such as postpartum atonic uterine bleeding, hydatidiform mole and fetal death in advanced pregnancy.

Flunoprost carries a fluorine in the 9-position for chemical stabilization, the pharmacological profile is dominated by potent PGE_2 as well as $PGF_{2\alpha}$ properties. Receptor affinity in human uterine particles (PGE_2) and in rabbit corpora lutea ($PGF_{2\alpha}$) increased by a factor of 5 and 2, respectively. In abortifacient activity in pregnant guinea pigs it was 100 times more potent than PGE_2. In contrast to $PGF_{2\alpha}$, flunoprost exhibited potent bronchio- and venoconstricting properties. The constrictor response in isolated portal veins of the rat was several orders of magnitude higher than $PGF_{2\alpha}$. Venoconstriction mediated by flunoprost should be investigated further therapeutically in indications such as nasal decongestion and allergic rhinitis.

A further compound which retains the 9β-halogen to provide chemical stability is nocloprost which exhibits potent antiulcer and cytoprotective activity. The particular advantage of nocloprost compared with other PGE_2 analogues under clinical investigation stems from its very low bioavailability when given orally. Kinetic studies in animals and man showed that after

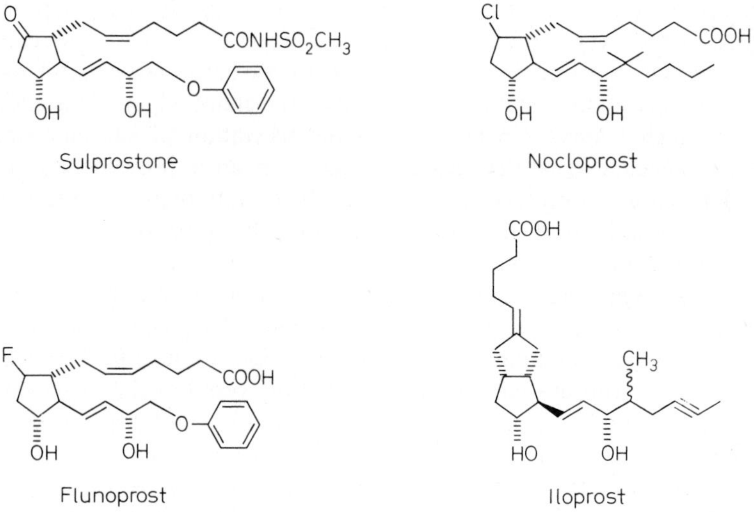

Fig. 1. Stable prostaglandin and prostacyclin analogues.

oral application, less than 5% of the total dose appeared in the blood-stream. The predominant proportion is extensively metabolized in the gut and in the liver which restricts the pharmacological effect of nocloprost to the stomach and the upper part of the intestine. Effective doses for antiulcer activity in the rat are in the range of about 10 ng/kg p.o. for nocloprost as well as for the standard compound 16,16-dimethyl-PGE$_2$. Diarrheal activity, however, was observed with 1,600 μg/kg of nocloprost (vs. 20 μg/kg of 16,16-dimethyl-PGE$_2$), which shifts the therapeutic index up by a factor of about 80 [3]. Phase I studies in healthy volunteers have been completed, and clinical trials will prove its therapeutic value.

An area of more recent clinical interest was established through the use of prostacyclin in situations of thromboembolic disorders. Prostacyclin is for the most part formed in vascular endothelial cells from membrane phospholipids. Upon stimulation, specific receptors mediate an increase in intracellular cAMP causing a multitude of different biological effects. The most prominent properties are the inhibition of platelet aggregation and the relaxation of smooth muscle cells resulting in vasodilatation. Cytoprotective and organ-protecting effects of prostacyclin are less well understood. We have concentrated on a chemically stable prostacyclin analogue, ilo-prost, with a biological profile very similar to the natural prostacyclin [4]. Iloprost inhibits the aggregation of platelets and the secretion of procoagulant activity from the thrombocytes such as ADP and serotonin and indirectly interferes with the generation of TXA$_2$. Iloprost inhibits an intricate interplay with endothelial cells, adhesion, aggregation and probably migration and oxygen radical production of white blood cells. The rescue of blood vessels from noxious factors such as mediator release and radical formation maintains the barrier function of the endothelium. This effect on the endothelial lining – in addition to the relaxation of vascular smooth muscle – appears to be the principal reason for an improvement in organ blood flow, for cytoprotective and eventually for antiatherosclerotic activity.

In vivo inhibition of platelet aggregation by iloprost was assessed in electrically damaged arterioles from the mesenteric region of anesthetized rats. Under light microscopy, an electrode was placed on an arteriole and transient formation of a white body was induced by endothelial damage through electrical stimulation. After topical application of 0.3 μM ADP, a firm platelet thrombus completely occluded the vessel at the damaged site. After infusion of prostacyclin or iloprost, higher concentrations of ADP were required to cause formation of platelet thrombi than under control conditions. An effective dose to reopen a vessel at a 10-fold higher concentration of ADP was obtained at 0.64 $\mu g/kg \cdot min$ for PGI_2 and at 0.25 $\mu g/kg \cdot min$ for iloprost [5].

In a further model, microvascular and permeability effects of iloprost were assessed in the Syrian hamster check pouch preparation using intra-vital microscopy with video-imaging and contrast enhancement with intra-venous FITC-labeled Dextran 70. At a nonhypotensive dose of iloprost a dilatation of arterioles and veins was observed together with an increase in the density of perfused capillaries. Local application of histamine, serotonin or bradykinin elicited an impressive increase of the venular leakage of fluorescein-labeled dextran, an effect which vanished after 20 min infusion of the vehicle only. When iloprost was infused prior to the stimulus, the leakage was almost totally prevented or markedly attenuated [6]. The beneficial effects of iloprost on the microvasculature may be explained by an improvement of tissue perfusion and by a functional antagonism of mediator-induced tissue edema and vasospasm.

The effects on the microvasculature taken together with the antiplatelet and vasodilating properties of prostacyclin and iloprost prompted the evaluation of the clinical usefulness of iloprost. The therapeutic efficacy was investigated in clinical studies in patients with peripheral arterial occlusive disease (PAOD), with diabetic angiopathy and with rest pain. Patients were selected according to age, sex and risk factors and were divided at random into either the placebo or the treatment group. The individually tolerated dose of iloprost was assessed in the first 3 days of treatment by a stepwise titration of the dose which was then held constant throughout the study. The dose never exceeded 2 ng/kg $\cdot$ min and lasted for 6 h a day for a full length of 28 or 14 days. At the end of the study the number of responders

Table 1. Clinical results with iloprost in severe limb ischemia

Efficacy criterion	Diagnosis	Iloprost			Placebo		
		resp.	nonresp.	total	resp.	nonresp.	total
Trophic lesions	PAOD (study 1)	32[1] (61.5)	20	52 (100)	8 (17)	39	47 (100)
		↑ └── p < 0.05 ──┘ ↑					
	diabet. angiop. (study 2)	31 (62)	19	50 (100)	12 (23.5)	39	51 (100)
		↑ └── p < 0.05 ──┘ ↑					
Ischemic rest pain (study 3)		30 (62.5)	18	48 (100)	23 (42.6)	31	54 (100)
		↑ └── p < 0.05 ──┘ ↑					
	PAOD	19 (63.3)	11	30 (100)	18 (53)	16	34 (100)
	diabet. angiop.	11 (61.1)	7	18 (100)	5 (25)	15	20 (100)
		↑ └── p<0.05 ──┘ ↑					

[1] Number and percentage (in parentheses) of patients.

was evaluated according to the patients who showed partial healing of large ulcerations at the end of 28 days' treatment or complete healing of ulcerations during or at the end of the treatment and according to patients free of pain without any analgesics for a minimum of 5 consecutive days including the 3 days after completion of the therapy (table 1).

Sixty-two percent of the patients with PAOD and diabetic angiopathy responded to the treatment with iloprost and showed significant healing of ulcers and/or relief of rest pain. About 20% of the patients with trophic lesions and 42% with rest pain responded to placebo. Intensive hospital care involving the treatment of concomitant diseases and a 2-weak infusion of saline solution causing hemodilution can explain the placebo response. In a 1 year follow-up study the long-term benefit for trophic lesions was determined. In 88% of the patients with PAOD who responded positively to iloprost, the clinical improvement was generally maintained. The prognosis for nonresponders, however, was not encouraging. Subsequent conventional therapies had been ineffective in these patients and within 4 months 75% had to be amputated. There was a long-term benefit for 59% of the diabetics who had shown an initial response to iloprost. However, the clinical condition in another 35% deteriorated and some of the patients had to undergo amputation during the follow-up treatment. Conventional therapies were apparently not able to stabilize the initially improved clinical status. A second treatment phase with iloprost should be initiated in this group of patients.

The efficacy of short-term applied intravenous iloprost (4 ng/kg · min for 3–5 days) was measured in patients with Raynaud's disease. They fulfilled clinical, serological and capillaroscopic criteria for systemic sclerosis. Healing of skin lesions and the subjective improvement in vasospastic pain attacks, that is reduction in frequency, duration and intensity, were considered positive responses.

Twenty-five of 37 patients showed healing of skin lesions. Pain attacks were reduced in 30 of 38 patients, while only 5 (21%) of 24 patients improved under placebo. Although treatment with iloprost lasted only for 3–5 days, clinical benefit was observed for approximately 6 weeks. Further controlled studies in Raynaud patients are therefore anticipated.

All patients treated with iloprost reported a similar pattern of side effects. About 70% experienced flush symptoms and headaches. Nausea occurred in 30% and vomiting in 16% of the patients. Sedation, apathy, restlessness or sudden sweating were uncharacteristic and may have been related to the clinical condition being treated or to concomitant diseases. A reduction in side effects was achieved by individual dose titration during the first days of treatment up to a level causing the first symptoms of flush and headache.

The clinical experience up to now suggests that iloprost is clinically effective in severe disorders such as PAOD and Raynaud's disease. Side effects like flush and headache can be dealt with, and the clinical use of iloprost can be regarded as safe.

In conclusion, it could be shown that labile arachidonic acid metabolites can be converted through chemical modification into stable compounds with greatly differing pharmacological profiles. Some of these substances hold a considerable potential as beneficial drugs in the clinic.

References

1   Coleman, R.A.: Methods in prostanoid receptor classification; in Benedetto, Prostaglandins and Related Substances – A Practical Approach, pp. 267–303 (IRL Press, Oxford 1987).

2   Hess, H.-J.; Bindra, J.S.; Constantine, J.W.; Elger, W.; Loge, O.; Schillinger, E.; Losert, W.: A tissue-selective prostaglandin E$_2$ analog with potent antifertility effects. Experientia *33*: 1076–1077 (1977).

3   Losert, W.F.; Loge, O.; Radüchel, B.; Skuballa, W.: Nocloprost. Drugs Future *13*: 926–930 (1988).

4   Schillinger E.; Krais, T.; Lehmann, M.; Stock, G: Iloprost; in Scriabine, New Cardiovascular Drugs, pp. 209–231 (Raven Press, New York 1986).

5   Müller, B.; Stürzebecher, S.; Witt, W.: Antithrombotic action of the stable prostacyclin analogue iloprost in rat mesenteric arterioles. Microcirc. Clin. Exp. *3:* 330 (1984).

6   Müller, B.; Schmidtke, M.; Witt, W.: Action of the stable prostacyclin analogue iloprost on microvascular tone and permeability in the hamster cheek pouch. Prostaglandins *29:* 187–198 (1987).

E. Schillinger, PhD, Research Laboratories of Schering AG, Müllerstrasse 170–178, D–1000 Berlin 65 (FRG)

Zor U, Naor Z, Danon A (eds): Leukotrienes and Prostanoids in Health and Disease.
New Trends Lipid Mediators Res. Basel, Karger, 1989, vol 3, pp 100–105

Leukotrienes and Asthma

O. Cromwell, A.J. Wardlaw, A.B. Kay

Department of Allergy and Clinical Immunology, The National Heart and Lung
Institute, London, UK

Clinically, asthma is regarded as a disorder of widespread airway
obstruction which reverses either spontaneously or with appropriate treatment. It is further characterized by bronchial hyperresponsiveness which
can be precipitated by the inhalation of certain inflammatory mediators as
well as specific allergens, and which relates closely to the presence and
severity of the asthma. The pathophysiological features of the disease
include airway obstruction, increased vascular permeability, mucus hyper-
secretion, inflammatory cell infiltration into bronchial and parenchymal
tissue and damage to the respiratory epithelium. The mechanism linking
these profound inflammatory features of the disease and the fundamental
abnormalities of impaired lung function and hyperresponsiveness is the
release of biologically active chemical mediators by various cell types into
the airways. Prominent among these are histamine, prostaglandins,
leukotrienes and platelet-activating factor. It is possible to learn a great
deal about the pharmacological contribution of various mediators by the
study of the disease during its natural course, but in order to obtain some
insight into the cellular and molecular mechanisms it has proved useful to
work with models of the disease both in man and experimental animals. In
allergic asthmatics, inhalation of specific allergen provokes two forms of
bronchoconstrictor response, an early reaction which is rapid in onset and
reaches a maximum some 15–30 min after challenge and a late reaction
which occurs in a proportion of patients and peaks between 6 and 9 h after
the challenge. Mediator measurements and changes in inflammatory cell
numbers and activation state in blood and bronchoalveolar lavage fluid
recovered during the course of these reactions have contributed to our
understanding of the mechanisms of these. One important fact which has
emerged from these studies is that asthma is not attributable to a single
mediator or indeed the mediators derived from one cell type but rather it
is a complex orchestration involving the direct and indirect actions of many

mediators, synergistic interactions, amplification mechanisms involving cytokines and neural and neuropeptide control mechanisms.

Leukotriene Inhalation

Inhalation of aerosolized leukotriene C_4 (LTC_4) and leukotriene D_4 (LTD_4) by normal nonasthmatic volunteers has shown that they are approximately 3,800 and 5,900 times more potent than histamine, respectively [1, 2]. In the case of asthmatic subjects, the rank order of potency of the leukotrienes is the same, but the subjects were generally more responsive [3]. Leukotriene E_4 (LTE_4) was the least potent of the leukotrienes tested being 137 and 177 times more potent than histamine in asthmatic and normal subjects, respectively [4]. Treatment with the oral leukotriene antagonist L-649,923 prior to inhalation of LTD_4 antagonized the bronchoconstriction causing almost a 4-fold shift of the dose-response curve [5]. Whilst the latter report demonstrates that it is possible to achieve leukotriene antagonism in vivo, a study to test the efficacy of L-649,923 in blocking an allergen-induced asthmatic response concluded that the drug afforded only a very small degree of protection [6]. The conclusion in this case was that either the drug was not sufficiently potent to diminish the effect of endogenous LTD_4 or alternatively that LTD_4 does not play a major role in the asthmatic response. The latter view is supported to some extent by the results of an attempted pharmacological dissection of the immediate asthmatic reaction [7]. Treatment of atopic asthmatics with terfenadine (180 mg, a selective histamine H_1 receptor antagonist) 3 h prior to allergen challenge produced a 50% inhibition of the early asthmatic response. Administration of flurbiprofen (a potent cyclooxygenase inhibitor, 150 mg daily for 3 days) caused a 30% reduction in the early asthmatic response, thus implicating cyclooxygenase products such as PGD_2 in the mechanisms of the bronchoconstriction. The balance of the response, some 20% after deducting the contributions from histamine (50%) and cyclooxygenase metabolites (30%), is presumably accounted for by mediators such as leukotrienes and platelet-activating factor.

There is no direct evidence from inhalation studies and studies using specific leukotriene antagonists to suggest that these mediators are involved in either the bronchoconstrictor or inflammatory components of the late asthmatic response. However, in vitro studies and those from animal models indicate that LTC_4 and LTD_4 are potent mediators in the induction of vascular permeability and also act as potent mucus secretagogues, pharmacological actions that are both considered to be important in the inflammatory component of asthma [for review, cf. 8].

Plasma Leukotriene Measurements

We have attempted to measure changes in the concentrations of plasma sulphidopeptide leukotrienes during both the natural course of the disease and following exercise and allergen provocation. An initial problem with all these studies, which used a double antibody radioimmunoassay to measure the leukotrienes, was to remove nonspecific immunoreactivity from the sample extracts. This was achieved by methanol extraction of the samples and fractionation on reverse-phase C18 silica extraction cartridges washed sequentially with chloroform and petroleum ether and eluted with methanol. This method achieved a mean LTC_4 recovery of $59.5 \pm 3.9\%$ (SE, n = 29) and an LTB_4 recovery of $100 \pm 10.3\%$ (SE, n = 10). Levels of LTC_4 immunoreactivity measured in normal nonasthmatic subjects and patients with mild asthma or acute severe asthma were not significantly different and ranged between 0.2 and 1.0 pmol/ml. The samples, which have been collected in the presence of metabolic inhibitors, were pooled and further analysed by reverse-phase high-performance liquid chromotography and found to contain predominantly LTC_4 and LTE_4 in the ratio of approximately 1:1.2. Plasma LTB_4 concentrations in normal subjects, asthmatics and patients with cystic fibrosis again were not significantly different and ranged between 0.24 and 0.95 pmol/ml. Concentrations of serum neutrophil chemotactic activity (NCA), a mast cell-associated activity which has been shown to be raised in association with both early and late asthmatic reactions, were significantly raised in the patients with acute severe asthma and provided a positive control in these studies [9]. The relevance of this observation is endorsed by the fact that levels of both NCA and histamine decrease during the course of successful treatment.

A group of 6 asthmatic subjects submitted to a 6-min exercise challenge and were monitored for 1 h thereafter. They produced a mean fall in FEV_1 (forced expiratory volume in 1 s) of 29%, and this was paralleled by significant increases in concentrations in plasma histamine and serum NCA. However, there were no significant changes in either LTC_4 immunoreactivity or LTB_4. We have obtained similar results in a group of 9 allergic asthmatics challenged with specific allergen. There are several explanations as to why it is not possible to detect increases in plasma leukotriene concentrations in the asthmatic subject. Firstly, the leukotrienes may not be making a significant contribution to the asthmatic response. Secondly, any changes in plasma levels would be a consequence of spillover from events taking place in the lung, and any leukotriene generated may have been diluted systemically or rapidly metabolized. Finally, in view of the fact that leukotrienes are relatively much more potent than histamine and PGD_2 as bronchoconstrictors, it is possible that the amounts produced in the lung which are sufficient to contribute to the asthmatic response are below the sensitivity of the detection method.

Bronchoalveolar Lavage Studies

The measurements of mediator concentrations in bronchoalveolar lavage (BAL) fluid are potentially more rewarding than measurements in plasma because sampling is carried out in the immediate vicinity of the reactions in the lung. A study that involved instilling specific allergen into a lung segment of each of 5 patients with allergic asthma resulted in the detection of a 150-fold increase in PGD_2, <8 pg/ml as against a mean of 322 pg/ml. Substantial increases in 15-hydroxyeicosatetraenoic acid were also observed but no 5-lipoxygenase products were detected [10]. We have measured the concentrations of LTC_4 and LTB_4 in BAL fluid from 16 atopic asthmatics (8 symptomatic, 8 asymptomatic) and 14 nonasthmatic control subjects. The amount of LTC_4 immunoreactivity and LTD_4 detected in the symptomatic asthmatics was significantly higher than in the controls (LTB_4: 0.58 ± 0.06 vs. 0.36 ± 0.05 pmol/ml, $p < 0.05$; LTC_4: 0.36 ± 0.1 vs. 0.12 ± 0.02 pmol/ml, $p < 0.01$). Reverse-phase HPLC analysis of the LTC_4 immunoreactivity suggested that LTE_4 accounts for most of this with $LTE_4 > LTD_4 > LTC_4$ in the mean relative proportions 3.3:2.4:1, respectively. These results are in accord with the findings of Lam et al. [11].

In a separate study we have attempted to ascertain whether or not leukotrienes are contributing to events in the late asthmatic response. BAL was performed in a total of 14 asthmatic subjects divided into two groups, those who experienced only an early asthmatic response following antigen inhalation and those who experienced both early and late asthmatic reactions. Lavage was performed on two separate occasions, once after a control diluent challenge and once after specific antigen inhalation. There were no consistent changes in LTB_4 concentrations in either group of patients but a small significant increase in LTC_4 was observed in the group experiencing both early and late phase reactions. When viewed in the context of the changes seen with individual patients from the other group, the changes in LTC_4 could not be considered impressive. Macrophages accounted for the majority of its cells recovered in the lavage on both occasions but there were significant increases in the numbers of lymphocytes, neutrophils and eosinophils in those patients experiencing a late asthmatic response thus implicating these cells in the inflammatory component of this reaction and indicating a potential source of LTC_4 (eosinophils).

Leukotrienes in Perspective

Our attempts to measure leukotrienes in association with asthmatic responses have not led to demonstrations of dramatic changes in the levels of these mediators. However, BAL studies in symptomatic asthmatics and asthmatics experiencing an antigen-induced late phase response suggest that both LTC_4 and LTB_4 may be making at least a minor contribution to the events occurring in the lung. These results taken together with the fact that the leukotriene antagonist L-649,923 has only a trivial effect in antagonizing antigen-induced bronchoconstricton [6] and the fact that most of the early asthmatic response can be accounted for by histamine and cyclooxygenase products [7] suggest that LTC_4 may not be making a major contribution as a bronchoconstrictor agonist. Alternatively, the leukotrienes may be contributing to the inflammatory events associated with the late asthmatic response and chronic ongoing asthma by promoting vascular permeability and tissue edema and acting as mucus secretagogues. However, one point that emerges quite clearly from these studies is that the measurements of 5-lipoxygenase metabolites as a means of monitoring the asthmatic response does not provide a feasible means of assessing the efficacy of potential leukotriene antagonists.

References

1   Weiss JW, Drazen JM, Coles N, et al: Bronchoconstrictor effects of leukotriene C in humans. Science 1982;216:196–198.
2   Weiss JW, Drazen JM, McFadden ER, et al: Airway constriction in normal humans produced by inhalation of leukotriene D. Potency, time course and effect of aspirin therapy. JAMA 1983;269:2814–2817.
3   Adelroth E, Morris MM, Hargreave FE, et al: Airway responsiveness to leukotrienes C_4 and D_4 and to methacholine in patients with asthma and normal controls. N Engl J Med 1986;315:480–484.
4   O'Hickey SP, Arm JP, Spur BW, et al: Relative responsiveness to leukotriene E_4, methacholine and histamine in normal and asthmatic subjects (abstract). J Allergy Clin Immunol 1988;81:1919.
5   Barnes NC, Piper PJ, Costello J: The effect of an oral leukotriene antagonist L649,923 on histamine and leukotriene D_4-induced bronchoconstricton in normal man. J Allergy Clin Immunol 1987;79:816–821.
6   Britton JR, Hanley SP, Tattersfield AE: The effect of an oral leukotriene D_4 antagonist L-649,923 on the response to inhaled antigen in asthma. J Allergy Clin Immunol 1987;79:811–816.
7   Holgate ST, Twentyman OP, Rafferty P, et al: Primary and secondary effector cells in the pathogenesis of asthma. Int Arch Allergy Appl Immunol 1987;82:498–506.
8   Drazen JM, Austen KF: Leukotrienes and airway responses. Am Rev Respir Dis 1987;136:985–998.

9   Buchanan DR, Cromwell O, Kay AB. Neutrophil chemotactic activity (NC) in acute severe asthma (status asthmaticus). Am Rev Respir Dis 1987;136:1397–1402.

10   Murray J, Tonnel AB, Brash AR, et al: Release of prostaglandin D_2 into human airways during acute antigen challenge. N Engl J Med 1986;315:800–804.

11   Lam S, Chan H, LeRiche JC, et al: Release of leukotrienes in patients with bronchial asthma. J Allergy Clin Immunol 1988;81:711–717.

O. Cromwell, MD, Department of Allergy and Clinical Immunology, The National Heart and Lung Institute, Dovehouse Street, GB-London 3W3 LY (UK)

Zor U, Naor Z, Danon A (eds): Leukotrienes and Prostanoids in Health and Disease.
New Trends Lipid Mediators Res. Basel, Karger, 1989, vol 3, pp 106–109

Role of Sulfidopeptide Leukotriene and Histamine Receptors in Human Airway Hypersensitization in vitro

S. Nicosia[a], *T. Vigano*'[a], *D. Oliva*[a], *M.T. Crivellari*[a], *M.R. Accomazzo*[a], *G.E. Rovati*[a], *A. Verga*[b], *M. Mezzetti*[c], *G.C. Folco*[d]

[a]Institute of Pharmacological Sciences and [c]4th Surgical Clinic, University of Milan; [b]Bayer Italia, Garbagnate, Milan; [d]Institute of Pharmacology and Pharmacognosy, University of Parma, Italy

Introduction

A hallmark of human asthma is hyperresponsiveness of the airways, which represents an exaggerated bronchoconstrictor response to a range of stimuli, both pharmacological and physical. The mechanisms responsible for this phenomenon are still unknown, although many hypotheses for bronchial hyperreactivity have been proposed [1]. Among the bronchoconstrictor stimuli, histamine and prostaglandins (PGs), in particular PGD_2, are supposed to play a primary role in human airway hypersensitivity [2, 3]. On the contrary, the hyperresponsiveness of asthmatic patients to sulfidopeptide leukotrienes (LTC_4, LTD_4, and LTE_4) still represents a controversial question [4, 5].

The aim of the present work was to evaluate whether such hypersensitization was associated with alterations in binding of ^3H-LTC_4, ^3H-LTD_4 and ^3H-mepyramine (H_1 antagonist).

Methods

Macroscopically normal human lung parenchyma from individuals was minced and divided into two aliquots. After washing, the minced tissue was incubated for 3 h at 37 °C in Tyrode's buffer, pH 7.4, with (passively sensitized, S) or without (control, C) hyperimmune serum (5 µg/ml IgE). At the end of the incubation, lung fragments (C and S) were extensively washed and an aliquot of both C and S was withdrawn for binding assay (see below). The C and S specimens were then challenged using an anti-human IgE antibody (Dr. S. Ahlstedt, Pharmacia, Uppsala) at 37 °C for 15 min. Aliquots of supernatant solutions were then taken for quantitative measurement of PGD_2 by enzyme immunoassay (EIA) [6], or biologically assayed for LTC_4-like activity [7]. Binding of LTC_4, LTD_4 and histamine was performed in crude membrane preparations obtained from C and S fragments before immunological challenge. Binding was assessed at 25 °C for 40 min in the presence of 2 nM ^{3}H-LTD$_4$ [8] or 5 μM ^{3}H-mepyramine [9] or at 0 °C for 20 min in the presence of 50 nM ^{3}H-LTC$_4$ [10]. Tissue bound IgE content was assayed by radioimmunoassay (RIA) [11].

Results and Discussion

The biosynthesis of LTC$_4$-like activity and PGD$_2$ from endogenous arachidonic acid in normal (C) and passively sensitized (S) human lung parenchyma from 23 individuals has been evaluated upon immunological challenge with anti-human IgE antibodies. The results, reported in figure 1, indicate that there is no increase in the release of either PGD$_2$ or LTC$_4$-like activity in the passively sensitized samples as compared to the non-sensitized ones.

Passive sensitization, however, markedly increased the content of tissue bound IgE, by approximately 10-fold (386 ± 134 in C fragments vs. 3,524 ± 609 U IgE/g tissue). The apparent discrepancy between the marked increase in tissue bound IgE and the lack of effect on the release of eicosanoids upon passive sensitization could be explained on the basis of the hypothesis that unsensitized human lung already possesses a sufficient amount of bound IgE which causes mast cell degranulation and the consequent massive release of allergic mediators, when challenged with anti-IgE antibody.

Concerning the problem whether modifications in sulfidopeptide leukotrienes and histamine receptors might be involved in human hyper-responsiveness, binding of ^{3}H-LTC$_4$, ^{3}H-LTD$_4$ and ^{3}H-mepyramine was performed on membranes from C and S fragments. Table 1 shows that no significant change in binding parameters of either LTC$_4$ or LTD$_4$ was observed in the different experimental conditions; on the contrary, passive sensitization caused a small, but statistically significant ($p < 0.02$) increase in ^{3}H-mepyramine binding. The increase (3.3–156%) occurred in 18 out of 23 specimens.

In conclusion, our present results indicate that, after passive sensitization of human lung with hyperimmune serum, despite a marked increase in tissue bound IgE, no modifications in the release of allergic mediators (PGD$_2$ and sulfidopeptide leukotrienes) upon immunological challenge could be revealed. In the sensitized tissue, neither ^{3}H-LTC$_4$ nor ^{3}H-LTD$_4$ binding revealed any modification upon sensitization, while ^{3}H-mepyramine binding shows a moderate but significant increase. This different behavior of leukotrienes with respect to histamine receptors might be in agreement with the observation that LTC$_4$ and LTD$_4$ are unique broncho-constrictors as suggested by Adelroth et al. [12].

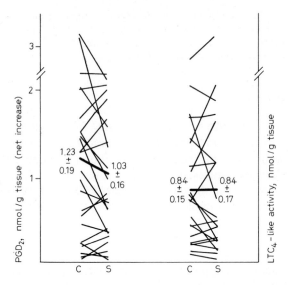

Fig. 1. Release of PGD_2 and LTC_4-like activity from normal (C) and passively sensitized (S) human lung parenchyma, following anti-IgE challenge. The data are expressed as mean $\pm SE$ values.

Table 1. Binding of ^3H-LTC_4, ^3H-LTD_4 and ^3H-mepyramine in control (C) and passively sensitized (S) human lung parenchyma

	^3H-LTC_4		^3H-LTD_4		^3H-mepyramine	
	C	S	C	S	C	S
pmol/mg protein mean	10.91	10.60	0.0713	0.0689	64.45	74.67
$\pm SD$	± 7.74	± 7.76	± 0.042	± 0.043	± 43.14	± 46.04
Responses/total[1]	8/20		11/23		18/23	

[1] Responses/total: number of specimens which responded with an increase in binding when challenged, over total number of specimens assayed.

Finally, in order to establish whether passive sensitization with hyperimmune serum could represent a good in vitro model for hyperresponsiveness on human airways, it would be important to investigate whether a direct correlation exists between the increase in ^3H-mepyramine binding and the in vitro hyperresponsiveness to the contractile effect of histamine in sensitized human lung parenchyma.

References

1   Simonsson BG: Bronchial hyperreactivity. Eur J Respir Dis 1980;61(suppl 108):21–28.
2   Anderton RC, Cuff MT, Frith PA, et al: Bronchial responsiveness to inhaled histamine and exercise. J Allergy Clin Immunol 1979;63:315–321.
3   Hardy CC, Robinson C, Tattersfield AE, et al: The bronchoconstrictor effect of inhaled prostaglandin D_2 in normal and asthmatic men. N Engl J Med 1984;311:209–213.
4   Griffin M, Weiss JW, Leitch AG, et al: Effects of leukotriene D on the airways in asthma. N Engl J Med 1983;308:436–439.
5   Bisgaard H, Groth S, Madsen F: Bronchial hyperreactivity to leukotriene D_4 and histamine in exogenous asthma. Br Med J 1985;290:1468–1471.
6   Pradelles P, Grassi J, Maclouf J: Enzyme immunoassay of eicosanoids using AchE as label: an alternative to RIA. Anal Biochem 1985;57:1170–1173.
7   Ferreira SH, Costa FS: A laminar flow technique with much increased sensitivity for the detection of smooth muscle stimulating substances. Eur J Pharmacol 1976;39:379–381.
8   Lewis MA, Mong S, Vessella RL, et al: Identification and characterization of leukotriene D_4 receptors in adult and fetal human lung. Biochem Pharmacol 1985;34:4311–4317.
9   Casale TB, Rodbard D, Kaliner M: Characterization of histamine H_1 receptors on human peripheral lung. Biochem Pharmacol 1985;34:3285–3292.
10 Rovati GE, Oliva D, Sautebin L, et al: Identification of specific binding sites for leukotriene C_4 in membranes from human lung. Biochem Pharmacol 1985;34:2831–2837.
11 Ceska M, Lundkvist U: A new and simple radioimmunoassay method for the determination of IgE. Immunochemistry 1972;9:1021–1030.
12 Adelroth E, Morris MM, Hargreave FE, et al: Airway responsiveness to leukotrienes C_4 and D_4 and to methacoline in patients with asthma and normal controls. N Engl J Med 1986;315:480–484.

S. Nicosia, MD, Institute of Pharmacological Sciences, Via Balzaretti 9,
I-20133 Milano (Italy)

Zor U, Naor Z, Danon A (eds): Leukotrienes and Prostanoids in Health and Disease.
New Trends Lipid Mediators Res. Basel, Karger, 1989, vol 3, pp 110–113

Formation of Lipoxygenase Products of Arachidonate Metabolism in Disaggregated Lung Cells

Michael K. Bach, John R. Brashler, Herman W. Smith

Hypersensitivity Diseases Research, The Upjohn Company, Kalamazoo, Mich., USA

Several lines of evidence suggest that 15-HETE may play an important proinflammatory role in the lung: (1) Bronchoalveolar lavage fluids from both normal and asthmatic humans contain high concentrations of 15-HETE [1]. (2) The passive administration of nanogram amounts of 15-HETE to dog tracheas in vivo caused increased mucus production, the accumulation of inflammatory cells, and an increase in loss of water from the lungs [2]. Cultured tracheal epithelial cells produce large amounts of 15-HETE [3, 4]. This paper describes our initial attempts to find the cellular source of 15-HETE in the lung and the ability of lymphokines to induce its selective formation.

Methods and Materials

Disaggregation of minced human lung specimens by digestion with collagenase and chymopapain was adapted from the method of Holgate et al. [5] and, where indicated, cells were further fractionated by centrifugation (1,500 g for 20 min) on discontinuous Percoll gradients. Challenge was in the presence of 17 μM arachidonate and 10 μM A23187, usually for 5 min. Challenge with cytokines, which required longer incubations, was carried out in laminin-coated tubes in the presence of 2.7×10^{-5} M PhCl28A [6]. Eicosanoid concentrations were estimated using selective radioimmunoassays (RIA) [7] and the identity of the radio-immunochemically determined 15-HETE with authentic 15-HETE was confirmed by combined HPLC and RIA.

Results and Discussion

The disaggregated human lung cells produced, on average, 0.5 ng LTB_4, 21 ng LTC_4, 23 ng LTD_4 and 54 ng 15-HETE per 10^6 cells. LTE_4 production was not detectable (<0.1 ng). Preincubation with arachidonate, which was concentration-independent, potentiated the production of 15-HETE much more than it potentiated the formation of LTD_4 which is consistent with the reported observation that the 15-lipoxygenase (15-LO) has a lower affinity for substrate than does the 5-lipoxygenase (5-LO) and that, generally, arachidonate must be supplied exogenously to induce the formation of this mediator [8, 9]. The K_m for arachidonate for the formation of sulfidopeptide leukotrienes and 15-HETE, based on a double reciprocal (Lineweaver Burk) plot of the results, was 9 and 21 μM, respectively.

To achieve a linear relationship between eicosanoid production and cell number, a prerequisite for identifying the cells responsible for the eicosanoid formation, the ratio of the ionophore concentration to the cell concentration must be kept constant [10]. Making the assumption that substantially all the synthetic capacity for a given mediator resides in a single cell type, two different methods were used to quantitate the synthetic capacity of different cells. The first method compared the fraction of the total number of cells of a given type which were present in each of the Percoll fractions to the fraction of the total eicosanoid synthetic capacity which was present in the same fraction. If a given cell type is truly responsible for the formation of the eicosanoids, these two numbers should be closely correlated; this was the case for the comparison with respect to the distribution of eosinophils but not to the distribution of any other cell type. The second method compared the specific synthetic capacity (ng/10^6 cells) for the different cell types in each of the Percoll fractions. These values should be highest for the cells which are truly responsible for the production and more importantly, they should be the most nearly constant when compared from one Percoll fraction to the next (i.e., the standard error for the mean synthetic capacity of that cell type in all the Percoll fractions should be the smallest). Both these predictions were also true in the case of the eosinophils, suggesting that the eosinophils can account for most of the 5- and 15-LO metabolites which are produced in the lung cell suspensions. The possibility remains, however, that another cell type, possibly a pneumocyte, may contribute to production of these mediators albeit at a much lower synthetic rate than the eosinophils.

We examined the ability of interleukin-1 (IL-1β) and tumor necrosis factor alpha (TNF-α) to elicit the production of lipoxygenase products in the disaggregated lung cells [11, 12]. PhCl28A, a potent inhibitor of the prostaglandin-15-dehydrogenase (27 μM) [9], was added to these incubations to minimize metabolism of 15-HETE to the 15-keto-eicosatetraenoic acid by prostaglandin-15-dehydrogenase [13] during the longer incubation periods which were required. There was a dose-dependent, selective, though very modest, stimulation of 15-HETE formation when 0.56–57 nM recombinant human IL-1β was added to incubations in the presence of a low concentration of arachidonate. Stimulation by similar concentrations of TNF-α was smaller and was not dose-dependent. However, there was synergistic stimulation of 15-HETE formation in cells which had been preincubated with arachidonate and were then exposed to combinations of these cytokines in laminin-coated culture tubes [14], which was as high as 38% of the ionophore-induced formation.

A possible explanation for the requirement for high cytokine concentrations in these experiments is the short incubation periods which had to be used. To address this, we have recently begun to culture the disaggregated lung cells overnight. In preliminary results, overnight culture of the eosinophil-enriched Percoll fraction in the presence of GM-CSF resulted in good retention of cell viability. Eicosanoid production was at least as high as that of the freshly isolated cells but the effect of the inclusion of cytokines during the overnight incubations has not yet been examined.

References

1   Murray JJ, Tonnel AB, Brash AR, et al: Release of prostaglandin D_2 into human airways during acute antigen challenge. N Engl J Med 1986;315:800–804.
2   Johnson HG, McNee ML, Sun FF: 15-Hydroxyeicosatetraenoic acid is a potent inflammatory mediator and agonist of canine tracheal mucus secretion. Am Rev Respir Dis 1985;131:917–922.
3   Holtzman MJ, Aizawa H, Nadel JA, et al: Selective generation of leukotriene B_4 by tracheal epithelial cells from dogs. Biochem Biophys Res Commun 1983;114:1071–1076.
4   Hunter JA, Finkbeiner WE, Nadel JA, et al: Predominant generation of 15-lipoxygenase metabolites of arachidonic acid by epithelial cells from human trachea. Clin Res 1985;33:78A.
5   Holgate ST, Burns GB, Robinson C, et al: Anaphylactic- and calcium-dependent generation of prostaglandin D_2 (PGD_2), thromboxane B_2, and other cyclooxygenase products of arachidonic acid by dispersed human lung cells and relationship to histamine release. J Immunol 1984;133:2138–2144.
6   Bakhle YS, Pankhania, JJ: Inhibitors of prostaglandin dehydrogenase (PhCl28A and PhCk61A) increase output of prostaglandins from rat isolated lung. Br J Pharmacol 1987;92:189–196.
7   Bach MK, Brashler JR, White GJ, et al: Experiments on the mode of action of piripost (U-60,257), an inhibitor of leukotriene formation in cloned mouse mast cells and in rat basophil leukemia cells. Biochem Pharmacol 1987;36:1461–1466.
8   Soberman RJ, Harper TW, Betteridge D, et al: Characterization and separation of the arachidonic acid 5-lipoxygenase and linolenic acid ω-6-lipoxygenase (arachidonic acid 15-lipoxygenase) of human polymorphonuclear leukocytes. J Biol Chem 1985;260:4508–4515.
9   Fruteau de Laclos B, Braquet P, Borgeat P: Characteristics of leukotriene (LT) and hydroxyeicosatetraenoic acid (HETE) synthesis in human leukocytes in vitro. Effect of arachidonic acid concentration. Prostaglandins Leukotrienes Med 1984;13:47–52.

10   Bach MK, Brashler JR: Ionophore A23187-induced production of slow reacting sub-
      stance of anaphylaxis (SRS-A) by rat peritoneal cells in vitro: Evidence for production
      by mononuclear cells. J Immunol 1978;120:998–1005.
11   Smith RJ, Epps DE, Justen JM, et al: Human neutrophil activation with interleukin-1.
      A role for intracellular calcium and arachidonic acid lipoxygenation. Biochem Pharma-
      col 1987;36:3851–3858.
12   Silberstien DS, David JR: Tumor necrosis factor enhances eosinophil toxicity to
      Schistosoma mansoni larvae. Proc Natl Acad Sci USA 1986;83:1055–1059.
13   Bergholte JM, Soberman RJ, Hayes R, et al: Oxidation of 15-hydroxyeicosatetraenoic
      acid and other hydroxyfatty acids by lung prostaglandin dehydrogenase. Archs Biochem
      Biophys 1987;257:444–450.
14   Nathan CF: Neutrophil activation on biological surfaces. Massive secretion of hydrogen
      peroxide in response to products of macrophages and lymphocytes. J Clin Invest
      1987;80:1550–1560.

Michael K. Bach, PhD, MD, Distinguished Scientist, Hypersensitivity Diseases
Research, The Upjohn Company, Kalamazoo, MI 49001 (USA)

Zor U, Naor Z, Danon A (eds): Leukotrienes and Prostanoids in Health and Disease.
New Trends Lipid Mediators Res. Basel, Karger, 1989, vol 3, pp 114–118

Interleukin-1 and Tumor Necrosis Factor – Mediators of the Acute Phase Response and Possible Modes for Their Manipulation

S. Endres,[a,1] *C.A. Dinarello*[b,2]

[a]Medizinische Klinik Innenstadt der Universität München, FRG, and
[b]Division of Geographic Medicine and Infectious Diseases, Department of Medicine,
Tufts University, New England Medical Center, Boston, Mass., USA

Biological Activities of Interleukin-1

Aside from its pyrogenic effects, interleukin-1 (IL-1) influences a wide array of biological functions [reviewed in 1]. Originally described in purified material from monocyte supernatants, most effects have been confirmed using recombinant IL-1 [2].

IL-1 is a cofactor in activating T lymphocytes inducing synthesis of interleukin-2 and interleukin-2 receptors. It increases antibody synthesis from B cells and acts in synergy with interferons in the enhancement of natural killer cell activity. IL-1 has been found to be identical to hemo-poietin-1, which was recognized by its ability to synergize with colony-stimulating factors in affecting hematopoietic stem cell proliferation [3], and it induces the release of granulocytes from the bone marrow.

IL-1 triggers several changes in metabolism. Hepatic synthesis of the acute phase proteins (serum amyloid A, C-reactive protein, fibrinogen, α_2-macroglobulin), participating in different stages of host defense, is increased. Concurrently, synthetic capacity is economized by a drop in albumin synthesis and more amino acids are liberated by induction of skeletal muscle proteolysis. Fibroblast collagen production is enhanced, possibly contributing to the process of wound healing. IL-1 has also been identified as an osteoclast-activating factor.

IL-1 has profound effects on vascular endothelial cell function. It activates human endothelial cells in vitro to synthesize and release prostaglandins I_2 and E_2 [4]. While these two arachidonic acid metabolites

[1] S.E. is supported by grant En-169/1 of the Deutsche Forschungsgemeinschaft.
[2] The authors thank Reza Ghorbani for his assistance in these studies.

increase blood flow, IL-1 also orchestrates a cascade of cellular and biochemical events that lead to vascular congestion, clot formation and cellular infiltration.

IL-1 induces other lymphokines (interleukin-2, interleukin-3, interleukin-6, granulocyte macrophage colony-stimulating factor (GM-CSF), interferon-β, -γ, tumor necrosis factor (TNF), lymphotoxin) and can serve as an autocrine signal promoting the expression of its own gene [5].

Modulation of Cytokine Synthesis

To date, the only known pharmacologic agents to reduce cytokine synthesis are corticosteroids and ciclosporin A. We have recently completed a study to investigate whether dietary supplementation with n-3 fatty acids contained of fish oils, affects the synthesis of IL-1 and TNF [6]. The main forms of n-3 fatty acids are eicosapentaenoic acid and docosahexaenoic acid. These fatty acids are scarce in a normal western diet, but make up an appreciable part of the lipid intake in diets rich in cold water fish and seal meat [7]. Epidemiological studies have shown that a high intake of n-3 fatty acids correlates with a low incidence of atherosclerosis and inflammatory diseases such as asthma and type I diabetes mellitus [8]. In healthy volunteers [9] and in patients [10], n-3 dietary supplementation results in decreased neutrophil chemotaxis and inhibition of leukotriene B$_4$ (LTB$_4$) generation by neutrophils and monocytes. In clinical studies, dietary supplementation with n-3 fatty acids has led to improvements in patients with rheumatoid arthritis [11]. Prospective studies in animals fed n-3 fatty acids support these clinical observations [12–14].

Since IL-1 and TNF are principal mediators of inflammation, reduced production of these cytokines may contribute to the amelioration of inflammatory symptoms in patients taking n-3 supplementation. Therefore, we decided to investigate the effects of n-3 fatty acids on the synthesis of the cytokines IL-1 and TNF.

Nine healthy volunteers added 16 g of fish oil concentrate (MaxEPA®) per day to their normal diet. This supplement, taken as capsules, was given for a period of 6 weeks. In vitro IL-1 and TNF production was assessed during four phases of the study: before beginning the n-3 fatty acid supplementation, after 6 weeks of the supplementation; and during the 10th and 20th week after cessation of the diet. To ascertain reproducibility, the assay was performed on 3 different days within 1 week for each of the four study phases.

In vitro production of IL-1 and TNF was determined by incubating peripheral blood mononuclear cells for 24 h with different stimuli. At the

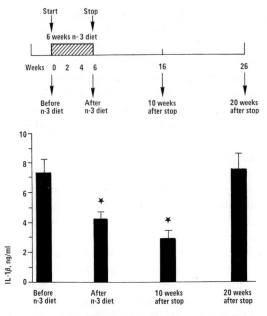

Fig. 1. Production of IL-1β stimulated with endotoxin. Mononuclear cells were incubated for 24 h with 1 ng/ml of endotoxin. IL-1β was determined by RIA. The bars represent the mean values for 9 volunteers with error bars as the standard error of the mean. *Significant difference from the baseline (before n-3 diet) at p < 0.05.

end of the incubation, the cells were lysed by freeze-thawing to obtain total, that is cell-associated plus secreted cytokine. IL-1β [15], IL-1α [16] and TNF [17] were measured by specific radioimmunoassays.

Figure 1 illustrates the production of IL-1β during the course of the study. Our findings demonstrate that n-3 fatty acid supplementation reduced the ability of IL-1β production induced by endotoxin. The effect was most pronounced 10 weeks after stopping the supplementation and suggests prolonged incorporation of n-3 fatty acids into a pool of circulation MNC. The capacity of the MNC from these donors to synthesize IL-1β returned to the presupplement level 20 weeks after ending the supplementation. Similar results were observed when we measured IL-1α and TNF.

There are several clinical implications of the present findings. There is a pathophysiologic rationale for therapeutic trials with n-3 fatty acids in certain diseases with documented involvement of inflammatory cytokines such as IL-1 and TNF. For example, rheumatoid arthritis [18], psoriasis and type I diabetes mellitus are likely candidates.

Our findings may also have relevance for atherogenesis. As described above, IL-1 and TNF induce the expression of intercellular adhesion molecule-1 on the surface of vascular endothelial cells, stimulate the production of platelet-activating factor, plasminogen activator inhibitor and procoagulant activity in endothelial cells, and induce smooth muscle cell proliferation [19]. Furthermore, IL-1 and TNF are generated in the vessel wall itself, by endothelial and smooth muscle cells [20]. Thus, since the vessel wall forms both a cardinal target tissue and a source of cytokines, a suppressive effect on their generation by a n-3 supplemented diet may contribute to its protective effect against atherosclerosis.

In a subsequent study we have investigated the effect of oral aspirin on the production of IL-1 and TNF in a protocol indentical to the one described above [21]. Preliminary data show that aspirin appears to *increase* the ex vivo production of IL-1β by mononuclear cells. This is particularly remarkable since in most other biological systems, e.g. the inhibition of platelet function, n-3 fatty acids and cyclooxygenase inhibitors like aspirin, induce a parallel effect. Aspirin enhances IL-1β production probably via decreased formation of PGE_2 which is an inhibitor of IL-1β production.

References

1   Dinarello CA: Interleukin-1 and its biologically related cytokines. Adv Immunol 1989;44:153–205.
2. Dinarello CA, Cannon JG, Mier JW, et al: Multiple biological activities of human recombinant interleukin-1. J Clin Invest 1986;77:1734–1739.
3   Moore MA, Warren DJ: Synergy of interleukin-1 and granulocyte colony stimulating factor: in vivo stimulation of stem-cell recovery and hemopoietic regeneration following 5-fluorouracil treatment of mice. Proc Natl Acad Sci USA 1987;84:7134–7138.
4   Rossi V, Breviario F, Ghezzi P, et al: Interleukin-1 induces prostacyclin in vascular cells. Science 1985;229:1174–1176.
5   Dinarello CA, Ikejima T, Warner SJC, et al: Interleukin-1 induces interleukin-1. I. Induction of circulating interleukin-1 in rabbits in vivo and in human mononuclear cells in vitro. J Immunol 1987;316:379–385.
6   Endres S, Ghorbani R, Kelley VE, et al: The effect of dietary supplementation with n-3 polyunsaturated fatty acids on the synthesis of interleukin-1 and tumor necrosis factor by mononuclear cells. N Engl J Med 1989;320:265–271.
7   Bang HO, Dyerberg J, Hjørne N: The composition of food consumed by Greenland Eskimos. Acta Med Scand 1976;200:69–73.
8   Kromann N, Green A: Epidemiological studies in the Upernavik District, Greenland. Acta Med Scand 1980;208:401–406.
9   Lee TH, Hoover RL, Williams JD, et al: Effect of dietary enrichment with eicosapentaenoic and docosahexaenoic acids on in vitro neutrophil and monocyte leukotriene generation and neutrophil function. N Engl J Med 1985;312:1217–1224.

10   Payan DG, Wong MYS, Chernov-Rogan T, et al: Alterations in human leukocyte
      function induced by ingestion of eicosapentaenoic acid. J Clin Immunol 1986;6:402–410.
11   Kremer JM, Jubiz W, Michalek A, et al: Fish-oil fatty acid supplementation in active
      rheumatoid arthritis. A double-blinded, controlled, crossover study. Ann Intern Med
      1987;106:497–502.
12   Prickett JD, Robinson DR, Steinberg AD: Dietary enrichment with the polyunsaturated
      fatty acid eicosapentaenoic acid prevents proteinuria and prolongs survival in
      NZB × NZW F_1 mice. J Clin Invest 1981;68:556–559.
13   Kelley VE, Ferretti A, Izur A, et al: A fish oil diet rich in eicosapentaenoic acid reduces
      cylooxygenase metabolites, and suppresses lupus in mrl-lpr mice. J Immunol
      1985;134:1914–1919.
14   Weiner BH, Ockene IS, Levine PH, et al: Inhibition of atherosclerosis by cod-liver oil in
      a hyperlipidemic swine model. N Engl J Med 1986;315:841–846.
15   Endres S, Ghorbani R, Lonnemann G, et al: Measurement of immunoreactive inter-
      leukin-1β from human mononuclear cells: optimization of recovery, intrasubject consis-
      tency and comparison with interleukin-1α and tumor necrosis factor. Clin Immunol
      Immunopathol 1988;49:424–438.
16   Lonnemann G, Endres S, Van der Meer JWM, et al: A radioimmunoassay for human
      interleukin-1α: measurement of IL-1α produced in vitro by human blood mononuclear
      cells stimulated with endotoxin. Lymphokine Res 1988;7:75–85.
17   Van der Meer JWM, Endres S, Lonnemann G, et al: Concentrations of immunoreactive
      human tumor mecrosis factor alpha produced by human mononuclear cells in vitro. J
      Leukocyte Biol 1988;43:216–223.
18   Krane SM, Dayer JM, Simon LS, et al: Mononuclear cell conditioned medium contain-
      ing mononuclear cell factor (MCF), homologous with interleukin-1, stimulates collagen
      and fibronectin synthesis by adherent rheumatoid synovial cells: effects of prostaglandin
      E_2 and indomethacin. Collagen Relat Res 1985;5:99–117.
19   Libby P, Ordovas JM, Auger KR, et al: Endotoxin and tumor necrosis factor induce
      interleukin-1 gene expression in adult human vascular endothelial cells. Am J Pathol
      1986;124:179–186.
20   Warner SJC, Auger KR, Libby P: Human interleukin-1 induces interleukin-1 gene
      expression in human vascular smooth muscle cells. J Exp Med 1987;165:1316–1321.
21   Endres S, Cannon JG, Ghorbani R, et al: In vitro production of IL-1β, IL-1α, tumor
      necrosis factor and IL-2 in a large cohort of human subjects: distribution, effect of
      cyclooxygenase inhibition and evidence of independent gene regulation (submitted).

Charles A. Dinarello, MD, Division of Geographic Medicine and Infectious
Diseases, Department of Medicine, Tufts University-New England Medical Center,
750 Washington Street, Boston, MA 02111 (USA)

Zor U, Naor Z, Danon A (eds): Leukotrienes and Prostanoids in Health and Disease.
New Trends Lipid Mediators Res. Basel, Karger, 1989, vol 3, pp 119–123

DuP 654 – A New Topical Anti-Inflammatory

N.R. Ackerman, D.G. Batt, W. Galbraith, K.R. Gans, R.R. Harris [1]

E.I. du Pont de Nemours & Co., Inc., Medical Products Department,
Experimental Station, E-400/2273, Wilmington, Del., USA

The common inflammatory diseases of the skin are characterized by the cardinal signs of inflammation, namely, heat, redness, edema and pain. Histologically, most inflammatory lesions are characterized by dilation of the blood vessels and accumulation of polymorphonuclear leukocytes (PMNs). A subset of the leukotrienes, which are products of the 5-lipoxygenase (5-LO) pathway of arachidonic acid (AA) metabolism, have been demonstrated as playing a role in these skin inflammatory conditions [1]. They are, for example, elevated in psoriasis [2]. The sulfidopeptide leukotrienes, LtC_4 and LtD_4, enhance vascular permeability and may be involved in the edematous phase of an inflammatory reaction [3]. LtB_4 has a potent chemoattractant activity; it may in part be responsible for the PMN infiltration in inflammatory skin conditions [4].

Based on the hypothesis that products of the 5-LO pathway are involved in perpetuating skin inflammatory lesions, inhibitors of this enzyme could prove beneficial in diseases such as psoriasis and atopic dermatitis [5]. In this report, the topical anti-inflammatory activity of DuP 654, a potent inhibitor or the 5-LO pathway, will be reported. DuP 654, 2-(phenylmethyl)-1-naphthol, is synthesized by the base-catalyzed condensation of benzaldehyde with tetralone. It is a weak acid which is insoluble in water and freely soluble in ethanol and other organic solvents.

Materials and Methods

Tetradeconoyl phorbol acetate, phenidone and dexamethasone were purchased from Sigma; Rev 5901A was a gift from Revlon and Lonapalene was a gift from Syntex.

[1] We would like to acknowledge P. Sayers, S. Lundy, R. Dowling, R. Collins and D. Pedicord for excellent technical assistance.

[14]C-labeled AA was purchased from NEN. CF$_1$ and Balb/C mice were purchased from Charles River and were used in a temperature- and humidity-controlled animal facility with food and water ad libitum.

AA-Induced Edema

The method used is an adoption from procedures published earlier [6, 7]. Male 18–20 g CF$_1$ mice were used in these experiments. AA from NuChekPrep was prepared daily in acetone at a concentration of 100 mg/ml. To challenge the ears, 10 µl of AA solution was placed on the inner surface of the ear. Test compounds were dissolved in acetone and applied in 10 µl to both ears just prior to challenge with AA. One hour after challenge, the animals were euthanized by cervical dislocation and both ears removed; 6-mm discs were then taken using a skin biopsy punch and the mass determined. In the control animal, the difference between the weights of the solvent only or solvent + AA ear punches was determined and termed the swelling number. In a like manner, the swelling numbers for the compound-treated animals were also determined. ED$_{50}$ were determined by linear regression.

Myeloperoxidase Assay

To determine the relative numbers of PMNs in the inflamed site, the specific enzyme marker for these cells, myeloperoxidase, was employed [8]. In preliminary experiments, the number of PMNs were determined both histologically and enzymatically (with the myeloperoxidase assay). Both techniques gave a similar profile on the infiltration of PMNs. Groups of animals were treated as above with DuP 654, 10 µg/ear, and AA, 1 mg. At 1, 2, 4, 6 and 8 h postchallenge, the animals were sacrificed, ears removed and a punch taken for weight determination. The tissue was then homogenized and the myeloperoxidase levels determined using O-diansidine (Sigma Chemical Co.) and H$_2$O$_2$. Protein was determined by OD$_{280/260}$ using bovine serum albumin as a standard.

Contact Sensitivity to Dinitrofluorobenzene

Balb/C mice were sensitized on days 0 and 1 with 25 µl 0.5% 2,4-dinitrofluorobenzene (DNFB) (Kodak Chemicals) in acetone-olive oil 4:1 on shaved abdomens. On day 5, DuP 654 or control drug (dexamethasone,

Table 1. Effects of selected agents on the edematous phase of an AA-induced inflammatory response[1]

Compound	Inhibition of ear edema (ED_{50}, μg/ear)
Indomethacin	700
Phenidone	35
DuP 654	11
RS-43179 (Lonapalene)	1,050
Rev 5901-A	103
Dexamethasone	>10

[1] AA, 1 mg, was applied to the ears of male CF_1 mice. Test compounds were applied immediately prior to the AA. One hour after challenge the animals were sacrificed, the ears removed and 6-mm discs obtained. Differences in weight of drug- and vehicle-treated ears were determined and the ED_{50}s determined.

1 μg/ear) was applied to both ears and the right ear was challenged with 10 μl of 0.4% DNFB in the acetone-olive oil vehicle. The reaction in the ears was determined by the thickness of ears as measured by a thickness gauge (Mitutoyo Tokyo No. 7326; 0–0.0050 inch).

In vivo Activity

Topical application of AA to the ears of mice leads to a rapid (1 h) edematous response with a longer term influx of PMNs. When DuP 654 was coapplied with the AA, a dose-dependent inhibition of the edema was observed. It had an ED_{50} (i.e., dose which reduced the edema by 50%) of 11 μg/ear. It was approximately 10- and 100-fold more potent than REV 5901 and Lonapalene, respectively (table 1).

PMN infiltration was observed 4–8 h after the application of AA. When DuP 654 10 μg was coadministered with the AA, the PMN influx was reduced by over 75% (fig. 1). This effect lasted throughout the 6-hour experimental period. Phenidone, 100 μg/ear, demonstrated comparable activity.

The delayed-type hypersensitivity (DTH) response to DNFB closely approximates contact dermatitis seen clinically. For this reason, activity in this model is seen as particularly important. DuP 654 had an ED_{50} of 0.9 μg/ear. Dexamethasone had an ED_{50} of 0.7 μg/ear, while phenidone had an ED_{50} of 15 μg/ear. Rev 5901 and Lonapalene were inactive in this model.

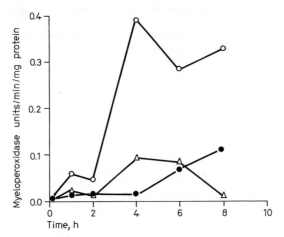

Fig. 1. Myeloperoxidase in AA-treated ears. AA, 1 mg (○), was applied to the ears of male CF_2 mice. Groups of animals were treated as above with DuP 654, 10 μg/ear (△), or phenidone, 100 μg/ear (●). At varying times after challenge the animals were sacrificed, ears removed and a 6-mm punch obtained. The tissue was homogenized and the myeloperoxidase levels determined by the method of Bradley et al. [8].

Discussion

In experiments not presented, DuP 654 was shown to be a potent inhibitor of the 5-LO pathway. Using the rat basophil leukemia (RBL)-derived enzyme, DuP 654 had an IC_{50} (i.e., concentration which prevented 5-HETE production by 50%) of 2.4×10^{-8} M. It also prevented LtC_4 production by zymosan-stimulated resident mouse peritoneal macrophages with an IC_{50} of 3.1×10^{-7} M.

The high level of activity of DuP 654 in the AA-induced ear model can probably be accounted for based on the aforementioned lipoxygenase inhibitory activity. The edematous response and the cell influx can be correlated with elevations in 5-LO products [7]. In studies not reported, treatment with DuP 654, 10 μg/ear, led to greater than 90% reductions of the 5-LO products 5-HETE and LtB_4.

Perhaps the most surprising finding was the high level of potency and efficacy seen in the DTH model. This is primarily a mononuclear cell-dependent system with virtually no literature basis for a significant leukotriene involvement. One mediator of potential importance in this model is interleukin-1 (IL-1). This has been reported to be 'proinflammatory' in a variety of model systems [9] and is present in elevated amounts in skin inflammatory conditions [10]. Recently, inhibitors of the lipoxygenase pathway have been reported to reduce the release of this mediator from monocytes [11]. Studies are in progress to determine if blockade of IL-1 release contributes to the mechanism of action of DuP 654.

References

1   Camp RDR, Mallet AI, Cunningham FM, et al: The role of chemoattractant lipoxygenase products in the pathogenesis of psoriasis. Br J Dermatol 1985;113(suppl 28):98–103.
2   Brain SD, Camp RDR, Kobza Black A, et al: Leukotrienes C_4 and D_4 in psoriatic skin lesions. Prostaglandins 1985;29:611–619.
3   Camp RDR, Couts AA, Greaves MW, et al: Response of human skin to the intradermal injection of leukotrienes C_4, D_4 and B_4. Br J Pharmacol 1983;80:497–502.
4   Dowd PM, Kobza Black A, Woolard PW, et al: Cutaneous responses to 12-hydroxy-5,8,10,14-eicosatetraenoic acid (12-HETE) and 5,12-dihydroxyeicosatetraenoic acid (leukotriene B_4) in psoriasis and normal human skin. Arch Dermatol Res 1987;279:427–434.
5   Ackerman NR, Arner EC, Mackin WM, et al: Anti-inflammatory consequences of 5-lipoxygenase inhibition. First International Conference on Leukotrienes and Prostanoids in Health and Disease, Tel Aviv 1985; Zor et al (eds): Advances in Prostaglandin, Thromboxane and Leukotriene Research. New York, Raven Press, 1986, pp 47–62.
6   Young JM, Spires DA, Bedord CJ, et al: The mouse ear inflammatory response to topical arachidonic acid. J. Invest Dermatol 1984;82:367–371.
7   Opas EE, Bonney RJ, Humes JL: Prostaglandin and leukotriene synthesis in mouse ears inflamed by arachidonic acid. J. Invest Dermatol 1985;84:253–256.
8   Bradley PP, Priebat DA, Christensen RD, et al: Measurement of cutaneous inflammation: estimation of neutrophil content with an enzyme marker. J Invest Dermatol 1982;78:206–209.
9   Larrick JW, Kunkel SL: The role of tumor necrosis factor and interleukin-1 in the immunoinflammatory response. Pharm Res 1988;5:129–139.
10   Camp RDR, Fincham NJ, Cunningham FM, et al: Psoriatic skin lesions contain biologically active amounts of an interleukin-1-like compound. J Immunol 1986;137:3469–3474.
11   Dinarello CA, Bishai I, Rosenwasser LJ, et al: The influence of lipoxygenase inhibitors in the in vitro production of human leukocyte pyrogen and lymphocyte-activating activity (interleukin-1). Int J Immunopharmacol 1984;6:43–50.

Neil R. Ackerman, PhD, E.I. du Pont de Nemours & Co., Inc., Medical Products Department, Experimental Station, E400/2273, PO Box 80400, Wilmington, DE 19880–0400 (USA)

Zor U, Naor Z, Danon A (eds): Leukotrienes and Prostanoids in Health and Disease.
New Trends Lipid Mediators Res. Basel, Karger, 1989, vol 3, pp 124–129

Pemedolac: A Novel Nonopiate Analgesic

Barry M. Weichman, Thuy T. Chau

Division of Immunopharmacology, Wyeth-Ayerst Research, Princeton, N.J., USA

Introduction

Among the pharmacological properties shared by the nonsteroidal anti-inflammatory drugs (NSAIDs) are anti-inflammatory actions, analgesic effects, and a gastroirritant side effect liability. Mechanistically, each of these actions may result from the inhibition of prostaglandin biosynthesis either at the inflammatory lesion, at an undefined site in pain transmission, or in the gastrointestinal mucosa, respectively [1–3]. In general, NSAIDs exhibit each of these three properties over a similar dose range.

Pemedolac (cis-1-ethyl-1,3,4,9-tetrahydro-4-(phenylmethyl)pyrano-[3,4-b]indole-1-acetic acid; fig. 1) is structurally derived from the NSAID, etodolac [4]. However, despite its ability to inhibit prostaglandin biosynthesis, pemedolac differs in pharmacologic profile from the NSAIDs. In experimental animal models, pemedolac selectively displays analgesic properties, whereas the anti-inflammatory and gastroirritant actions are noted at much higher doses.

Effect on Prostaglandin Biosynthesis

In vitro, pemedolac inhibited A23187-induced thromboxane (Tx) B_2 and prostaglandin (PG) E_2 biosynthesis from oyster glycogen-elicited rat peritoneal neutrophils with IC_{50} values of 0.02 and 0.03 μM, respectively. Pemedolac did not alter 5-HETE or leukotriene (LT) B_4 generation at concentrations of 1 and 10 μM, although a small increase in the lipoxygenase metabolites was noted at 0.1 μM.

Injection of 0.6% acetic acid into the mouse peritoneal cavity produced a rapid generation of PGI_2 (quantitated by radioimmunoassay of 6-keto-$PGF_{1\alpha}$), which functioned in a synergistic fashion with the acetic

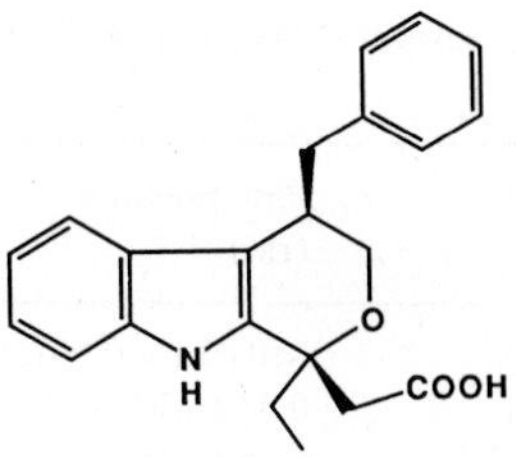

Fig. 1. Structure of pemedolac.

acid to induce a writing response, a quantitative measure of pain [5]. Pemedolac dose dependently inhibited the acetic acid-induced biosynthesis of 6-keto-PGF$_{1\alpha}$ with an ID$_{50}$ of 0.44 mg/kg p.o. [6]. Inhibition of 6-keto-PGF$_{1\alpha}$ was correlated with an inhibition of the writing response for pemedolac, indomethacin, ibuprofen, and etodolac. Thus, pemedolac functions in this mouse analgesic model similarly to an NSAID.

Analgesic Profile

Pemedolac displayed potent analgesic actions in mouse, rat, and dog models, with its potency defined by ED$_{50}$ values of 2.0 mg/kg p.o. or less. Pemedolac inhibited the writhing response induced by acetylcholine (ED$_{50}$ = 0.3 mg/kg), acetic acid (ED$_{50}$ = 1.2 mg/kg), and phenylbenzoquinone (ED$_{50}$ = 2.0 mg/kg) in the mouse. This differential potency in the three writhing models was also observed for the peripherally acting analgesics, whereas the opiates (morphine and codeine) exhibited a similar potency in the three models. In these mouse models, pemedolac was similar in potency with indomethacin, piroxicam and zomepirac (table 1). In the Randall-Selitto paw pressure model in rats, hyperalgesia was induced with brewers' yeast, Freund's complete adjuvant (FCA), or PGE$_2$. Pemedolac increased the pain threshold of the hyperalgesic paw in a concentration-dependent manner; the ED$_{50}$ values were ≤ 0.03 mg/kg p.o., but the slope of the dose-response curves were very shallow. In contrast, NSAIDs such as piroxicam, indomethacin and etodolac had steeper dose-response curves with ED$_{50}$ values of 0.35, 0.40 and 4.5 mg/kg p.o., respectively, with FCA as the hyperalgesic stimulus. In the rat acetic acid writhing model, pemedolac had an ED$_{50}$ of 2.0 mg/kg relative to indomethacin's ED$_{50}$ of 3.6 mg/kg. In the dog urate synovitis paw pressure model, pemedolac and indomethacin were equipotent with ED$_{50}$ values of 0.17 and 0.20 mg/kg p.o., respectively.

The analgesic properties of pemedolac were independent of an opiate mechanism. Prior treatment of mice or rats with naloxone (1 mg/kg s.c.)

Table 1. Comparison of pemedolac with reference NSAIDs in several animal analgesic models (ED$_{50}$: mg/kg p.o. (95% CL))

Drug	PBQ writing[1] (mouse)	AA writing[1] (rat)	Paw pressure[2] (rat)
Pemedolac	2.0 (0.5–8.3)	2.0 (0.8–4.5)	<0.03 (<0.001–0.38)
Indomethacin	1.5 (0.9–2.8)	3.6 (2.2–6.1)	0.40 (0.17–0.94)
Piroxicam	3.4 (1.6–6.9)	7.0 (1.4–36)	0.35 (0.18–0.66)
Zomepirac	4.7 (2.6–8.4)	NT	0.60 (0.30–1.2)
Ibuprofen	32 (22–49)	55 (24–124)	65 (33–126)
Naproxen	57 (29–112)	4.5 (2.1–9.8)	14 (6.2–30)
Diflunisal	133 (78–227)	NT	1.1 (0.5–2.2)

[1] Writhing was induced in Swiss albino mice (15–25 g) by the i.p. injection of 0.02% 2-phenyl-1,4-benzoquinone (0.15 ml/10 g body weight) or in male Sprague-Dawley rats (140–190 g) by i.p. injection of 1% acetic acid (0.5 ml/100 g body weight). Writhing was determined for the 0–15 min and 5–20 min period for the PBQ and acetic acid studies, respectively. Drug or vehicle was administered 60 min prior to the PBQ or acetic acid. The ED$_{50}$ represents the dose of drug at which writhing was inhibited by 50%.

[2] Freund's complete adjuvant (0.1 ml of 5 mg/ml *Mycobacterium butyricum*) was injected into the hindpaw. At 24 h, rats received either drug or vehicle, and then 1 h later, the pain threshold of the injected paw was determined. The number of drug-treated animals having a threshold at least 50% greater than the vehicle-treated animals was determined. The ED$_{50}$ is the dose at which this criterion was established in 50% of the rats.

did not antagonize the analgesic activity of pemedolac, whereas it completely blocked that of morphine. Pemedolac (10 μM) did not compete for binding to the opiate receptor in rat brain homogenates nor did it interact with α- or β-adrenergic, serotonergic, histaminergic or muscarinic receptors in various isolated tissue preparations at 1 μM. At a dose of 50 mg/kg p.o., pemedolac was inactive in the mouse tail-flick test and exhibited a weak effect in the mouse hot-plate assay. Further, tolerance did not develop in mice or rats upon repeated administration of pemedolac.

Separation of Analgesic Actions from Anti-inflammatory and Gastroirritant Effects

Inasmuch as pemedolac was capable of inhibiting prostaglandin biosynthesis and was structurally derived from the NSAID, etodolac, it was anticipated that it would share the anti-inflammatory and gastroirritant properties common to the NSAIDs. However, these latter activities were only noted at doses much higher than the analgesic doses making pemedolac's pharmacologic profile distinct from the NSAIDs. Pemedolac was

Table 2. Comparison of pemedolac with NSAIDs in models of acute inflammation in rats

Drug	Carrageenan paw edema[1] ED_{50} (95% CL)	Preventative adjuvant edema[2] ED_{50} (95% CL)	RPAR pleurisy[3] ED_{30} (95% CL)
Pemedolac	48% inh at 100 mpk	37% inh at 25 mpk	46 (16–131)
Piroxicam	1.3 (0.2–9.0)	3.2 (1.4–7.2)	0.9 (0.3–2.5)
Indomethacin	2.9 (1.1–7.9)	1.0 (0.4–2.7)	2.5 (1.4–4.5)
Naproxen	2.9 (0.12–73)	31 (9–107)	3.9 (2.1–7.4)
Ibuprofen	21 (8–57)	51 (14–180)	NT

[1] λ-Carrageenan (0.1 ml of 2% suspension in saline) was injected intradermally into the left hindpaw of male Sprague-Dawley rats (150–200 g). Drug or vehicle was administered 60 min prior to the carrageenan. Paw edema was defined by the change in paw volume at 3 h after carrageenan. The ED_{50} represents the dose of drug (mg/kg p.o.) at which edema was inhibited by 50% relative to vehicle-treated rats.
[2] Freund's complete adjuvant (0.1 ml of 5 mg *Mycobacterium butyricum*/ml mineral oil) was injected intradermally in the left hindpaw of male Sprague-Dawley rats (180–200 g) on day 0. Drug or vehicle was administered on days 0, 1 and 2. Paw edema was the difference in paw volumes on day 0 (before FCA) and on day 3 (24 h after last drug dose). The ED_{50} represents the dose of drug at which paw edema was inhibited by 50% relative to vehicle treated rats.
[3] A reverse passive Arthus reaction (RPAR) was induced in the pleural cavity of male Lewis rats (150–200 g) by injection of 5 mg bovine serum albumin (BSA) i.v. followed 30 min later by intrapleural injection of 1 mg anti-BSA. Drug or vehicle was administered 60 min prior to the anti-BSA. The volume of fluid in the pleural cavity was measured at 4 h after anti-BSA. The ED_{30} represents the dose of drug at which fluid accumulation was inhibited by 30% relative to vehicle-treated rats.

compared to reference NSAIDs in three standard models of acute inflammation in rats. Pemedolac inhibited carrageenan-induced paw edema (48% inhibition at 100 mg/kg p.o.), fluid accumulation in the pleural cavity following a reverse passive Arthus reaction (ED_{30} = 46 mg/kg p.o.), and the day 3 primary inflammation of the FCA-injected hindpaw (37% inhibition at 25 mg/kg/day p.o. × 3 days). In each of these inflammation models, indomethacin, piroxicam, naproxen, and ibuprofen exhibited anti-inflammatory activity at doses similar to their respective analgesic doses (table 2).

Gastroirritation was noted in 50% of the rats (UD_{50}) at 107 mg/kg p.o. for a single dose of pemedolac given to fasted rats and at 140 mg/kg/day p.o. × 4 days when administered daily to fed rats. Thus, the gastroirritant properties of pemedolac were noted at doses similar to those required for its anti-inflammatory actions.

Two additional actions common to the NSAID class are antipyresis and inhibitory effects on platelet aggregation. In yeast-induced hyperthermic rats, pemedolac exhibited antipyretic actions over the 10–20 mg/kg p.o. dose range, whereas the drug did not affect body temperature in normothermic animals. Aggregation of human platelets was inhibited by high concentrations of pemedolac (e.g., collagen-induced aggregation decreased by 57% at 50 μM).

Site of Action

The mechanism underlying pemedolac's selective actions as an analgesic remains to be determined. Preliminary pharmacokinetic analysis indicated pemedolac to be rapidly absorbed ($t_{max} = 30$–60 min) with a $t_{1/2}$ in plasma ranging from 4 h in mice to 6–7 h in rats. In the animal models, pemedolac displayed a functional analgesic $t_{1/2}$ of 10 h in the mouse PBQ writhing assay when administered at 10 mg/kg p.o. (5 × ED_{50}). Similarly, analgesic activity has been noted through 16 h in the rat paw pressure model.

Two lines of evidence suggest pemedolac does not function centrally. Fist, drug distribution studies in rats with [^{14}C]-pemedolac did not detect significant concentrations of pemedolac in the brain nor was drug accumulation noted in other tissues. Second, intracerebroventricular administration of the sodium salt of pemedolac to mice resulted in a dose-dependent inhibition of the writing response. However, the i.v./i.c.v. dose ratio for pemedolac was only 2 (morphine ratio = 20), suggesting potential drug leakage into the periphery. This latter point was confirmed by the observation that i.c.v. pemedolac inhibited the acetic acid-induced biosynthesis of 6-keto-PGF$_{1\alpha}$ in the mouse peritoneal cavity.

To summarize: pemedolac displays analgesic activity at doses lower than its anti-inflammatory and gastroirritant actions, and as such, appears distinguished from the typical NSAID. The ultimate clinical utility of pemedolac will depend upon the confirmation of its unique pharmacologic profile in man and the determination of its analgesic efficacy. Initial clinical and animal toxicological studies have confirmed the safety of pemedolac.

References

1   Vane JR: Inhibition of prostaglandin synthesis as a mechanism of action of aspirin-like drugs. Nature New Biol 1971;231:232–235.
2   Ferreira SH: Prostaglandin hyperalgesia and the control of inflammatory pain; in Bonta IL, Bray MA, Parnham MJ (eds): The Pharmacology of Inflammation. Handbook of Inflammation. New York, Elsevier Science, 1985, vol 5, pp 107–116.

3   Robert A: Cytoprotection by prostaglandins. Gastroenterology 1979;77:761–767.
4   Humber LG: Etodolac: The chemistry, pharmacology, metabolic disposition, and clinical profile of a novel antiinflammatory pyranocarboxylic acid. Med Res Rev 1987;7:1–28.
5   Collier HOJ, Dinneen LC, Johnson CA, et al: The abdominal constriction response and its suppression by analgesic drugs in the mouse. Br J Pharmacol Chemother 1968;32:295–310.
6   Berkenkopf JW, Weichman BM: Production of prostacyclin in mice following intraperitoneal injection of acetic acid, phenylbenzoquinone and zymosan: Its role in the writhing response. Prostaglandins 1988;36:693–709.

Barry M. Weichman, PhD, Division of Immunopharmacology, Wyeth-Ayerst Research, CN-8000, Princeton, NJ 08543-8000 (USA)

Zor U, Naor Z, Danon A (eds): Leukotrienes and Prostanoids in Health and Disease.
New Trends Lipid Mediators Res. Basel, Karger, 1989, vol 3, pp 130–134

Immunomodulatory Properties of Platelet-Activating Factor

Pierre Braquet, David Hosford

Institut Henri Beaufour, Le Plessis Robinson, France

Platelet-activating factor (PAF) is a phospholipid mediator of allergic reactions also involved in various acute inflammatory processes. There is now accumulating evidence that this autacoid plays a major role in the regulation of various compartments of the immune response [1], including modulation of interleukin-1 and 2 (IL-1, IL-2) and tumor necrosis factor (TNF) production, regulation of helper and suppressive T cell proliferation and control of natural killer (NK) cell activity. Until recently, most of these studies had been performed in vitro and the possible in vivo involvement of PAF remained uncertain. However, the prolongation of cardiac allograft survival in azathioprine-treated animals by the specific PAF antagonist BN 52021 provided the first evidence that PAF was indeed involved in the in vivo regulation of immune processes. In this short review we will summarize the effects of the autacoid on various immune functions.

In vitro Immunomodulatory Effects of PAF

When human peripheral blood mononuclear leukocytes (PBML) are cultured for 72 h in the presence of the mitogen phytohemagglutinin A (PHA), their proliferation is markedly impaired in the presence of PAF (10 nM to 1 μM). Addition of PAF to the lymphocytes at the beginning of the culture period induces maximal suppression. A similar effect is observed when PBML are depleted of monocytes [2]. The inhibition of PBML proliferation by PAF is prevented by indomethacin, indicating that in this system, PAF may exert its effect via the generation of various cyclooxygenase metabolites.

Indeed, prostaglandin E_2 (PGE_2) may exert a powerful suppressive effect on lymphocyte proliferation [2]. PAF has been also shown to significantly reduce IL-2 production by PBML. Inhibition of lymphoctye proliferation and IL-2 production induced by PAF are not observed in the presence of the PAF antagonist, BN 52021 ($10 \mu M$) [2, 3]. Recently, Ward et al. [4] reported that IL-2-induced proliferation of human T lymphoblasts is dose dependently antagonized by two structurally unrelated PAF antagonists, L-652,731 and CV-3988, but not BN 52021. In addition, PAF and two synthetic analogues of PAF have been shown to enhance the proliferation of IL-2-stimulated human lymphoblasts [4].

As demonstrated by coculture experiments, PAF induces significant suppressor cell activity at concentrations ranging from 0.1 pM to 10 nM [5]. The expression of suppressor cell function during the coculture is abrogated by indomethacin, suggesting that cyclooxygenase metabolites of arachidonic acid are involved in the suppressive effect of PAF.

PAF receptor antagonists such as BN 52021, WEB 2086 and CV-3988 induce significant suppressor cell activity, although to a lesser extent than PAF. Interestingly, when added concomitantly with PAF to PBML, BN 52021 partially inhibits the effect of the mediator [5]. When incubated with PAF for 24 h, monocyte-depleted lymphocyte populations show a 30% decrease in $CD4^+$ T cell number and a 50% increase in $CD8^+$ T cell number. At extremely high doses ($10 \mu M$) PAF suppresses $CD4^+$ T cell proliferation [6]. Maximal inhibition occurs when PAF is present during the first 24 h of the cell culture period and is not associated with an alteration in IL-2 production.

Stimulated macrophages release both PAF [7] and IL-1 [8]. The capability of IL-1 to participate in many aspects of the immune response suggests that PAF is also an important regulatory molecule of immunoinflammatory reactions. Depending on the dose of the mediator, opposite effects on lipopolysaccharide (LPS)-induced IL-1 production are observed [9]. PAF at 1 pM increases IL-1 production, whereas at 0.1 μM a decrease is observed. BN 52021 reverses by more than 70% the increase or inhibition of IL-1 production induced by PAF, while having no direct effect itself on lymphokine release.

In addition to modulating IL-1 and IL-2 synthesis, PAF can also influence TNF production. We have shown that PAF stimulates TNF production from peripheral blood derived monocytes and at picomolar concentrations amplifies LPS-induced TNF production, effects inhibited by various PAF antagonists [10]. PAF also acts synergistically with interferon-γ (INF-γ) to increase the monocyte cytotoxicity.

In addition to directly modulating cell activity, at very low concentrations both PAF and TNF can 'prime' cells such as neutrophils (PMN) to respond in an enhanced manner to subsequent agonistic stimuli that would otherwise be ineffectual. Amplified responses including aggregation, adhesiveness, superoxide production and elastase release have been reported using FMLP as the inducing agonist following priming with PAF [11]. Furthermore, we have recently shown that TNF can prime the PAF-induced superoxide generation by human PMN [12], the enhancing effect of the mediator being completely abolished by BN 52021, kadsurenone, BN 52111 and WEB 2086. The PAF antagonists also decrease by 50% the superoxide production elicited solely by TNF, indicating that TNF-induced superoxide generation is partially mediated by a mechanism involving endogenous PAF. These results indicate the importance of interactions between PAF and TNF in immunoinflammatory responses.

Addition of PAF (0.1 pM to 1 nM) increases NK cell activity [13]. On the contrary, significant inhibition of NK cell activity is observed at 10 μM PAF, possibly due to a toxic effect of the autacoid. BN 52021 blocks the suppression of NK cell activity induced by 10 μM PAF and increases the PAF-induced enhancement of NK function at 10 pM to 0.1 μM. Addition of PAF 18 h before or at the start of the NK assay induces a similar increase of NK activity suggesting an effect of the autacoid on target cells. Mandi et al. [14] have shown that high concentrations of BN 52021 (30–120 μM) induce an inhibition of human cytotoxicity. This inhibitory effect of BN 52021 on NK cell activity is similar regardless of the time of its addition to the culture, i.e. added to lymphocytes 60 min before mixing with target cells or added immediately at the start of the reaction. However, the inhibition is higher when the target cells are preincubated for 90 min prior to the beginning of the assay. Interestingly, preincubation of lymphocytes with BN 52021 for 18 h does not cause an inhibition of cytotoxicity.

Effects of PAF on ex vivo Cytokine Production

After 7 days' instillation with a total amount of 4.5 or 9 μg PAF into the rat via a subcutaneously implanted osmotic minipump, monocytes exhibit an increase in the LPS-induced IL-1 production. In contrast, when the rats are treated with 28 μg PAF, a marked decrease in the LPS-induced IL-1 production is observed [15]. Splenocytes from rats receiving 1 μg PAF exhibit an increased ex vivo capability to produce IL-2 after stimulation with Con A. Higher doses of PAF only produce minimal effects on Con A-induced IL-2 production. These alterations of IL-2 and IL-1 production evoked by PAF are not observed when the animals are concomitantly treated with BN 52021.

Conclusion

These studies indicate that PAF can modulate various immune processes both in vitro and ex vivo. Although its precise mechanism of action remains unclear, the data reviewed here support the concept that PAF is not only a mediator of acute allergic and inflammatory reactions but also contributes to a long-term process such as chronic diseases. Further studies are required to determine the role of endogenously produced PAF in the regulation of the immune response.

References

1   Braquet P, Rola-Pleszczynski M: Platelet-activating factor and cellular immune responses. Immunol Today 1987;8:345–352.
2   Rola-Pleszczynski M, Pignol B, Pouliot C, et al: Inhibition of human lymphocyte proliferation and interleukin-2 production by platelet-activating factor (PAF-acether): reversal by a specific antagonist, BN 52021. Biochem Biophys Res Commun 1987;142:754–760.
3   Pignol B, Henane S, Mencia-Huerta JM, et al: Platelet-activating factor (PAF) inhibits interleukin-2 (IL-2) production and proliferation of human lymphocytes (abstract). 10th Int Congr Pharmacology, Sydney 1987.
4   Ward SG, Lewis GP, Westwick J: A role of platelet-activating factor (PAF) in human lymphocyte proliferation; in Paubert-Braquet M, Braquet P, Demling R, et al (eds): Lipid Mediators in the Immunology of Shock. Nato ASI Series. New York, Plenum Press, 1987, pp 483–493.
5   Rola-Pleszczynski M, Pouliot C, Pignol B, et al: Platelet-activating factor induces human suppressor cell activity. Fed Proc 1987;46:743.
6   Dulioust A, Vivier E, Salem P, et al: Immunoregulatory functions of PAF. Effect of PAF on CD4+ cell proliferation. J Immunol 1988;140:240–245.
7   Braquet P, Touqui L, Shen TY, et al: Perspectives in platelet-activating factor research. Pharmacol Rev 1987;39:97–145.
8   Dinarello CA: Interleukin-1 and the pathogenesis of the acute-phase response. N Engl J Med 1984;311:1413–1419.
9   Pignol B, Henane S, Mencia-Huerta JM, et al: Effect of PAF-acether (platelet-activating factor) and its specific antagonist, BN 52021, on interleukin-1 (IL-1) synthesis and release by rat monocytes. Prostaglandins 1987;33:931–939.
10 Bonavida B, Braquet P: Effect of platelet-activating factor (PAF) on monocyte activation and production of tumor necrosis factor (TNF). Prostaglandins 1988;35:802A.
11 Vercellotti GM, Yin HQ, Gustafson KS, et al: Platelet-activating factor primes neutrophil responses to agonists: role in promoting neutrophil-mediated endothelial damage. Blood. 1988;71:1100–1106.
12 Paubert-Braquet M, Longchamp MO, Koltz PL, et al: Tumor necrosis factor (TNF) primes platelet-activating factor (PAF)-induced superoxide generation by human neutrophils (PMN): consequences in promoting PMN-mediated endothelial cell (EC) damage. Prostaglandins 1988;35:803A.
13 Rola-Pleszczynski M, Tarcotte S, Cagnon L, et al: Enhancement of natural killer cell activity by platelet-activating factor; in Braquet P (ed): New Trends in Lipid Mediators

Research, vol 1: Platelet-Activating Factor and Cell Immunology. Basel, Karger, 1988, pp 89–98.

14   Mandi Y, Farkas G, Koltai M, et al: The effect of BN 52021, a PAF-acether antagonist, on natural killer activity; in Braquet P (ed): New Trends in Lipid Mediators Research, vol. 1: Platelet-Activating Factor and Cell Immunology. Basel, Karger, 1988, pp 76–88.

15   Pignol B, Henane S, Sorlin B, et al: Effect of long-term in vivo treatment with platelet-activating factor on interleukin-1 and interleukin-2 production by rat splenocytes; in Braquet P (ed): New Trends in Lipid Mediators Research, vol 1: Platelet-Activating Factor and Cell Immunology. Basle, Karger, 1988, pp 38–43.

Pierre Braquet, PhD, DSc, Institut Henri Beaufour, 17, avenue Descartes, F–92350 Le Plessis Robinson (France)

Zor U, Naor Z, Danon A (eds): Leukotrienes and Prostanoids in Health and Disease.
New Trends Lipid Mediators Res. Basel, Karger, 1989, vol 3, pp 135–142

Leukotrienes and Pulmonary Inflammation

M.A. Bray

Research Department, Ciba-Geigy AG, Basel, Switzerland

Introduction

The involvement of inflammation and inflammatory mediators in the pathogenesis and maintenance of pulmonary disease is now generally accepted and research in this area has been concentrated on the role of inflammation in the more chronic aspects of asthma, in particular, the hyperreactivity and lung tissue damage seen in many cases of this disease. In this context the role of the leukotrienes (LTs) is being widely studied and the consequences of the inflammatory effects of these mediators need to be taken into account when examining their importance in chronic pulmonary disease, particularly with reference to the assessment of the possible therapeutic activity of inhibitors of 5-lipoxygenase or LT antagonists.

In this brief review I shall discuss the effects of the LTs on lung tissue in terms of their ability to induce the signs and symptoms of an inflammatory response in the lung following an allergenic stimulus.

Leukotriene Formation in the Lung

Considerable evidence is now available to demonstrate that both peptido-LTs and LTB_4 are generated in the lungs of animals following either antigen challenge or challenge with other stimuli. For example, LT levels have been shown to be elevated during the late bronchial response following allergen challenge in susceptible sheep [1]. Whilst there is comparatively little data on the presence of LTs in the lung of patients undergoing an asthma attack or following antigen challenge, raised levels of LTs have been detected in the blood of children during acute asthma attacks [2]. Also, challenge of human lung tissue in vitro with specific antigen following passive sensitization causes release of both peptido-LTs and LTB$_4$ [3] and the release of, at least, the peptido-LTs has been correlated with bronchial contraction using allergen challenged isolated human lung tissue [4]. Measurement of LT levels in broncho-alveolar lavage (BAL) fluid or nasal washouts have also been used to demonstrate the presence of LTs in lung disease, for example, raised levels of LTB$_4$ have been detected in the BAL fluid of infants with chronic obstructive pulmonary disease (COPD) [5], whilst LTE$_4$ and 20-OH-LTB$_4$ have been detected in BAL fluid from asthma patients [6]. Also, nasal washout fluids from allergic rhinitis patients following challenge contain raised levels of peptido-LTs [7]. The cells responsible for endogenous pulmonary generation of the LTs in the noninflamed lung could include the tissue mast cell, alveolar macrophage (AM) and the pulmonary epithelium. Several studies have shown that challenge of human lung mast cells in vitro with anti-IgE causes the release of LTs although the LTB$_4$ levels are low [8, 9] and a recent study has been demonstrated that mast cells cultured for longer periods release both LTC$_4$ and LTB$_4$ after challenge with anti-IgE [10].

Human lung macrophages in BAL fluid generate primarily LTD$_4$ in culture [11] and this study also demonstrated an increased generation of LTD$_4$ from macrophages derived from allergic asthmatics when compared to healthy subjects. On the other hand, stimulation of AM with the calcium ionophore A23187 (CaI) in vitro leads to the formation primarily of LTB$_4$ [12]. Animal studies have shown that AM release LTs following challenge with IgE immune complexes (mainly LTC$_4$) in both mice and rats [13–15]. Such immune complex induced stimulation presumably acts via binding to low affinity IgE receptors on the target AM. Heat aggregated IgG stimulates the release of LTB$_4$ from human AM when incubated in the presence of gamma-interferon (to enhance the expression of IgG Fc receptors) [16] and it will be of interest to see whether aggregated IgE or IgE immune complexes have similar properties as aggregated IgE has been shown to stimulate LTC$_4$ and LTB$_4$ release from human blood monocytes to the same extent as aggregated IgG [17]. Finally, data from studies of canine pulmonary epithelial cells in culture have shown that these cells have the capacity to generate LTB$_4$ when incubated in the presence of arachidonic acid [18].

Inflammatory cells recruited into the lung following antigen challenge such as neutrophils, eosinophils, platelets, monocytes (macrophages) and lymphocytes may also contribute to pulmonary LT generation. A number of stimuli cause neutrophils to releaese, almost exclusively, LTB$_4$ [reviewed in 19] whilst human eosinophils generate primarily peptido-LTs although they can also form LTB$_4$ following stimulation with CaI [20, 21]. Human blood monocytes have the capacity to generate all the LTs and platelets have also been shown to release LTB_4 and LTC_4 when stimulated with CaI plus arachidonic acid.

Leukotrienes and Pulmonary Oedema

Pulmonary oedema has been correlated with the development of bronchial hyperreactivity in animal models, for instance, guinea pigs inhaling smoke develop oedema and the peak of this response correlates with the maximal increase in histamine hyperreactivity [22]. It has been speculated that the thickening of the bronchial tree seen as a clinical feature of the asthmatic lung could be contributed to by a chronic increase in pulmonary fluid and cell extravasation as well as smooth muscle hyperplasia [22].

The direct effects of the peptido-LTs on pulmonary hemodynamics have been widely studied and they appear to be potent vasoconstrictors although there is considerable species variation [23]. In man, application of LTD_4 to the nasal mucosa results in an increase in blood flow but no increase in nasal resistance or fluid secretion [24]. These data are in contrast to animal studies where the peptido-LTs appear capable of causing oedema in the lung of dogs [25], sheep [26] and rabbits [27]. LTB_4 has been shown to be a stimulator of peripheral vascular permeability in animals and this response is dependent on the presence of circulating neutrophils [28]; however, LTB_4 does not appear to modify vascular tone. Studies on isolated perfused guinea pig lungs indicate that LTB_4 can cause extravascular fluid accumulation in the absence of blood elements as infusion of this LT causes a gradual increase in lung weight [29]. Staub et al. [30] noted a small increase in extravascular fluid and protein following LTB_4 infusion in sheep but concluded that the effect could be due to passive extravasation occurring during the passage of neutrophils across the vascular endothelium and alveolar epithelium rather than via a direct effect of the LT.

Leukotrienes and Pulmonary Mucus Transport and Secretion

Leukotrienes have been shown to be able to alter mucus velocity in both animals and man [31, 32], possibly by lowering ciliary beat frequency [33]. The ability of LTs to stimulate the production of mucus in the lung has been demonstrated using both animal and human tracheal tissue explants [34–37]. Studies in our laboratories using porcine trachea have demonstrated that nanogram quantities of both LTD_4 and LTC_4 rapidly stimulate fluid secretion from tracheal mucus glands; however, both LTE_4 and LTB_4 were inactive at doses up to $1.0 \, \mu M$ [Subramanian et al., submitted].

Leukotrienes and Inflammatory Cell Accumulation and Activation

Most available data points towards LTB_4 as being the LT with the capacity to recruit cells to sites of inflammation [19, 38]. Intravenous infusion of LTB_4 into sheep has been shown to cause a rapid accumulation of ^{111}In-labelled neutrophils into the lung [39] and in dogs aerosol application of LTB_4 into the lung can cause neutrophil accumulation and hyperreactivity to subsequent challenges with acetylcholine [40]. This airway hyperreactivity correlates with the accumulation of neutrophils into the BAL fluid and increased thromboxane B_2 formation. Very little data are available regarding the chemotactic activity of the peptido-LTs; however, one report describes the LTD_4 and LTE_4 induced accumulation of eosinophils into the conjunctiva of the eyes of guinea pigs following topical application [41].

Whilst it is tempting to speculate that LTB_4 or the peptido-LTs may be responsible for, at least, part of the acute inflammatory cell accumulation seen in asthma, there are other candidate mediators such as the platelet-activating factor (PAF), the colony-stimulating factors (CSFs) and high-molecular-weight chemotactic factors from mast cells. LTB_4 may also contribute to pulmonary monocyte accumulation as LTB_4 is a potent macrophage and lymphocyte chemotactic factor in vitro, particularly for T cells [42].

Activation of acute inflammatory cells in the lung may cause release of products such as toxic oxygen metabolites and destructive lysosomal enzymes which are contributors to the irreversible tissue damage seen as a feature of chronic asthma. Neutrophils generate elastase and other proteases whilst eosinophils can generate a cationic peptide and major basic protein capable of damaging pulmonary epithelium. LTB_4 is a relatively weak stimulator of neutrophil degranulation and superoxide anion generation [43]. Also, both LTB_4 and LTD_4 are able to stimulate the AM respiratory burst [44]; however, the effect is weak when compared to zymosan or phorbol myristate acetate.

Leukotrienes and the Induction of Allergic Inflammation

B-cell activation is modulated by T cells and accessory cells which are involved in antigen processing and presentation. Specific T-cell subpopulations can both enhance (T-helper; T_h) or suppress (T-suppressor; T_s) B-cell responses to antigen principally via the secretion of interleukins. Recent data from studies in the mouse have demonstrated that T_h cells are able to specifically enhance the formation of IgE by specific B cells via the action of interleukin-4 (IL-4, BSF-1) [45–47]. The possibility that a similar

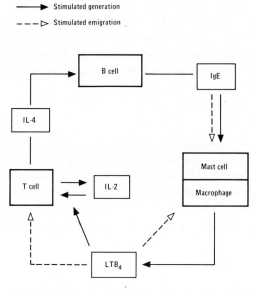

Fig. 1. Possible role of LTB_4 in enhancing local allergic reactions.

up-regulation of B-cell IgE formation occurs in asthmatics is indicated by the ability of supernatant fluids from cultures of T cells derived from the blood of atopic individuals to specifically enhance IgE formation by normal blood B cells or an EBV-transformed IgE-producing human B-cell lines [48].

Recruitment of both T_h cells and T_s cells into the lung has been demonstrated to occur in allergic asthma [49]. LTB_4 has been shown to stimulate the movement of T cells, particularly those of the T_h phenotype [42], and activation of both mast cells by IgE and macrophages/monocytes with IgE or IgG immune complexes leads to the generation of LTB_4.

A number of studies have demonstrated that LTB_4 is able to modulate interleukin generation. Thus, for example, LTB_4, in the presence of indomethacin, stimulates both interleukin-2 and gamma-interferon production by mitogen-activated human T cells in vitro [50] and interleukin-1 from human monocytes [51, 52].

Taking these data together it is possible to speculate that LTB_4 could form part of a positive feedback loop contributing to the antigen hypersensitivity characteristic of many patients with allergic diseases. Thus, LTB_4 generated from mast cells or macrophages following stimulation with IgE could help to recruit T cells to the site of antigen challenge and subsequently enhance the activation of these cells and the secretion of lymphokines which, in turn, would stimulate further IgE formation from the specific B cells (fig. 1). In this context it will be of particular interest to see whether LTB_4 is able to stimulate the formation of IL-4 from T_h cells in the same way as has been shown for interleukin-2 and gamma-interferon.

Conclusions

The data reviewed above has demonstrated that the LTs have properties consistent with their being considered as pulmonary inflammatory mediators. Thus it is premature to consider the LTs simply as mediators of acute bronchospasm in asthma and other effects of the LTs such as those outlined above should be taken into account when drugs aimed at modifying the actions of these potent lipid mediators are tested for efficacy in pulmonary disease.

References

1   Delehunt JC, Perruchoud AP, Yerger L, et al: The role of SRS-A in the late bronchial response following antigen challenge in allergic sheep. Am Rev Resp Dis 1984;130:748–754.

2   Schwartzberg SB, Shelov SP, Van Praag D: Blood leukotriene levels during acute asthma attack in children. Prostaglandins Leukotrienes Med 1987;26:143-155.

3   Salari H, Borgeat P, Fournier M, et al: Studies on the release of leukotrienes and histamine by human lung parenchymal and bronchial fragments upon immunologic and nonimmunologic stimulation. J Exp Med 1985;162:1904–1915.

4   Dahlen S-E, Hansson G, Hedqvist P, et al: Allergen contraction of lung tissue from asthmatics elicits bronchial contraction that correlates with the release of leukotrienes. Proc Natl Acad Sci USA 1983;80:1712–1716.

5   Westcott JY, Stenmark KR, Murphy RC: Analysis of leukotriene B_4 in human lung lavage by HPLC and mass spectrometry. Prostaglandins 1986;31:227–237.

6   Lam S, Chan H, Leriche JC, et al: Release of leukotrienes in patients with bronchial asthma. J Allergy Clin Immunol 1988;81:711–717.

7   Creticos PS, Peters SP, Adkinson NF Jr, et al: Peptide leukotriene release after antigen challenge in patients sensitive to ragweed. N Engl J Med 1984;310:1626–1630.

8   Peters SP, MacGlashan DW, Schulman ES, et al: Arachidonic acid metabolism in purified human lung mast cells. J Immunol 1984;132:1972–1979.

9   Peters SP, Freeland HS, Kelly SJ, et al: Is leukotriene B_4 an important mediator in human IgE-mediated allergic reactions? Am Rev Respir Dis 1987;135:S42–S45.

10 Levi-Schaffer F, Austen KF, Caulfield JP, et al: Co-culture of human lung-derived mast cells with mouse 3T3 fibroblasts: morphology and IgE-mediated release of histamine, prostaglandin D_2, and leukotrienes. J Immunol 1987;139:494–500.

11 Damon M, Chavis C, Crastes de Paulet A, et al: Arachidonic acid metabolism in alveolar macrophages. A comparison of cells from healthy subjects, allergic asthmatics and chronic bronchitis patients. Prostaglandins 1987;34:291–309.

12   Martin TR, Altman LC, Albert RK, et al: Leukotriene B_4 production by the human alveolar macrophage: A potential mechanism for amplifying inflammation in the lung. Am Rev Respir Dis 1984;129:106–111.

13   Rouzer CA, Scott WA, Hamill AL, et al: Secretion of leukotriene C and other arachidonic acid metabolites by macrophages challenged with immunoglobulin E immune complexes. J Exp Med 1982;156:1077–1086.

14   Rankin JA, Hitchcock M, Merrill W, et al: IgE-dependent release of leukotriene C_4 from alveolar macrophages. Nature 1982;297:329–331.

15   Rankin JA, Hitchcock M, Merrill WW, et al: IgE immune complexes induce immediate and prolonged release of leukotriene C_4 (LTC_4) from rat alveolar macrophages. J Immunol 1984;132:1993–1999.

16   Rankin JA, Schrader CE, Lewis RA: Leukotriene B_4 release by human alveolar macrophages incubated with IgG: Effect of gamma-interferon (abstract). Clin Res 1986;34:504A.

17   Ferreri NR, Howland WC, Spiegelberg HL: Release of leukotrienes C_4 and B_4 and prostaglandin E_2 from human monocytes stimulated with aggregated IgG, IgA, and IgE. J Immunol 1986;136:4188–4193.

18   Holtzman MJ, Aizawa H, Nadel JA, et al: Selective generation of leukotriene B_4 by tracheal epithelial cells from dogs. Biochem Biophys Res Commun 1983;114:1071–1076.

19   Bray MA: Pharmacology and pathophysiology of leukotriene B_4. Br Med Bull 1983;39:249–254.

20   Weller PF, Lee CW, Foster DW, et al: Generation and metabolism of 5-lipoxygenase pathway leukotrienes by human eosinophils: Predominant production of leukotriene C_4. Proc Natl Acad Sci USA 1983;80:7626–7630.

21   Henderson WR, Harley JB, Fauci AS: Arachidonic acid metabolism in normal and hypereosinophilic syndrome human eosinophils: generation of leukotrienes B_4, C_4, D_4 and 15-lipoxygenase products. Immunology 1984;51:679–686.

22   Hogg JC, Pare PD, Moreno R: The effect of submucosal edema on airways resistance. Am Rev Respir Dis 1987;137:S54–S56.

23   Garcia JGN, Noonan TC, Jubiz W et al: Leukotrienes and the pulmonary microcirculation. Am Rev Respir Dis 1987;136:161–169.

24   Bisgaard H, Olsson P, Bende M: Effect of leukotriene D_4 on nasal mucosal blood flow, nasal airway resistance and nasal secretions in humans. Clin Allergy 1986;16:289–297.

25   Shapiro JM, Mihm FG, Trudell JR, et al: Leukotriene D_4 increases extravascular lung water in the dog. Circ Shock 1987;21:121–128.

26   Noonan TC, Kern DF, Malik AB: Pulmonary microcirculatory responses to leukotrienes B_4, C_4 and D_4 in sheep. Prostaglandins 1985;30:419–434.

27   Farrukh IS, Sciuto AM, Spannhake EW, et al: Leukotriene D_4 increases pulmonary vascular permeability and pressure by different mechanisms in the rabbit. Am Rev Respir Dis 1986;134:229–232.

28   Wedmore CV, Williams TJ: Control of vascular permeability by polymorphonuclear leukocytes in inflammation. Nature 1981;289:646–650.

29   Noonan TC, Selig WM, Kern DF, et al: Mechanism of peptidoleukotriene-induced increases in pulmonary transvascular fluid filtration. J Appl Physiol 1986;61:1928–1934.

30   Staub NC, Schultz EL, Koike K, et al: Effect of neutrophil migration induced by leukotriene B_4 on protein permeability in sheep lung. Fed Proc 1985;44:30–35.

31   Russi EW, Abraham WM, Chapman GA, et al: Effects of leukotriene D_4 on mucociliary and respiratory function in allergic and nonallergic sheep. J Appl Physiol 1985;59:1416–1422.

32   Ahmed T, Greenblat DW, Birch S, et al: Abnormal mucociliary transport in allergic
     patients with antigen-induced bronchospasm: role of slow reacting substance of anaphy-
     laxis. Am Rev Respir Dis 1981;124:110–114.
33   Bisgaard H, Pederson M: SRS-A leukotrienes decrease the activity of human respiratory
     cilia. Clin Allergy 1987;17:95–103.
34   Shelhamer JH, Marom Z, Sun F, et al: The effect of arachinoids and leukotrienes on the
     release of mucus from human airways. Chest 1982;81:36S–37S.
35   Marom Z, Shelhamer JH, Bach MK, et al: Slow-reacting substance, leukotrienes C_4 and
     D_4, increase the release of mucus from human airways in vitro. Am Rev Respir Dis
     1982;126:449–451.
36   Coles SJ, Neill KH, Reid LM, et al: Effects of leukotrienes D_4 and C_4 on glycoprotein
     and lysozyme secretion by human bronchial mucosa. Prostaglandins 1983;25:155–170.
37   Lundgren JD, Shelhamer JH, Kaliner MA: The role of eicosanoids in respiratory mucus
     hypersecretion. Ann Allergy 1985;55:5–13.
38   Bray MA: Leukotrienes in inflammation. Agents Actions 1986;19:87–99.
39   Malik AB, Perlman MB, Cooper JA, et al: Pulmonary microvascular effects of arachi-
     donic acid metabolites and their role in lung vascular injury. Fed Proc 1985;44:36–42.
40   O'Bryne PM, Leikauf GD, Aizawa H, et al: Leukotriene B_4 induces airway hyperrespon-
     siveness in dogs. J Appl Physiol 1985;59:1941–1946.
41   Spada CS, Woodward DF, Hawley SB, et al: Leukotrienes cause eosinophil emigration
     into conjunctival tissue. Prostaglandins 1986;31:795–809.
42   Jordan ML, Hoffman RA, Debe EF, et al: In vitro locomotion of allosensitized T
     lymphocyte clones in response to metabolites of arachidonic acid is subset specific. J
     Immunol 1986;137:661–668.
43   Serhan CN, Rodin A, Smolen JE, et al: Leukotriene B_4 is a complete secretagogue in
     human neutrophils: a kinetic analysis. Biochem Biophys Res Commun 1982;107:1006–
     1012.
44   Patterson NAM, McIver DJI, Schurch S: The effect of leukotrienes on porcine alveolar
     macrophage function. Prostaglandins Leukotrienes Med 1986;25:147–161.
45   Coffman RL, Carty J: A T cell activity that enhances polyclonal IgE production and its
     inhibition by interferon-gamma. J Immunol 1986;136:949–954.
46   Coffman RL, Ohara J, Bond MW, et al: B cell stimulatory factor-1 enhances the IgE
     response of lipopolysaccharide-activated B cells. J Immunol 1986;136:4538–4541.
47   Snapper CM, Finkelman FD, Paul WE: Differential regulation of IgG1 and IgE synthesis
     by interleukin-4. J Exp Med 1988;167:183–196.
48   Sherr EH, Stein LD, Dosch H-M, et al: IgE-enhancing activity directly and selectively
     affects activated B cells: evidence for a human IgE differentiation factor. J Immunol
     1987;138:3836–3843.
49   Gonzalez MC, Diaz P, Galleguillos FR, et al: Allergen-induced recruitment of bron-
     choalveolar helper (OKT4) and suppressor (OKT8) T-cells in asthma. Am Rev Respir
     Dis 1987;136:600–604.
50   Rola-Pleszczynski M, Chavaillaz P-A, Lemaire I: Stimulation of interleukin-2 and
     interferon-gamma production by leukotriene B_4 in human lymphocyte cultures.
     Prostaglandins Leukotrienes Med 1986;23:207–210.
51   Rola-Pleszczynski M, Lemaire L: Leukotrienes augment interleukin-1 production by
     human monocytes. J Immunol 1985;135:3958–3961.
52   Kunkel SL, Chensue SW, Spengler M, et al: Effects of arachidonic acid metabolites and
     their inhibitors on interleukin-1 production; in Kluger MJ, et al (eds): The Physiologic,
     Metabolic, and Immunologic Actions of Interleukin-1. New York, Liss, 1985, pp 297–307.

M.A. Bray, PhD, Ciba-Geigy AG, R 1056.215, CH–4002 Basel (Switzerland)

Zor U, Naor Z, Danon A (eds): Leukotrienes and Prostanoids in Health and Disease.
New Trends Lipid Mediators Res. Basel, Karger, 1989, vol 3, pp 143–148

Leukotrienes and Other Lipid Mediators in Psoriasis and Other Inflammatory Dermatoses

Clive B. Archer, Malcolm W. Greaves

Institute of Dermatology, St. Thomas's Hospital, London, UK

Human skin has been used for many years as a readily observable site for studying the pro-inflammatory properties of putative mediators of inflammation. In addition to measuring cutaneous responses such as weal and flare, biopsy of the skin over a period of time allows the examination of the histopathological effects of inflammatory mediators. By retrieving exudate samples from lesional and nonlesional skin in the inflammatory skin diseases, psoriasis and eczema (dermatitis) one can quantify the amounts of various mediators in vivo. These test systems provide a means for the study of potentially useful anti-inflammatory drugs, administered either systemically or topically. One may also gain insight into mechanisms of inflammation in other less accessible human organs, such as the lungs.

Here we review the evidence for the involvement of a number of lipid compounds in the pathogenesis of psoriasis and other inflammatory dermatoses. These lipids included the leukotrienes LTB_4, LTC_4 and LTD_4, the arachidonate lipoxygenase product 12-hydroxyeicosatetraenoic acid (12-HETE), the prostaglandins PGE_1, PGE_2, PGD_2 and prostacyclin (PGI_2), and the family of ether-linked phospholipids, platelet-activating factor (PAF, Paf-acether, AGEPC). Each mediator will be considered within the context of Dale's criteria [1] and the possible importance of inflammatory mediator interactions will be emphasized.

Leukotrienes and 12-Hydroxyeicosatetraenoic Acid

The arachidonate lipoxygenase products LTB_4, LTC_4, LTD_4 and 12-HETE induce inflammatory changes in human skin following local administration [2]. Like most inflammatory diseases, psoriasis is a chronic inflammatory disorder, and much attention has focussed on those compounds which can induce inflammatory cell accumulation, particularly the neutrophil chemoattractant, LTB_4.

Camp et al. [3] found that topical application of nanogram doses of LTB_4 under occlusion for 6 h to normal human skin induced erythema at 12 h, reaching a maximal response at 24 h. Biopsy of these sites at 24 h showed pockets of neutrophils in the epidermis, resembling the Munro microabscesses which may be seen histologically in psoriasis. 12-HETE produces similar findings in normal skin but at microgram doses [4]. Both LTB_4 and 12-HETE have been shown to induce these inflammatory cellular events in the normal appearing skin of patients with psoriasis, although the application of these compounds did not produce a psoriatic lesion [5, 6]. The 5-lipoxygenase products LTC_4 and LTD_4 produce dose-dependent weal and erythema responses in human skin following intradermal injection of nanogram amounts [7].

Biologically active amounts of LTB_4 have been found in extracts of scale and chamber fluid from abraded psoriatic skin lesions [8, 9], and elevated concentrations of LTB_4 and arachidonic acid were reduced by topical application of a potent corticosteroid, clobetesol propionate [10]. Increased levels of 12-HETE have also been observed in psoriatic lesional skin [11, 12]. Grabbe et al. [13], using relatively nonspecific HPLC ultraviolet absorbance profiles, reported the presence of significant amounts of 5-HETE in psoriatic scale extracts, but Camp et al. [12] found insignificant amounts of 5-HETE in lesional psoriatic scale and chamber fluid samples, analysed by gas chromatography-mass spectrometry. Elevated levels of LTC_4 and LTD_4 have also been found in skin chamber fluid from abraded psoriatic lesions [14].

LTB_4 immunoreactivity has been reported in suction blister fluid samples from areas of active atopic dermatitis (atopic eczema) [15], although the contribution made by 12-HETE or arachidonic acid in the assay is uncertain since both compounds cross-react with the LTB_4 antiserum [16]. Other forms of skin inflammation reported to be associated with increased concentrations of lipoxygenase products include allergic contact dermatitis [17], eosinophilic cellulitis [18], and ultraviolet B-irradiated and dithranol-treated human skin [19].

Before its withdrawal, benoxaprofen was convincingly shown to be an effective treatment for psoriasis [20, 21], but its action is probably independent of 5-lipoxygenase inhibition [22]. Lonapalene is a selective, potent inhibitor of human polymorphonuclear 5-lipoxygenase and has been shown to be clinically effective, as a topical agent, in psoriasis [23]. A recent study showed that topical lonapalene reduced the levels of LTB_4 in psoriatic lesional skin within 4 days of therapy, before the attainment of significant clinical improvement compared with the vehicle [24].

Prostaglandins

Intradermal injection of nanogram doses of the prostaglandins PGE_1, PGE_2, and PGD_2 in human skin produces erythema [25]. Prostacyclin (PGI_2) is also a potent cutaneous vasodilator, causing pronounced flushing following intravenous injection in man [26]. In human skin, intradermal injection of PGE_1, PGE_2 and PGD_2 induces weal responses [25], although in animal skin, prostaglandins have been reported to induce oedema formation only in the presence of other mediators of vascular permeability, such as bradykinin and histamine [27].

There is substantial evidence for the involvement of prostaglandins in inflammatory skin disorders [for review, see 28]. This evidence is based on measurement of cyclooxygenase products in inflamed skin, supported in some instances by the observed effects of cyclooxygenase inhibitors on the skin inflammation. It appears that in man, PGE_2 and perhaps prostacyclin play a role in the early stages of the inflammatory response of human skin to injury (e.g. UV erythema). PGD_2, the major mast cell-derived prostaglandin, may play a part in acute weal and flare responses, as seen in the urticarias.

Platelet-Activating Factor

Based on in vivo and in vitro studies, PAF seems likely to be a mediator of human inflammation that may be involved in the pathogenesis of inflammatory skin diseases, such as psoriasis. In man, intradermal injection of nanogram amounts of PAF elicits an acute weal and flare response which, depending on the dosage, may be succeeded by a late-onset area of erythema at the site of the resolved weal [29, 30]. As with LTB_4, PAF induces cellular accumulation in human skin [31, 32], with an increased number of eosinophils in atopics compared with nonatopics [33, 34].

PAF is rapidly metabolized and difficult to measure. However, PAF-like lipid has been demonstrated in the blood of patients with primary acquired cold urticaria [35]. PAF has also been retrieved from psoriatic scale [36]. Selective PAF antagonists are now becoming available for use as research and possible therapeutic tools. A specific PAF antagonist (BN52021) significantly suppressed mouse ear swelling when given orally, both during the elicitation phase of dinitrofluorobenzene (DNFB)-induced dermatitis and, to a lesser extent, in croton oil-induced dermatitis [37]. In human skin, weal and flare responses to PAF have been shown to be inhibited by the oral ginkgolide mixture (BN52063) [34, 38].

The role of PAF in acute and chronic inflammation should be considered within the context of other putative mediators of inflammation, and its relative importance will only be established when we are able to determine the clinical usefulness of specific PAF antagonists in human inflammatory diseases.

Inflammatory Mediator Interactions

We believe that no single putative mediator of inflammation is likely to produce all of the characteristic features of psoriasis and other inflammatory dermatoses. Studies in the skin of experimental animals have provided evidence for synergistic interaction between a number of inflammatory mediators, and a two-mediator concept of increased vascular permeability has been proposed, increased extravasation of plasma protein depending on the interaction between vasodilator and permeability factors [27].

In human skin, Archer et al. [39] showed that concurrent intradermal injection of PAF and the vasodilator PGE_2 leads to potentiation of acute weal and flare responses. Interestingly, synergistic interaction between LTB_4 and PGE_2 in human skin is of delayed-onset and is persistent [40]. The concept of inflammatory mediator interaction may be important in the pathogenesis of psoriasis and other inflammatory skin diseases and this is relevant to the development of new anti-inflammatory drugs.

In conclusion, human skin is a readily accessible and observable organ which can be used to study the actions and interactions of putative mediators of inflammation. The inflammatory skin diseases, such as psoriasis, provide interesting models of inflammation which can be studied in the assessment of potentially useful anti-inflammatory agents.

References

1   Dale, H.H.: Progress in autopharmacology. A survery of present knowledge of the chemical regulation of certain functions by natural constituents of the tissues. Bull. Johns Hopkins Hosp. *53:* 297–347 (1934).
2   Camp, R.D.R.; Greaves, M.W.: Inflammatory mediators in the skin. Br. Med. Bull. *43:* 401–414 (1987).
3   Camp, R.D.R.; Russell Jones, R.; Brian, S.D. et al.: Production of intraepidermal micorabscesses by topical application of leukotriene B_4. J. Invest. Dermatol. *82:* 202–204 (1984).
4   Dowd, P.M.; Kobza Black, A.; Woollard, P.M. et al.: Cutaneous responses to 12-hydroxy-5,8,10,14-eicosatetraenoic acid (12-HETE). J. Invest. Dermatol. *84:* 537–541 (1985).

5   Wong, E.; Camp, R.D.R.; Greaves, M.W.: The responses of normal and psoriatic skin
    to single and multiple topical applications of leukotriene B₄. J. Invest. Dermatol. *84:*
    421–423 (1985).

6   Dowd, P.M.; Kobza Black, A.; Woolard, P.M., et al.: Cutaneous responses to 12-
    hydroxy-5,8,10,14-eicosatetraenoic acid (12-HETE) and 5,12-dihydroxyeicosatetraenoic
    acid (leukotriene B₄) in psoriasis and normal human skin. Arch. Dermatol. Res. *297:*
    427–434 (1987).

7   Camp, R.D.R.; Coutts, A.A.; Greaves, M.W., et al.: Responses of human skin to
    intradermal injection of leukotrienes C₄, D₄ and B₄. Br. J Pharmacol. *80:* 497–502
    (1983).

8   Brain, S.D.; Camp, R.D.R.; Dowd, P.M., et al.: The release of leukotriene B₄-like
    material in biologically active amounts from the lesional skin of patients with psoriasis.
    J. Invest. Dermatol. *83:* 70–73 (1984).

9   Brain, S.D.; Camp, R.D.R.; Cunningham, F.M., et al.: Leukotriene B₄-like material in
    scale of psoriatic skin lesions. Br. J. Pharmacol. *83:* 313–317 (1984).

10 Wong, E.; Barr, R.; Cunningham, F.M., et al.: Topical steroid treatment reduces
    arachidonic acid and leukotriene B₄ in lesional skin of psoriasis. Br. J. Clin. Pharmacol.
    22: 627–633 (1986).

11 Hammarstrom, S.; Hamberg, M.; Samuelsson, B., et al.: Increased concentrations of
    nonesterified arachidonic acid, 12L-hydroxy-5,8,10,14-eicosatetraenoic acid, prostaglan-
    din E₂ and prostaglandin F₂ in epidermis of psoriasis. Proc. Natl. Acad. Sci. USA *72:*
    5130–5135.

12 Camp, R.D.R.; Mallet, A.I.; Woollard, P.M., et al: The identification of hydroxy fatty
    acids in psoriatic skin. Prostaglandins *26:* 431–437 (1983).

13 Grabbe, J.; Czarnetzki, B.M.; Rosenbach, T., et al.: Identification of chemotactic
    lipoxygenase products of arachidonic acid metabolism in psoriatic skin. J. Invest.
    Dermatol. *82:* 477–479 (1984).

14 Brain, S.D.; Camp, R.D.R.; Kobza Black, A., et al: Leukotrienes C₄ and D₄ in psoriatic
    skin lesions. Prostaglandins *29:* 611–619 (1985).

15 Ruzicka, T.; Simmet, T.; Peshar, B.A., et al.: Leukotrienes in skin of atopic dermatitis.
    Lancet *i:* 222–223 (1984).

16 Greaves, M.W.; Barr, R.M.; Camp, R.D.R.: Leukotriene B₄-like immunoreactivity and
    skin disease. Lancet *ii:* 160 (1984).

17 Barr, R.M.; Brain, S.; Camp, R.D.R., et al.: Human allergic and irritant contact
    dermatitis; levels of arachidonic acid and its metabolism in involved skin. Br. J.
    Dermatol. *111:* 23–28 (1984).

18 Wong, E.; Greaves, M.W.; O'Brien, T.: Increased concentrations of immunoreactive
    leukotrienes in cutaneous lesions of eosinophilic cellulitis. Br. J. Dermatol. *110:* 653–656
    (1984).

19 Kobza Black, A.; Barr, R.M.; Wong, E., et al.: Lipoxygenase products of arachidonic
    acid in human inflamed skin. Br. J. Clin. Pharmacol. *20:* 185–190 (1985).

20 Allen, B.R.; Littlewood, S.M.: Benoxaprofen: effect on cutaneous lesions of psoriasis.
    Br. Med. J. *285:* 1241–1243 (1983).

21 Kragballe, K.; Herlin, T.: Benoxaprofen improves psoriasis. A double-blind study. Arch.
    Dermatol. *119:* 548–552 (1983).

22 Masters, D.J.; McMillan, R.M.: 5-Lipoxygenase from human leucocytes (abstract). Br.
    J. Pharmacol. *81:* 70 (1984).

23 Lassus, A.; Forstrom, S.: A dimethoxynaphthalene derivative (RSO42179 gel) compared
    with 0.025% fluocinolone acetonide gel in the treatment of psoriasis. Br. J. Dermatol.
    113: 103–106 (1985).

24   Kobza Black, A.; Camp, R.D.R.; Cunningham, F.M., et al.: The clinical and pharmacological effect of lonapalene (RS-43179), a 5-lipoxygenase inhibitor, applied topically in psoriasis (abstract). Br. J. Dermatol. *119:* suppl. 33, p. 33 (1988).

25   Camp, R.D.R.: Prostaglandins, hydroxy fatty acids, leukotrienes and inflammation of the skin. Clin. Exp. Dermatol. *7:* 435–444 (1982).

26   O'Grady, J.; Warrington, S.; Moti, M.J., et al.: Effects of intravenous prostacyclin infusions in healthy volunteers – some preliminary observations; in Vane, Bergstrom, (eds): Prostacyclin. Raven Press, New York, 1979, p. 409.

27   Williams, T.J.; Peck, M.J.: Role of prostaglandin-mediated vasodilation in inflammation. Nature *270:* 530–532 (1977).

28   Greaves, M.W.: Pharmacology and significance of non-steroidal anti-inflammatory drugs in the treatment of skin diseases. J. Am. Acad. Dermatol. *16:* 751–764 (1987).

29   Archer, C.B.; Page, C.P.; Paul, W., et al: Inflammatory characteristics of platelet activating factor (PAF-acether) in human skin. Br. J. Dermatol. *110:* 45–50 (1984).

30   Basran, G.S.; Page, C.P.; Paul, W., et al.: Platelet activating factor: a possible mediator of the dual response to allergen. Clin. Allergy *14:* 75–79 (1984).

31   Archer, C.B.; Page, C.P.; Morley, J., et al: Accumulation of inflammatory cells in response to intracutaneous platelet activating factor (PAF-acether) in man. Br. J. Dermatol. *112:* 285–290 (1985).

32   Michel, L.; Mencia-Huerta, J.M.; Benveniste, J., et al.: Biologic properties of LTB_4 and PAF-acether in vivo in human skin. J. Invest. Dermatol. *88:* 675–681 (1987).

33   Henocq, E.; Vargaftig, B.B.: Accumulation of eosinophils in response to intercutaneous PAF-acether and allergens in man. Lancet *i:* 1378–1379 (1986).

34   Markey, A.C.; Barker, J.N.; Archer, C.B., et al.: The clinical and histological effects of platelet activating factor (PAF-acether) in atopic skin, and the effect on responses of the specific PAF-acether antagonist, BN52063 (abstract). Br. J. Dermatol. (in press).

35   Grandel, K.E.; Farr, R.S. Wanderer, A.A., et al.: Association of platelet-activating factor with primary acquired cold urticaria. New Engl. J. Med. *313:* 405–409 (1985).

36   Mallet, A.I.; Cunningham, F.M.: Structural identification of platelet activating factor in psoriatic scale. Biochem. Biophys. Res. Commun. *126:* 192–198 (1985).

37   Csato, M.; Czarnetzki, B.M.: Effect of BN52021, a platelet activating factor antagonist, on experimental murine contact dermatitis. Br. J. Dermatol. *118:* 475–479 (1988).

38   Chung, K.F.; McCusker, M.; Page, C.P., et al.: Effect of a ginkgolide mixture (BN 52063) in antagonising skin and platelet responses to platelet activating factor in man. Lancet *i:* 248–251 (1987).

39   Archer, C.B.; Frohlich, W.; Page, C.P., et al.: Synergistic interaction between prostaglandins and Paf-acether in experimental animals and man. Prostaglandins *27:* 495–501 (1984).

40   Archer, C.B.; Page, C.P.; Juhlin, L., et al.: Delayed onset synergism between leukotriene B_4 and prostaglandin E_2 in human skin. Prostaglandins *33:* 799–805 (1987).

C.B. Archer, MD, Institute of Dermatology, St. Thomas's Hospital,
GB-London SE1 7EH (UK). Present address: Department of Dermatology,
Bristol Royal Infirmary, Bristol (UK)

Zor U, Naor Z, Danon A (eds): Leukotrienes and Prostanoids in Health and Disease.
New Trends Lipid Mediators Res. Basel, Karger, 1989, vol 3, pp 149–153

Regulation of the Leukotriene B_4 Receptor by Protein Kinase C-Dependent and Independent Mechanisms

Daniel W. Goldman

Johns Hopkins University School of Medicine, Department of Medicine,
Division of Clinical Immunology, Baltimore, Md., USA

Introduction

Leukotriene B_4 (LTB_4), a product of the 5-lipoxygenation of arachidonic acid, has been established as an inflammatory mediator by virtue of its capacity to recruit and activate neutrophils both in vitro and at sites of inflammation in vivo [1, 2]. The human neutrophil expresses two classes of receptors for LTB_4 that transduce LTB_4 binding into the generation of the complex set of intracellular signals necessary for the full expression of neutrophil function [3, 4]. In addition to promoting neutrophil activation, these intracellular signals also act as feedback inhibitors at an early stage in the signal transduction process and thus mitigate the extent to which the neutrophil responds to LTB_4 and to other inflammatory mediators. In this paper the effects of inflammatory mediators and of protein kinase C (PKC) activators on LTB_4 receptor expression and on changes in the neutrophil responsiveness to LTB_4 will be described and the role of PKC in mediating these changes in LTB_4 receptor expression and responsiveness will be discussed.

Regulation of LTB_4 Receptor Expression

LTB_4 binds to two classes of receptors on the surface of the human neutrophil. The high affinity class of receptors binds LTB_4 with a dissociation constant (K_d) of approximately 0.5 nM. Binding to the high affinity set of receptors initiates increases in chemotactic and chemokinetic migration, increases in aggregation and adherence of neutrophils to surfaces, and stimulates a rapid and transient increase in the concentration of cytosolic calcium. The low affinity class of receptors binds LTB_4 with a K_d of between 60 to 100 nM. Binding of LTB$_4$ to the low affinity class of receptors initiates increases in granular enzyme release, increases in super-oxide generation, and under appropriate conditions, increases in cytosolic calcium concentrations [4].

A variety of preincubation conditions have been reported which decrease the expression of high affinity receptors for LTB$_4$ with concomi-tant decreases in responsiveness to LTB$_4$ [3, 5–9]. Of particular interest is the decrease in receptor expression after preincubation with f-met-leu-phe (fMLP), a chemotactic formyl-methionyl peptide. In human neutrophils, fMLP induces a rapid increase in the level of diacylglycerol with a parallel translocation and activation of PKC to the plasma membrane [10, 11]. Since direct activation of PKC by phorbol esters markedly decreases the expression of receptors for LTB$_4$ [7, 8], a more detailed examination of fMLP and phorbol ester induced changes in receptor expression was performed to assess the role of PKC in the regulation of LTB$_4$ receptor.

Activation of PKC in human neutrophils with phorbol esters or with an exogenously added diacylglycerol, oleoylacetylglycerol (OAG), progres-sively inhibits binding of [^{3}H]-LTB$_4$ to human neutrophils in a dose-depen-dent manner. The activation of PKC by OAG selectively inhibited binding by inducing a 5-fold decrease in the affinity of the high affinity receptors for LTB$_4$ with no change in the expression of low affinity receptors [7, 9]. More extensive activation of PKC by phorbol esters ultimately leads to the complete loss of high affinity receptor expression [7–9]. Preincubation of neutrophils with the chemotactic peptide fMLP also inhibits [^{3}H]-LTB$_4$ binding to the neutrophil by inducing a 4- to 5-fold decrease in affinity of the high affinity receptors for LTB$_4$. Inhibitors of the 5-lipoxygenase pathway and inhibitors of oxidative metabolism do not block the shift in receptor affinity induced by fMLP and OAG, thus ruling out the possibility that the endogenous generation of LTB$_4$ or of reactive oxygen metabolites are responsible for the changes in high affinity receptor expression [7, 9].

Regulation of Neutrophil Responsiveness to LTB$_4$

An early event in the signal transduction pathway initiated by LTB$_4$ and by other chemotactic factors is a rapid and transient increase in the concentration of cytosolic calcium. Activation of PKC by increasing con-centrations of OAG will progressively inhibit the capacity of LTB$_4$ and fMLP to increase cytosolic calcium. Preincubation of neutrophils with a maximal concentration of OAG alters LTB$_4$-elicited increases in cytosolic calcium in two ways. Firstly, OAG desensitizes the responsiveness of the neutrophil towards LTB$_4$ such that 1,000–10,000-fold higher concentrations of LTB$_4$ are needed to produce an increase in cytosolic calcium comparable to that observed in control neutrophils. Neutrophils pretreated with OAG are also desensitized towards fMLP, but to a lesser degree; up to 100-fold higher concentrations of fMLP are necessary to elicit an increase in cytosolic calcium comparable to control. Secondly, the neutrophils are deactivated with respect to LTB$_4$ and fMLP. A maximal concentration of OAG inhibits by approximately 80–90% the maximal increase in cytosolic calcium elicited by LTB$_4$ but only inhibits by 25% the maximal response elicited by fMLP. Thus, activation of PKC by OAG preferentially desensitizes and deactivates the LTB$_4$-elicited increases in cytosolic calcium [7, 9].

Preincubation of neutrophils with fMLP mimics the effects of PKC activation on LTB$_4$-elicited increases in cytosolic calcium. Preincubation of neutrophils with 3 nM fMLP and 100 nM fMLP desensitizes by 10- and 1,000-fold, respectively, LTB$_4$-elicited increases in cytosolic calcium; whereas, preincubation with 1 μM fMLP completely eliminates the capacity of 1 μM LTB$_4$ to elicit an increase in cytosolic calcium. Preincubation of the neutrophils with 1 μM fMLP also eliminates the capacity of fMLP to elicit an increase in cytosolic calcium, consistent with an fMLP-induced down-regulation of fMLP receptor expression [12]. When C5a is used as a stimulus, the effects of fMLP are similar to the effects of PKC activation on fMLP-induced increases in cytosolic calcium. Preincubation of neutrophils with 1 μM fMLP produces only a 10-fold shift in the dose-response curve for C5a and inhibits by approximately 30–40% the maximal increase in cytosolic calcium that can be elicited by C5a. Activation by fMLP thus appears to preferentially desensitize and deactivate the responsiveness of the neutrophil towards LTB$_4$ by progressively uncoupling the LTB$_4$ receptor from the signal transduction pathways.

Effects of PKC Inhibitors

To confirm the role of PKC as a mediator of the fMLP effects on LTB$_4$ receptor expression and on coupling of the LTB$_4$ receptor to changes in cytosolic calcium, the effects of a potent PKC antagonist, staurosporine [13], was examined. Staurosporine inhibited phorbol ester induced increases in superoxide generation and in degranulation. However, staurosporine markedly potentiated the capacity of fMLP to elicit an increase in superoxide generation and in degranulation. Staurosporine blocked the phorbol ester induced decreases in LTB$_4$ receptor expression and LTB$_4$-elicited increases in cytosolic calcium. However, fMLP-induced inhibition of LTB$_4$ binding and of neutrophil responsiveness to LTB$_4$ were not blocked by the addition of 200 nM staurosporine. The staurosporine data suggest that

Table 1. Regulation of LTB$_4$ receptor expression and receptor-mediated increases in cytosolic calcium in human neutrophils by diacylglycerol, an activator of PKC, and by fMLP

	Regulatory factor	
	diacylglycerol	fMLP
High-affinity receptors for LTB$_4$		
Receptor density	increased	no change
Affinity for LTB$_4$	decreased	decreased
Increases in cytosolic calcium		
Stimulus		
LTB$_4$	desensitization	desensitization
	deactivation	deactivation
fMLP	partial desensitization	desensitization
	partial deactivation	deactivation
C5a	partial desensitization	partial desensitization
	partial deactivation	partial deactivation
Staurosporine effects	antagonist	no effect

activation of PKC by fMLP is not responsible for the fMLP-induced changes in LTB$_4$ receptor expression and LTB$_4$-induced increases in cytosolic calcium.

Conclusions

The effect of fMLP and activators of PKC on neutrophil changes on the LTB$_4$ receptor and its associated signal transduction mechanisms are summarized in table 1. Despite their similar effects, the finding that fMLP alters neutrophil responsiveness by a mechanism other than activation of PKC indicates the alternative biochemical events need to be carefully examined in order to obtain a more complete understanding of the mechanisms which regulate LTB$_4$ receptor expression and the responsiveness of the neutrophil to inflammatory stimuli.

References

1   Bray MA: Leukotrienes in inflammation. Agents Actions 1986;19:87–99.
2   Goetzl EJ, Payan DG, Goldman DW: Immunopathogenetic roles of leukotrienes in human diseases. J Clin Immunol 1984;4:79–84.
3   Goldman DW, Geotzl EJ: Heterogeneity of human polymorphonuclear leukocyte receptors for leukotriene B$_4$. J Exp Med 1984;159:1027–1041.

4   Goldman DW, Gifford LA, Marotti T, et al: Molecular and cellular properties of human
     PMN leukocyte receptors for leukotriene B$_4$. Fed Proc 1987;46:200–203.
5   Goldman DW, Chang FH, Gifford LA, et al: Inhibition by pertussis toxin of chemotac-
     tic factor-induced calcium mobilization and function in human polymorphonuclear
     leukocytes. J Exp Med 1985;162:145–156.
6   Goldman DW, Gifford LA, Olson DM, et al: Transduction by leukotriene B$_4$ receptors
     of increases in cytosolic calcium in human polymorphonuclear leukocytes. J Immunol
     1985;135:525–530.
7   Goldman DW: Activation of protein kinase C (PKC) decreases leukotriene B$_4$ (LTB$_4$)
     receptor expression on human neutrophils (N). Fed Proc 1987;46:606.
8   O'Flaherty JT, Redman JF, Jacobson DP: Protein kinase C regulates leukotriene B$_4$
     receptors in human neutrophils. FEBS Lett 1986;206:279–282.
9   Goldman DW: Regulation of the receptor system for leukotriene B$_4$ on human
     neutrophils. Ann NY Acad Sci 1988;524:187–195.
10   Pike MC, Jakoi LC, McPhail LC, et al: Chemoattractant-mediated stimulation of the
     respiratory burst in human polymorphonuclear leukocytes may require appearance of
     protein kinase activity in the cells' particulate fraction. Blood 1986;67:909–913.
11   Preiss JE, Bell RM, Niedel JE: Diacylglycerol mass measurements in stimulated HL-60
     phagocytes. J Immunol 1987;138:1542–1545.
12   Donabedian H, Gallin JI: Deactivation of human neutrophil chemotaxis by chemoat-
     tractants: effect on receptors for the chemotactic factor f-Met-Leu-Phe. J Immunol
     1981;127:839–844.
13   Tamaoki T, Nomoto H, Takahashi I, et al: Staurosporine, a potent inhibitor of
     phospholipid/Ca^{++} dependent protein kinase. Biochem Biophys Res Commun
     1986;135:397–402.

Daniel W. Goldman, PhD, The Johns Hopkins University School of Medicine,
Department of Medicine, Division of Clinical Immunology, 5601 Loch Raven
Boulevard, Baltimore, MD 21239 (USA)

Zor U, Naor Z, Danon A (eds): Leukotrienes and Prostanoids in Health and Disease.
New Trends Lipid Mediators Res. Basel, Karger, 1989, vol 3, pp 154–160

Lipoxygenase Products in Radiation Injury and Radioprotection[1,2]

Thomas L. Walden, Jr., Nushin K. Farzaneh, Lisa Richards

Radiation Biochemistry Department, Armed Forces Radiobiology Research Institute, Bethesda, Md., USA

Introduction

Metabolites of arachidonic acid are important biological mediators in the sequelae of radiation injury and in recovery from radiation damage [1]. Interestingly, the cytoprotective properties of these same metabolites can be exploited for effective radioprotection [1–6]. The roles of the cyclo-oxygenase products in radiation injury and protection have recently been reviewed [1–3]. The availability of lipoxygenase products and methods for detection are permitting delineation of the roles of the lipoxygenase pathway in radiobiology. These roles are similar to those already identified for prostaglandins [1, 7, 8]. Since alterations in their synthesis and release following irradiation are dependent on the cell type, not all lipoxygenase metabolites will become elevated, nor will there be a common time point or pattern for elevation. In a similar manner, not all lipoxygenase products are radioprotective.

Production of Leukotrienes Postirradiation

Radiation exposure induces alterations in both prostaglandin synthesis and metabolism [reviewed in 1]. Alterations in tissue prostaglandin concentrations have been linked to mediation of radiation injury in several species

[1] Supported by the Armed Forces Radiobiology Research Institute (AFRRI), Defense Nuclear Agency, under Research Work Unit 000152.

[2] The views presented in this paper are those of the authors; no endorsement by the Defense Nuclear Agency has been given or should be inferred. Any AFRRI research included in this review was performed according to the principles enunciated in the 'Guide for the Care and Use of Laboratory Animals,' prepared by the Institute of Laboratory Animal Resources, National Research Council.

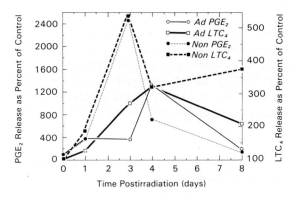

Fig. 1. Elevation of leukotriene C_4 and prostaglandin E_2 in mouse peritoneal macrophages after irradiation. Peritoneal macrophages were obtained by lavage from B6D2F1 female mice on the indicated days post 10 Gy ^{60}Co γ-irradiation. The peritoneal exudates were divided into adherent and nonadherent populations. They were placed in tissue culture medium and the PGE_2 and LTC_4 released into the medium was quantitated by radioimmunoassay. Drawn from data in table 3 of Steel et al. [8].

[1], including man [9]. In addition, prostaglandin synthesis is important for hematopoietic recovery processes. Information concerning the status of lipoxygenase products is now available.

In vivo cobalt-60 γ-irradiation with 10 Gy increases the in vitro release of LTC_4, and PGE_2 by murine peritoneal macrophages during the first 8 days after irradiation [8]. Peak release of LTC_4 into the medium occurs on day 4 postirradiation (fig. 1). At this time, release by adherent cell populations is 3 times greater than the release by control peritoneal macrophages. Release remains elevated in macrophages obtained from animals 8 days postirradiation. Release of LTC_4 by nonadherent peritoneal exudate cells is also elevated after irradiation and peaks on day 4. PGE_2 is also increased postirradiation and peaks at the same times as LTC_4 release. The macrophages obtained during this period are larger in size and contain numerous vacuoles from phagocytosis. It is not clear whether the increased release of LTC_4 and PGE_2 result from direct radiation stimulation or from indirect activation through phagocytic activity. Irradiation does not induce maximal arachidonic acid release since adherent and nonadherent peritoneal exudate cells still respond to stimulation by the calcium ionophore, A23187. LTC_4 release by macrophages obtained postirradiation can be further stimulated with A23187 to 6- to 8-fold greater release than produced by unstimulated-irradiated cells.

Exposure of murine peritoneal macrophages to 254 nm UV light also produces an increase in LTC_4 and LTB_4 synthesis [10]. One hour after exposure of macrophages to $0.14 \, W/m^2$ for 30 s, LTC_4 production increased from undetectable levels in control cells to 21 $ng/10^6$ cells $\pm$ 3.5 ng (SEM, n = 4). LTB_4 production also increased by a smaller amount. A23187 was more effective in stimulating LTC_4 and LTB_4 production than UV light exposure. Cell toxicity to UV exposure was higher than for exposure to A23187, and this may have affected the overall production of leukotrienes since arachidonic acid release was higher in the UV-irradiated cell population.

Radioprotection Induced by Leukotrienes

Prostaglandins induce cytoprotection to a number of biological insults including chemicals [11], physical agents [12], and radiation [1]. Leukotrienes and some other lipoxygenase products are also able to induce similar protective responses when given prior to irradiation [5–7, 13]. Radioprotection induced by leukotrienes has been observed for cells in culture [2, 13], hematopoietic stem cells in vivo and in vitro [5], intestinal crypt stem cells [3], and for animal survival [6]. The results for the most effective lipoxygenase product, LTC_4, are summarized in table 1. The leukotriene must be present prior to irradiation for protection.

Pretreatment of mice with 10 μg of LTC_4 or LTD_4 resulted in increased numbers of intestinal crypt colonies in the intestinal epithelium 4 days after receiving 15 Gy of cesium-137 γ-irradiation [3]. LTC_4 provided a 2.8-fold increase in crypt colonies, and LTD_4 provided a 60% increase; however, pretreatment with LTE_4 was not radioprotective. Both values are in comparison to animals receiving vehicle alone.

Addition of 2.5 μM LTC_4 to the medium of V79A03 Chinese hamster fibroblasts prior to irradiation increases reproductive survival measured by colony formation. Two to three times more cells survive after irradiation to form colonies when pretreated with LTC_4 than with sham treatment. A specific binding site for $[^3H]$-LTC_4 on V79A03 cells has been identified that is associated with the ability of LTC_4 pretreatment to induce protection. Trypsinization destroys the specific binding. If cells are harvested with trypsin, replated 4 h prior to irradiation, and treated with LTC_4 for 2 h prior to irradition, no protection is observed. If this same experiment is repeated using cells harvested by scraping to preserve the binding site, LTC_4 pretreatment induces radioprotection. This indicates the binding site is necessary for LTC_4 to induce protection. Both tissue culture medium and V79A03 cells can metabolize LTC_4 to LTD_4 and LTE_4, but the latter two compounds were not as effective radioprotectants. The LTC_4 binding site on the outer surface of the V79A03 cell is specific for LTC_4 and does not

Table 1. Dose reduction factors (DRF) for leukotriene-induced radioprotection

System	Pretreatment time	Dose	DRF	Reference
In vitro				
V79A03	2 h	2.5 μM	1.2[1]	7
In vivo (mouse)				
Stem cells				
CFU-S	15 min	0.4 mg/kg	1.55	5
CM-CFC	15 min	0.4 mg/kg	2.01	5
Animal survival				
Male mice	5 min	0.2 mg/kg	1.46	6

[1] Estimate for in vitro protection to V79A03 cells from preliminary survival curves [Walden, unpubl.].

bind LTD_4, LTE_4, or glutathione [Walden and Fitz, unpubl.]. Irradiation of the V79A03 cell with doses less than 50 Gy does not significantly affect the ability of this site to bind LTC_4 (fig. 2). A 100-Gy radiation dose decreases binding activity by 37%.

Leukotrienes induce radioprotection of mouse hematopoietic stem cells in vivo in a time- and dose-dependent manner [5]. Pretreatment with doses as low as 1 μg of LTC_4/mouse given 5 min prior to radiation exposure can increase the number of endogenous spleen colonies (hematopoietic stem cells) at 10 days postirradiation by 10-fold. Optimal protection is provided by 10 μg of LTC_4/mouse, or 400 μg LTC_4/kg of body weight. The effectiveness of a given radioprotective agent is measured in terms of its dose reduction factor (DRF). The DRF is the ratio formed by the cell or animal survival after irradiation that is observed in the presence of a given protective agent, divided by the survival of the sham-treated irradiated group. In order of effectiveness, $LTC_4 \gg B_4 > E_4 > D_4$. The mean lethal radiation dose (D_o) for pluripotent mouse hematopoietic stem cells receiving a placebo or sham-protective treatment is 0.85 Gy, and for cells in mice receiving 10 μg of LTC_4 at 10 min prior to irradiation is 1.32 Gy. The DRF for LTC_4 pretreatment is 1.32/0.85, or 1.55. The DRF for granulocyte-macrophage progenitor stem cells in mice receiving LTC_4 pretreatment is 2.01 [5]. Leukotriene C_4 pretreatment is also effective for enhancing animal survival. Treatment with 200 μg of LTC_4/kg body weight (5 μg/mouse) increases the $LD_{50/30}$ from 8.35 Gy in sham-treated mice to 12.55 Gy, providing a DRF of 1.45. All irradiated animals that received radiation doses greater than 9.25 Gy and did not receive a LTC_4 pretreatment, died

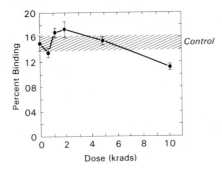

Fig. 2. Effect of irradiation on specific binding of LTC$_4$ by V79A03 cells in vitro. V79A03 Chinese hamster fibroblasts were grown as described [7] on α-minimal essential medium in 25-cm^2 flasks and irradiated with 250 KVP X-radiation at 15 mA and a dose rate of 1.69 Gy/min. Three hours postirradiation, cells were harvested with 0.02% EDTA in Dulbecco's phosphate-buffered saline (PBS), and washed 3 times with PBS to remove the EDTA. Binding assays contained 1×10^6 cells, 20 fmol [^{3}H]-LTC$_4$ (specific acitivity 40 Ci/mmol; DuPont-New England Nuclear, Boston, Mass), 10 mM serine-borate, pH 7.35, in 50 μl of PBS. Assays were incubated for 15 min on ice, and the nonspecific binding was determined in parallel assay tubes containing 200 ng of unlabeled LTC$_4$. Assays were stopped by centrifugation at 13,000 g for 1 min to pellet the cells, and supernatant containing the unbound label was removed. Assays were performed in quadruplicate, and standard error bars are provided. The shaded area represents the binding range for unirradiated cells. A one-way analysis of variance indicated a significant difference in the mean between radiation doses (SL = 0.001).

within 30 days after radiation exposure [6]. All mice pretreated with LTC$_4$ survived at least 30 days after radiation exposure. In terms of animal survival, LTC$_4$ is the most effective of the leukotrienes [Walden, unpubl. observation].

Many radioprotectants contain thiol moieties and are thought to act by thiol scavenging of free radicals produced by irradiation. Leukotriene protection by this mechanism is unlikely since the effective concentrations of leukotrienes are much smaller than the 200–1,400 mg/kg body weight required for most thiol-containing radioprotective agents. In addition, LTB$_4$ [5] and LTA$_4$ [14] are radioprotective but do not contain thiol groups. Most eicosanoid activities are initiated by interaction with specific receptors [15]. The basis for leukotriene-induced radioprotection is unknown, but is thought to be mediated through the action of specific leukotriene receptors. The ability to induce radioprotection is not a ubiquitous property of the lipoxygenase products. Lipoxin B$_4$ induces radioprotection of mouse hematopoietic stem cells [14], but lipoxin A$_4$ does not [5].

Conclusions

The synthesis of lipoxygenase products following exposure to ionizing and nonionizing radiation has important implications for radiation injury and radiotherapy, and for skin exposure to sunlight. These compounds are inflammatory agents and may contribute to the radiation-induced inflammation and tissue injury. Further work is required to determine where leukotrienes participate in radiation injury, and to delineate points following radiation exposure at which leukotriene synthesis is a normal physiological or recovery process. Studies which attempt to relate specific radiation-induced physiological responses to eicosanoid release must consider both pathways, cyclooxygenase and lipoxygenase. Blocking only the cyclooxygenase pathway may lead to misinterpretation since leukotriene synthesis can be stimulated by irradiation or by the shunting of arachidonic acid metabolism through the lipoxygenase pathway. The radioprotective properties of leukotrienes raises the possibility that tumors which synthesize leukotrienes may have increased resistance to therapy.

References

1   Walden, T.L., Jr.: A paradoxical role for eicosanoids: Radioprotectants and radiosensitizers; in Walden, T.L., Jr., Hughes, H.N. (eds): Prostaglandin and Lipid Metabolism in Radiation Injury, pp. 263–271 (Plenum Press, New York 1987).
2   Steel, L.K.; Catravas, G.N.: Protection against ionizing radiation with eicosanoids; in Polagar, P. (ed): Eicosanodis and Radiation, pp. 79–87 (Kluwer, Boston 1988).
3   Hanson, W.R.: Radiation protection by exogenous arachidonic acid and several metabolites; in Walden, T.L. Jr., Hughes, H.N. (eds): Prostaglandin and Lipid Metabolism in Radiation Injury, pp. 233–243 (Plenum Press, New York 1987).
4   Walden, T.L., Jr.; Patchen, M.L.; Snyder, S.L.: 16,16-Dimethyl-prostaglandin E_2 increases survival in mice following irradiation. Radiat. Res. 109: 440–448 (1987).
5   Walden, T.L., Jr.; Patchen, M.L.; MacVittie, T.J.: Leukotriene-induced radioprotection of hematopoietic stem cells in mice. Radiat. Res. 113: 388–395 (1988).
6   Walden, T.L., Jr.: Pretreatment with leukotriene C_4 enhances the whole-animal survival of mice exposed to ionizing radiation. Ann. N.Y. Acad. Sci. 524: 431–433 (1988).
7   Walden, T.L., Jr.; Holahan, E.V., Jr.; Catravas, G.N.: Development of a model system to study leukotriene-induced modification of radiosensitivity in mammalian cells. Prog. Lipid. Res. 25: 587–590 (1986).
8   Steel, L.K.; Hughes, H.N.; Walden, T.L., Jr.: Quantitative, functional and biochemical alterations in the peritoneal cells of mice exposed to whole-body gamma-irradiation. I. Changes in cellular protein, adherence properties and enzymatic activities associated with platelet-activating factor formation and inactivation, and arachidonic acid metabolism. Int. J. Radiat. Biol. 53: 943–964 (1988).
9   Tanner, N.S.; Stamford, I.F.; Bennett, A.: Plasma prostaglandins in mucositis due to radiotherapy and chemotherapy for head and neck cancer. Br. J. Cancer 43: 767–771 (1981).

10 Hardcastle, J.E.; Minoui, S.: Leukotriene synthesis by UV-irradiated macrophage cell cultures; in Walden, T.L., Jr., Hughes, H.N. (eds): Prostaglandin and Lipid Metabolism in Radiation Injury, pp. 173–178 (Plenum Press, New York 1987).

11 Ohno, T.; Ohtsuki, H.; Okabe, S.: Effects of 16,16-dimethyl-prostaglandin E_2 on ethalno-induced and aspirin-induced gastric damage in the rat. Gastroenterology *88:* 353–361 (1985).

12 Robert, A.; Nezamis, J.E.; Lancaster, C.; Hanchar, A.J.: Cytoprotection by prostaglandins in rats. Gastroenterology *77:* 433–443 (1979).

13 Walden, T.L., Jr., Kalinich, J.F.: Radioprotection by leukotrienes. Is there a receptor mechanism? Pharmacol. Ther. *39:* 379–384 (1988).

14 Walden, T.L., Jr.: Radioprotection of mouse hematopoietic stem cells by leukotriene A_4 and lipoxin B_4. J. Radiat. Res. *29:* 255–260 (1988).

15 Robertson, R.P.: Characterization and regulation of prostaglandin and leukotriene receptors: An overview. Prostaglandins *31:* 395–412 (1986).

Thomas L. Walden, Jr., PhD, Radiation Biochemistry Department, Armed Forces Radiobiology Research Institute, Bethesda, MD 20814-5145 (USA)

Brain

Zor U, Naor Z, Danon A (eds): Leukotrienes and Prostanoids in Health and Disease.
New Trends Lipid Mediators Res. Basel, Karger, 1989, vol 3, pp 161–165

Metabolism of Leukotrienes in Rat Brain Microvessels and Choroid Plexus

Jan Åke Lindgren, Irina Karnushina, Hans-Erik Claesson

Department of Physiological Chemistry, Karolinska Institutet, Stockholm, Sweden

Introduction

The central nervous system possesses the capacity to produce leukotrienes [1–6]. Furthermore, enhanced production of leukotrienes in gerbil brain after ischemia and reperfusion [3] has been demonstrated. Cysteinyl-containing leukotrienes have been reported to constrict cerebral arterioles [7] and increase the blood-brain barrier permeability, causing vasogenic brain edema [8]. In addition, LTC_4, in subnanomolar concentrations, may have neuroendocrine functions [9–11]. Thus, leukotrienes may be both physiological and pathophysiological mediators in the brain. Therefore, it can be assumed that the concentrations of these compounds are strictly controlled in the brain.

The present study shows that both cerebral microvessels and choroid plexus from rat efficiently transformed LTC_4 to LTD_4 and LTE_4. Although the microvessels converted exogenous LTA_4 to LTC_4, this was only observed in the presence of exogenously added glutathione. The results suggest that the blood-brain and blood-cerebrospinal fluid (CSF) barriers may constitute an important system for metabolism of leukotrienes in the brain.

Methods

Brain microvessels and choroid plexuses were purified from perfused brains of 8- to 10-week-old rats as described [12]. After 5 min pre-incubation of the tissue at 37 °C, leukotriene synthesis and metabolism was initiated by addition of synthetic leukotrienes or ionophore A23187. The incubations were terminated by addition of 3 volumes of cold ethanol. The samples were analyzed by reversed phase-high performance liquid chromatography (RP-HPLC). Identification and quantitation of the leukotrienes were performed using on-line UV spectroscopy and co-chromatography with synthetic standards [5].

Results

Leukotrienes were not produced by isolated brain microvessels after stimulation with ionophore A23187 in the absence or presence of arachidonic acid. In contrast, incubation of microvessels with LTA_4 for 5 min in the presence of glutathione led to the formation of LTC_4 and LTD_4. After incubation with $5-14\,\mu M$ LTA_4 for 5 min, 40–160 pmol of cysteinyl-containing leukotrienes/mg of protein were formed. In the absence of exogenously added glutathione, no production of LTC_4 and LTD_4 was observed.

Both microvessels and isolated choroid plexus efficiently converted LTC_4 to LTD_4 and LTE_4 (fig. 1, 2). In microvessels, the conversion rate of LTC_4 $(0.5\,\mu M)$ to LTD_4/LTE_4 was 100 pmol/mg protein × min. The conversion of LTC_4 to LTD_4 was rapid, while further metabolism of LTD_4 to LTE_4 was much slower. Choroid plexus converted about 40 pmol of LTC_4 $(0.5\,\mu M)$ to LTD_4/LTE_4 per milligram of protein × min.

In separate experiments, the capacity of the microvessels to metabolize LTC_4 was compared to the capacity of the total rat brain homogenate. In these experiments the conversion rates of LTC_4 $(0.5\,\mu M)$ to LTD_4 and LTE_4 were 0.6 and 68 pmol/mg protein × min in brain homogenate and microvessels, respectively. Thus, LTC_4 was metabolized about 100 times more efficiently by the microvessels. The total brain homogenate (containing 270 mg protein) converted 166 pmol LTC_4/min, while the microvessels (containing 1 mg protein), isolated from the same amount of brain homogenate, possessed about 40% of the total activity (68 pmol/min).

Discussion

The present report indicates that brain microvessels efficiently participate in the metabolism of LTC_4 in the central nervous system. In contrast to an earlier report [13], metabolism of LTC_4 was also observed in the choroid plexus. The results are in agreement with early findings demonstrating high activity of gamma-glutamyltranspeptidase (γ-GTP) in the endothelial cells of brain microvessels and epithelial cells of the choroid plexus [14, 15]. Conversion of LTA_4 to LTC_4 by brain microvessels was only observed in the presence of exogenous glutathione. This may indicate that the brain endothelium possesses less capacity to produce LTC_4 from LTA_4, as compared to endothelium from peripheral tissue [cf. 16, 17]. Preliminary results indicate that the choroid plexus has very limited capacity to transform LTA_4 to LTC_4 [Karnushina et al., unpubl.]. The transformation of LTC_4 to LTD_4 was rapid, in agreement with histochemical data

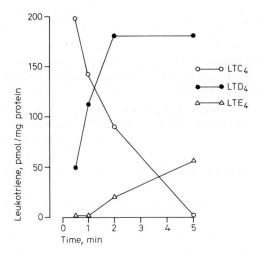

Fig. 1. Time courses of leukotriene formation by rat cerebral microvessels after addition of LTC$_4$ (0.5 μM).

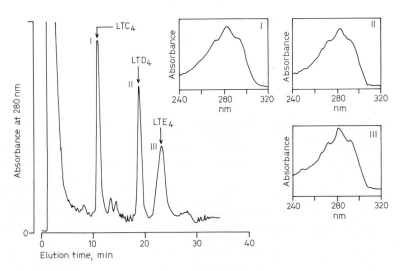

Fig. 2. RP-HPLC chromatogram and UV spectra of products formed by rat choroid plexus after incubation with LTC$_4$ (0.5 μM) for 10 min.

demonstrating that the γ-GTP activity is considerably higher in cerebral microvessels, as compared to microvessels from peripheral tissue [18]. The results indicate that brain microvessels and choroid plexus are mainly involved in leukotriene degradation, thereby protecting the central nervous system from these bioactive substances. Furthermore, the choroid plexus may be involved in the transport of leukotrienes out of the brain [19].

The cellular origin of leukotrienes in the brain is presently not known. Brain microvessels did not produce leukotrienes after incubation with ionophore A23187, indicating that this tissue lacks 5-lipoxygenase activity and is therefore not the primary source of leukotrienes in the brain. Further studies to elucidate the leukotriene-producing capacity in nerve and glial cells are in progress.

In conclusion, the present study shows that brain microvessels and choroid plexus readily metabolize leukotrienes. These compounds have been suggested to possess both pathophysiological and physiological roles in the central nervous system. Therefore, a potent system for degradation of leukotrienes in the central nervous system may be of great importance.

Acknowledgements

We thank Ms. Inger Forsberg, Ms. Barbro Näsman-Glaser and Ms. Monica Hendén for excellent technical assistance. This work was supported by grants and a fellowship (to I.K., visiting scientist from Biological Research Center, Szeged, Hungary) from the Swedish Medical Research Council (project No. 03X-06805, 03X071356 and 03W-08296) and Karolinska Institutets Research Funds.

References

1 Lindgren JÅ, Hökfelt T, Dahlen SE, et al: Leukotrienes in the rat central nervous system. Proc Natl Acad Sci USA 1984;81:6212–6216.
2 Dembinska-Kieć A, Simmet T, Peskar BA: Formation of leukotriene C_4-like material by rat brain tissue. Eur J Pharmacol 1984;99:57–62.
3 Moskowitz MA, Kiwak KJ, Hekimian K, et al: Synthesis of compounds with properties of leukotrienes C_4 and D_4 in gerbil brains after ischemia and reperfusion. Science 1984;224:886–889.
4 Wolfe LS, Pappius HM, Pokrupa R, et al: Involvement of arachidonic acid metabolites in experimental brain injury. Identification of lipoxygenase products in brain. Clinical studies on prostacyclin infusion in acute cerebral ischemia; in Hayaishi O, Yamamoto S (eds): Advances in Prostaglandin, Thromboxane and Leukotriene Research. New York, Raven Press, 1985, vol 15, pp 585–588.
5 Miyamoto T, Lindgren JÅ, Samuelsson B: Isolation and identification of lipoxygenase products from the rat central nervous system. Biochem Biophys Acta 1987;922:372–378.
6 Shimizu T, Takusagawa Y, Izumi T, et al: Enzymatic synthesis of leukotriene B_4 in guinea pig brain. J Neurochem 1987;48:1541–1546.

7   Rosenblum WI: Constricting effects of leukotrienes on cerebral arterioles of mice. Stroke 1985;16:262–263.

8   Black KL, Hoff JT: Leukotrienes increase blood-brain barrier permeability following intraparenchymal injections in rat. Ann Neurol 1985;18:349–351.

9   Hulting AL, Lindgren JÅ, Hökfelt T, et al: Leukotriene C_4 as a mediator of LH release from rat anterior pituitary cells. Proc Natl Acad Sci USA 1985;82:3834–3838.

10   Gerozissis K, Rougeot C, Dray F: Leukotriene C_4 is a potent stimulator of LHRH secretion. Eur J Pharmacol 1986;121:159–160.

11   Miyamoto T, Lindgren JÅ, Hökfelt T, et al: Regional distribution of leukotriene and monohydroxyeicosatetraenoic acid production in the rat central nervous system. FEBS Lett 1987;216:123–127.

12   Lindgren JÅ, Karnushina I, Claesson HE: Role of brain microvessels and choroid plexus in cerebral metabolism of leukotrienes. Ann NY Acad Sci, in press.

13   Spector R, Goetzl EJ: Leukotriene C_4 transport and metabolism in the central nervous system. J Neurochem 1986;46:1308–1312.

14   Albert Z, Orlowski M, Rzucidlo Z, et al: Studies on γ-glutamyltranspeptidase activity and its histochemical localization in the central nervous system of man and different animal species. Acta Histochem 1966;25:312–320.

15   Shine HD, Haber B: Immunochemical localization of γ-glutamyltranspeptidase in the rat CNS. Brain Res 1981;217:339–349.

16   Feinmark SJ, Cannon PJ: Endothelial cell leukotriene C_4 synthesis results from intercellular transfer of leukotriene A_4 synthesized by polymorphonuclear leukocytes. J Biol Chem 1987;261:16466–16472.

17   Claesson HE, Haeggström J: Human endothelial cells stimulate leukotriene synthesis and convert granulocyte released leukotriene A_4 into leukotrienes B_4, C_4, D_4, and E_4. Eur J Biochem 1988;173:93–100.

18   Orlowski M, Sessa G, Green JP: γ-Glutamyltranspeptidase in brain capillaries: possible site of a blood-brain barrier for amino acids. Science 1974;184:66–68.

19   Spector R, Goetzl EJ: Leukotriene C_4 transport by the choroid plexus in vitro. Science 1985;228:325–327.

J.Å. Lindgren, MD, Department of Physiological Chemistry, Karolinska Institutet,
Box 60400, S-104 01 Stockholm (Sweden)

Zor U, Naor Z, Danon A (eds): Leukotrienes and Prostanoids in Health and Disease.
New Trends Lipid Mediators Res. Basel, Karger, 1989, vol 3, pp 166–170

Formation of Cysteinyl-Leukotrienes by Human Intracranial Tumors

Th. Simmet[a], *W. Luck*[a], *M. Winking*[b], *W.K. Delank*[b], *B.A. Peskar*[a]

Departments of [a]Pharmacology and [b]Neurosurgery, Ruhr University, Bochum, FRG

Introduction

Brain tissue from various species has recently been demonstrated to possess the capacity for cysteinyl-leukotriene (LT) production [1–5]. In addition to the in vitro formation of cysteinyl-LT by rat brain tissue after stimulation with ionophore A23187 [1, 2], a number of pathophysiological stimuli such as concussive injury, subarachnoid hemorrhage, ischemic insult and seizures have been shown to initiate cysteinyl-LT formation in the gerbil brain in vivo [3, 6–8]. The findings that LTC_4 binding sites are present in rat and guinea pig brain and that LTC_4 and LTD_4 cause excitation of cerebellar Purkinje neurons suggest a putative role of cysteinyl-LT as neuromodulators [9–11]. In addition, at least in the periphery cysteinyl-LT are potent vasonconstrictors and seem to increase vascular permeability which is thought to be due to direct action of cysteinyl-LT on the endothelia of the postcapillary venule [11]. Only recently, we have shown that nonpathological human brain tissue has the capacity to synthesize cysteinyl-LT [5]. Here we demonstrate that human intracranial tumors release even larger amounts of cysteinyl-LT than nonpathological human brain tissue, which perhaps may be relevant for some pathological changes in brain function of brain tumor patients.

Materials and Methods

Fresh tissue from intracranial neoplasms was obtained from patients subjected to neurosurgical resection of such tumors. The study had been approved by the Ethic Committee of the Ruhr University. Immediately after resection the tumors were placed into ice-cold Tyrode solution and were rapidly transported to the laboratory. The tissues were manually cut into slices and washed extensively in ice-cold saline (0.9%). Aliquots of 100–150 mg of tissue slices were incubated at 37 °C in 2.0 ml of oxygenated Tyrode solution and were adapted to in vitro conditions for 60 min with buffer changes every 30 min. After an additional preincubation period of 30 min, the incubation buffer was removed and fresh buffer containing ionophore A23187 ($10 \mu M$) or its solvent DMSO was added and incubation was continued for a further 30 min. At the end of the incubation period the supernatants were collected, boiled for 4 min and proteins were finally pelleted by centrifugation. The supernatants were used for the radioimmunological determination of TXB_2 and LTC_4-like material [12, 13]. When enzyme inhibitors such as indomethacin ($2.8 \mu M$) or nordihydroguaiaretic acid (NDGA) ($10 \mu M$) were used, these compounds were present throughout the experiment.

Two-milliliter aliquots of whole human blood were allowed to clot in glass tubes at 37 °C for 30 min. After centrifugation, serum proteins were precipitated by addition of 3 volumes of ice-cold acetone. After evaporation the residues were resuspended in Tris-HCl 50 mM, pH 7.4, for radioimmunological determination of cysteinyl-LT.

Bioassay and reversed-phase HPLC was performed as previously described [5]. Urine samples were purified by reversed-phase HPLC prior to radioimmunological determination of LTE_4. All tumor tissues were examined histopathologically.

Results

Upon incubation in vitro, tissue slices from astrocytomas grade III, glioblastomas and meningiomas released LTC_4-like material spontaneously (12.8 ± 2.0 ng/g w.w./30 min, n = 10, 37.7 ± 9.0 ng/g w.w./30 min, n = 4, and 28.5 ± 7.7 ng/g w.w./30 min, n = 10, respectively) and even larger amounts after ionophore A23187 ($10 \mu M$) stimulation (165.2 ± 12.6 ng/g w.w./30 min, n = 10, 177.9 ± 52.3 ng/g w.w./30 min, n = 4, and 331.5 ± 52.3 ng/g w.w./30 min, n = 10, respectively). Astrocytomas which may occur at different malignancy grades seem to have a larger biosynthetic capacity for LTC_4-like material and TXB_2 with increasing malignancy (table 1) when the slices were incubated in the presence of ionophore A23187 ($10 \mu M$) (table 1). At grade IV, biosynthesis of LTC_4-like material and TXB_2 in the presence of ionophore A23187 ($10 \mu M$) was significantly higher ($p < 0.05$) than in the lower malignancy groups. By reversed-phase

Table 1. Formation of eicosanoids by astrocytoma tissue slices incubated in the presence of ionophore A23187 ($10 \mu M$)

	Malignancy grade (WHO)			
	I	II	III	IV
LTC_4-like material, ng/g w.w./30 min	33.1 (1)	28.9 ± 9.7 (3)	165.2 ± 12.6 (10)	436.1 ± 109.2 (4)
TXB_2, ng/g w.w./30 min	2.7 (1)	8.0 ± 0.8 (3)	23.3 ± 2.7 (10)	44.6 ± 5.6 (4)

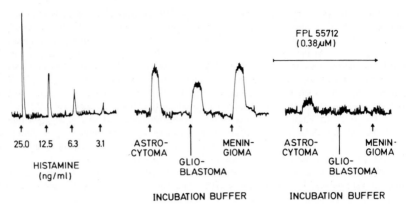

Fig. 1. Contractile activity of LTC$_4$-like material released from ionophore A23187 (10 μM)-stimulated human intracranial tumor tissue slices on the isolated guinea pig ileum and the effect of the SRS-A antagonist FPL 55712 (0.38 μM).

HPLC analysis, immunoreactivity from the three different tumor types was shown to coelute mainly with authentic LTD$_4$ and LTE$_4$ and to smaller amounts with LTC$_4$. The material obtained from these incubations was biologically active in the guinea pig ileum bioassay system (fig. 1), an effect which could be antagonized by the slow-reacting substance of anaphylaxis (SRS-A) antagonist FPL 55712. The identity of TXB$_2$ had been confirmed by thin-layer chromatography. During a 30-min incubation period the biosynthetic capacity of whole human blood for LTC$_4$-like material was 0.38 ± 0.09 and 14.1 ± 2.2 ng/ml serum in the absence and presence of ionophore A23187 (10 μM), respectively.

In 4 patients suffering from astrocytomas grade III or IV, we estimated urinary excretion of LTE$_4$ by radioimmunoassay after reversed-phase HPLC. Within 1 week after operation, excretion of LTE$_4$ decreased from 100% at the day of operation to $15.5 \pm 6.8\%$.

Discussion

The data presented clearly indicate that intracranial neoplastic tissues such as astrocytomas, glioblastomas and meningiomas have the capacity to synthesize large amounts of cysteinyl-LT. In fact, when one compares the spontaneous release of cysteinyl-LT of these tissues with previous data from our laboratory from nonpathological human gray and white matter [5], the neoplastic tissues release significantly ($p < 0.05$) larger amounts of cysteinyl-LT. Similarly, in the presence of ionophore A23187, much larger amounts of cysteinyl-LT were formed by the tumor tissues as compared to nonpathological human gray or white matter. In analogy, it has recently been shown that astrocytomas or glioblastomas have a higher biosynthetic capacity for various prostanoids than adjacent normal tissue [14]. Since the antibody used for cysteinyl-LT determination cross-reacts to 40% with LTD_4 and LTE_4, the pattern of cysteinyl-LT was further characterized by reversed-phase HPLC, demonstrating the formation of mainly LTD_4, LTE_4 and smaller amounts of LTC_4. A similar pattern has been observed in material from human gray matter while in human white matter LTC_4 appeared to be more rapidly metabolized to LTD_4 and LTE_4 [5]. In addition, the material was demonstrated to be biologically active in that it contracted the isolated guinea pig ileum as a standard bioassay tissue for SRS-A [15]. The specificity of this effect was confirmed by the antagonistic activity of the SRS-A antagonist FPL 55712 [16]. Because of the relatively low biosynthetic capacity of whole human blood for cysteinyl-LT as compared to the tumor tissue, contribution of blood-derived cysteinyl-LT due to possibly remaining blood contamination, can be considered negligible.

Astrocytomas offer the unique opportunity to study one tissue type at different grades of malignancy. Although we were able to get only a few tumors of grade (WHO) I, II and IV, there seems to be a correlation between increasing malignancy and increased cysteinyl-LT and TXB_2 production. This increased cysteinyl-LT and TXB_2 formation perhaps reflects the increased metabolic turnover of tumor cells. Whether this phenomenon has any pathophysiological consequence remains to be investigated. We hypothesized that under physiological conditions there should be very little formation of cysteinyl-LT in vivo. Therefore, if tumor tissue should have any basal secretion rate in vivo, there might appear some LTE_4, which is considered to be a main urinary metabolite of cysteinyl-LT in man [17], in the patient's urine. In fact, we found detectable amounts of LTE_4 in urine from patients suffering from astrocytomas grade III or IV and the urinary excretion of LTE_4 decreased by 84% within 1 week after operation as compared to excretion prior to operation, indicating that malignant astrocytomas perhaps produce cysteinyl-LT in vivo. This finding certainly deserves some further exploration. Thus, due to potential edemogenic and neuromodulatory effects, cysteinyl-LT formation by intracranial neoplasms might possibly contribute to complications encountered with intracranial tumor patients such as brain edema and seizure attacks.

References

1   Dembińska-Kieć A, Simmet Th, Peskar BA: Formation of leukotriene C_4-like material by rat brain tissue. Eur J Pharmacol 1984;99:57–62.

2   Lindgren JA, Hökfelt T, Dahlén SE, et al: Leukotrienes in the rat central nervous system. Proc Natl Acad Sci USA 1984;81:6212–6216.

3   Moskowitz MA, Kiwak KJ, Hekimian K, et al: Synthesis of compounds with properties of leukotrienes C_4 and D_4 in gerbil brain after ischemia and reperfusion. Science 1984;224:886–889.

4   Shimizu Th, Takusagawa Y, Izumi T, et al: Enzymic synthesis of leukotriene B_4 in guinea-pig brain. J Neurochem 1987;48:1541–1546.

5   Simmet T, Luck W, Delank WK, et al: Formation of cysteinyl-leukotrienes by human brain tissue. Brain Res 1988;456:344–349.

6   Kiwak KJ, Moskowitz MA, Levine L: Leukotriene production in gerbil brain after ischemic insult, subarachnoid hemorrhage, and concussive injury. J Neurosurg 1985;62:865–869.

7   Simmet Th, Seregi A, Hertting G: Formation of sulphidopeptide-leukotrienes in brain tissue of spontaneously convulsing gerbils. Neuropharmacology 1987;26:107–110.

8   Simmet Th, Seregi A, Hertting G: Characterization of seizure-induced cysteinyl-leukotriene formation in brain tissue of convulsion-prone gerbils. J Neurochem 1988;50:1738–1742.

9   Palmer MR, Mathews R, Murphy RC, et al: Leukotriene C elicits a prolonged excitation of cerebellar Purkinje neurons. Neurosci Lett 1980;18:173–180.

10 Palmer MR, Mathews WR, Hoffer BJ, et al: Electrophysiological response of cerebellar Purkinje neurons to leukotriene D_4 and B_4. J Pharmacol Exp Ther 1981;219:91–96.

11 Samuelsson B, Dahlén SE, Lindgren JA, et al: Leukotrienes and lipoxins: Structures, biosynthesis, and biological effects. Science 1987;237:1171–1176.

12 Anhut H, Bernauer W, Peskar BA: Radioimmunological determination of thromboxane release in cardiac anaphylaxis. Eur J Pharmacol 1977;44:85–88.

13 Aehringhaus U, Wölbling RH, König W, et al: Release of leukotriene C_4 from human polymorphonuclear leucocytes as determined by radioimmunoassay. FEBS Lett 1982;146:111–114.

14 Castelli MG, Butti G, Chiabrando C, et al: Arachidonic acid metabolic profiles in human meningiomas and gliomas. J Neuro-Oncol 1987;5:369–375.

15 Brocklehurst WE: The release of histamine and formation of slow-reacting substance of anaphylaxis (SRS-A) during anaphylactic shock. J Physiol (Lond) 1960;151:416–435.

16 Augstein J, Farmer JB, Lee TB, et al: Selective inhibitor of slow-reacting substance of anaphylaxis. Nature New Biol 1973;245:215–217.

17 Örning L, Kaijser L, Hammarström S: In vivo metabolism of leukotriene C_4 in man: urinary excretion of leukotriene E_4. Biochem Biophys Res Commun 1985;130:214–220.

Th. Simmet, MD, Ruhr-Universität Bochum, Medizinische Fakultät,
Abteilung für Pharmakologie und Toxikologie, Universitätsstrasse 150,
Postfach 10 21 48, D–4630 Bochum (FRG)

Zor U, Naor Z, Danon A (eds): Leukotrienes and Prostanoids in Health and Disease.
New Trends Lipid Mediators Res. Basel, Karger, 1989, vol 3, pp 171–174

The Role of Eicosanoids and Phospholipase A₂ in Brain Injury

E. Shohami[a], *Y. Shapira*[b], *G. Yadid*[a], *S. Yedgar*[c]

Departments of [a]Pharmacology, [b]Anesthesiology and [c]Biochemistry, The Hebrew University-Hadassah Medical School, Jerusalem, Israel

Introduction

The mechanisms underlying progressive necrosis after brain ischemia or trauma are not clearly understood. The pathological features of these conditions include edema, opening of the blood-brain barrier and inflammation.

Recent studies suggest that a number of biochemical and metabolic events contribute to the progressive and irreversible cell damage and to tissue hypoperfusion after cerebral ischemia or trauma. Eicosanoids might be involved in these mechanisms due to their effect on microcirculatory resistance on vascular permeability and on neural transmission.

In an attempt to evaluate the significance of the eicosanoids in cerebral ischemia, Gaudet et al. [4] first showed in gerbils a marked elevation in brain concentrations of PGs shortly after ischemia. In further studies of ischemia in rats [8], gerbils [6] and rabbits [5], we followed the eicosanoid production during the extended postischemic period (24 h up to 1 week). We found that ischemia and reperfusion produce differential changes in the eicosanoids production that are site-, time- and PG-specific. The eicosanoids produced after an ischemic episode may aggravate the initial insult, due to an excessive production of TXA_2 over PGI_2, or to edema formation, mediated by PGE_2 and PGI_2. On the other hand, PGD_2 and PGI_2 may exert self-protective mechanisms on the microcirculation as well as on brain metabolism and function.

Brain contusion induces metabolic and physiological responses, similar to those induced by ischemia. In order to study the effect of brain injury on brain eicosanoids and their relation to the development of neurological dysfunction, we have developed a model of closed head injury in rat [7]. In this model we followed the temporal changes in eicosanoids, specific gravity and neurological status of the rats after the trauma (up to 10 days) and we found increased production of the AA cascade metabolites as early as

15 min after trauma. 5-HETE synthesis increased 4-fold at the site of injury
and similarly, TXB_2 was elevated at the same time in both the injured and
noninjured zones [9, 10]. A unique pattern of concentration changes was
found up to 10 days after injury for the various eicosanoids. PGE_2,
6-keto-$PGF_{1\alpha}$ and TXB_2 levels peaked and declined back to normal at
different times after injury. The decrease in tissue-specific gravity, i.e.
edema formation, was significant at 15 min after trauma, it peaked at
24–48 h and returned to normal by 7–10 days.

In the present study we investigated the activity of PLA_2 in brain tissue
after head injury – both at the early (15 min and 4 h) and late (24 h)
post-trauma periods. The experiment was designed to study whether the
increase in PGs and 5-HETE formation after trauma is due, at least in part,
to the activation of PLA_2, which is a rate-limiting enzyme in the liberation
of AA from membrane phospholipids.

Materials and Methods

Head injury was induced in rats as described elsewhere [7] and they were sacrificed at
15 min, 4 and 24 h after injury. Cortical slices were taken from the injured zone, from the
corresponding region of the contralateral hemisphere and from the frontal lobe of both
hemispheres. These cortical slices were incubated in the presence of a fluorescent phospholipid analogue, 1-acyl-2-(N-4-nitrobenzo-2-oxa-1,3-diazole)amino caproylphosphatidylchlonine
(C_6-NBD-PC) which is a substrate for phospholipase A_2 (PLA_2) in intact cells. The action of
PLA_2 on this substrate produces only one fluorescent product which is separated from the
substrate as described in detail [11]. Thus, measurement of the fluorescence intensity of the
product yields the direct measure of PLA_2 activity [2].

Results and Discussion

The activity of brain PLA_2 at 15 min, 4 and 24 h after head trauma is
summarized in table 1, and is expressed as percent of activity in sham rats.
In slices taken from the injured zone of the left hemisphere, increased PLA_2
activity was found as early as 15 min after trauma ($p < 0.05$ as compared
to the same region of sham by ANOVA followed by Dunnett test). At the
same time, the frontal lobe had normal activity of PLA_2 and the activity in
the corresponding region in the contralateral hemisphere was somewhat
reduced. At 4 h after injury, in both the injured and the frontal zone of the
left, contused hemisphere, increased activity was found (although, due to
the highly scattered results, it was not statistically significant in the injured
zone), whereas in the right hemisphere, PLA_2 activity was normal in both
regions. Twenty-four hours after injury, a 2.5-fold increase in activity of

Table 1. Activity of PLA_2 (percent of sham) after head injury

Region	15 min (14)	4 h (7)	24 h (11)
Injured			
Left (traumatized)	170*	175	240*
Right (contralateral	65	130	95
Frontal			
Left	115	195	145
Right	85	130	125

Number of rats in parentheses.
*p < 0.05, ANOVA followed by Dunnett text.

PLA_2 was found at the site of injury, with only a slight increase in the remote region of the same hemisphere, and normal activity in the contralateral hemisphere.

There was a statistically significant correlation (cc = 0.469, p < 0.05, n = 21), using Pearson's coefficient correlation test, betwen PLA_2 activity and tissue release of PGE_2 whenever the latter was elevated, in the injured zone, as compared to sham.

The results presented here show increased activity of PLA_2 in cerebral cortical tissue already at 15 min, and up to 24 h after closed head injury. The parallel time course of increased PLA_2 activity and eicosanoid production in this model suggests that, at least in part, raised levels of free AA (the substrate for both the cyclooxygenase and lipoxygenase pathways) may be available due to a direct action of PLA_2 on membrane phospholipids. The correlation found between PLA_2 activity and PGE_2 levels after head injury may suggest that PLA_2 activation is indeed involved in the increased production of eicosanoids which was earlier reported in this model of head injury. Based on the present data, we cannot exclude the involvement of other phospholipases. However, the method employed in this study had been shown to be specific for the measurement of PLA_2 activity [2].

In our model of head injury, ischemic cell damage develops gradually after the impact [7]. It is hard to speculate whether the increased PLA_2 activity observed in the present study of 24 h is the result of local ischemia, or the products of the AA cascade are responsible for a delayed ischemia. However, the similar time course of PLA_2 activity and eicosanoids production strongly supports a role of this enzyme in the pathogenesis of traumatic brain injury. Previous studies by Wei et al. [12] demonstrate activation of PLC after brain injury, and in an ischemia model, Abe et al.

[1] also showed increased activity of PLC. On the other hand, other reports showed activation of PLA_2 during ischemia [3]. Thus, in the pathological conditions of ischemia and injury, most probably both enzymatic pathways are activated to yield the increased levels of free arachidonic acid, followed by its oxidation to eicosanoids.

References

1   Abe, K.; Kogure, K.; Yamamoto, H.: Imazawa, M.; Miyamoto, K.: Mechanism of arachidonic acid liberation during ischemia in gerbil cerebral cortex. J. Neurochem. *48:* 503–509 (1987).

2   Dagan, A.; Yedgar, S.: A facile method for direct determination of phospholipase A_2 activity in intact cells. Biochem. Int. *15:* 801–808 (1987).

3   Edgar, A.D.; Storsznzjder, J.; Horrocks, L.A.: Activation of ethanolamine phospholipase A_2 in brain during ischemia. J. Neurochem. *39:* 1111–1116 (1982).

4   Gaudet, R.J; Alam, I.; Levine, L.: Accumulation of cyclooxygenase products of arachidonic acid metabolism in gerbil brain during reperfusion after bilateral common carotid occlusion. J. Neurochem. *35:* 653–658 (1980).

5   Jacobs, T.P.; Shohami, E.; Baze, W.; Burgard, E.; Gunderson, C.; Hallenbeck, J.M.; Feuerstein, G.: Deteriorating stroke model: histopathology, edema and eicosanoid changes following spinal cord ischemia in rabbits. Stroke *18:* 741–750 (1987).

6   Kempski, O.; Shohami, E.; von Lubitz, D.; Hallenbeck, J.M.; Feuerstein, G.; Postischemic production of eicosanoids in gerbil brain. Stroke *18:* 111–119 (1987).

7   Shapira, Y.; Shohami, E.; Sidi, A.; Soffer, D., Freeman, S.; Cotev, S.: Experimental closed head injury in rats: mechanical, pathophysiologic and neurologic properties. *Crit. Care Med. 16:* 258–265 (1988).

8   Shohami, E.; Rosenthal, J.; Lavy, S.: The effect of incomplete cerebral ischemia on prostaglandin levels in rat brain. Stroke *13:* 494–498 (1982).

9   Shohami, E.; Shapira, Y., Sidi, A.; Cotev, S.: Head injury induces increased prostaglandin synthesis in rat brain. J. Cereb. Blood Flow Metab. *7:* 58–63 (1987).

10 Shohami, E.; Shapira, Y., Cotev, S.: Experimental closed head injury in rats: prostaglandin production in a noninjured zone. Neurosurgery *22:* 859–863 (1988).

11 Shohami, E.; Shapira, Y.; Yadid, G.; Reisfeld, N.; Yedgar, S.: Brain phospholipase A_2 is activated after experimental closed head injury in the rat (submitted, 1989).

12 Wei, E.P.; Lamb, R.G.; Kontos, H.A.: Increased phospholipase C activity after experimental brain injury. J. Neurosurg. *56:* 695–698 (1982).

E. Shohami, PhD, Department of Pharmacology, The Hebrew University, Hadassah Medical School, P.O. Box 1172, Jerusalem 91010 (Israel)

Zor U, Naor Z, Danon A (eds): Leukotrienes and Prostanoids in Health and Disease.
New Trends Lipid Mediators Res. Basel, Karger, 1989, vol 3, pp 175–182

Formation and Actions of Hepoxilin A₃ in Mammalian CA1 Neurons[1]

C.R. Pace-Asciak[a, b], *N. Gurevich*[c], *P.H. Wu*[c], *W-G. Su*[d],
E.J. Corey[d], *P.L. Carlen*[c]

[a]Research Institute, Hospital for Sick Children, Toronto; [c]The Playfair Neurosciences
Unit and Addiction Research Foundation, Departments of Medicine (Neurology),
Physiology and [b]Pharmacology, University of Toronto, Canada, and [d]Department of
Chemistry, Harvard University, Cambridge, Mass., USA

Arachidonic acid was tested electrophysiologically on rat hippocampal
CA1 neurons and found to cause a hyperpolarization which is sometimes
followed by a later depolarization, to augment the post-spike train long-
lasting after-hyperpolarization (AHP) and to increase the inhibitory postsynaptic potentials (IPSPs). These effects were mimicked by hepoxilin A₃.
Under identical conditions as the electrophysiological experiments, intact
hippocampal slices were incubated with arachidonic acid; hepoxilin A₃,
measured as the corresponding trioxilin methyl ester t-butyldimethylsilyl
ether derivative, was detected in the perfusate by GCMS. Our data show
that hepoxilin A₃ is released by the brain slices, its release is greatly
stimulated by the addition of exogenous arachidonic acid, and this enhanced release is blocked by the presence of the dual cyclooxygenase and
lipoxygenase blocker, BW-755C. These results demonstrate novel actions of
arachidonic acid on rat hippocampal neurons which are mediated through
its transformation into hepoxilin A₃. The hepoxilins may represent important second messengers of the arachidonic acid cascade in the mammalian
CNS.

Introduction

Brain is known to possess both cyclooxygenase and lipoxygenase
activities, although the physiological roles of metabolites of these pathways
(prostaglandins and the HPETEs) in the mammalian CNS are poorly

[1] Portions of these findings were presented at the Taipei Conference on Prostaglandin
and Leukotriene Research, Taipei, Taiwan, April 22–24, 1988.

understood [1–7]. Certain neurotransmitters augment the formation of 12- and 5-HPETE in rat brain pieces [8]. In Aplysia sensory neurons where both 12- and 5-HPETE are formed [9], it has recently been shown electrophysiologically that 12-HPETE but not 5-HPETE causes a membrane hyperpolarization and diminishes calcium spikes [10]. Since 12-HPETE is unstable in a biological system and is rapidly transformed into hydroxy epoxide metabolites, namely hepoxilins [11], the actions of 12-HPETE may not be intrinsic to this compound but to the hepoxilins. Indeed, the hepoxilins have been shown to mimic the actions of 12-HPETE in terms of insulin secretion [12] and calcium transport [13]. The hepoxilins are also formed in rat brain from both endogenous and exogenous arachidonic acid [14], and a specific epoxide hydrolase exists which degrades the hepoxilins [14]. Until now, actions of the hepoxilins have not been demonstrated in central mammalian neurons. The present experiments demonstrate the formation and electrophysiological actions of hepoxilin A_3 in intact rat brain hippocampal slices.

Materials and Methods

Electrophysiological Experiments

Male Wistar rats (150–170 g) were anesthetized with halothane, decapitated and coronal sections (400 μm) were cut with a vibratome in ice-cold artificial cerebrospinal fluid (ACSF). The slices were stored in oxygenated (95% O_2–5% CO_2) ACSF for at least 1 h at room temperature. The ACSF contained the following in mM: Na^+ 154; K^+ 4.25; Ca^{2+} 2; Mg^{2+} 2; Cl^- 131.5; HCO_3^- 26; HPO_4^- 1.25; SO_4^{2-} 2; dextrose 10. The pH of the medium was maintained at 7.4. Electrophysiological recordings were done in an interface-type chamber at 34 °C using standard techniques [15] with glass micropipettes filled with 3 M KCl or 3 M K acetate (resistances 60–150 MΩ). Data were stored on tape and recorded on chart paper. Orthodromic stimulation for the excitatory postsynaptic potentials (EPSPs) and inhibitory postsynaptic potentials (IPSPs) were done by mono- and bipolar tungsten electrodes. Electrophysiological recordings were obtained from hippocampal CA1 neurons using routine techniques [15].

Arachidonic acid (NuChek) was converted to the methyl ester derivative using diazomethane [16], and the methyl ester (AA-Me) was used after dissolving in DMSO at 1 mg/40 μl; hepoxilin methyl ester (HxA$_3$), a mixture of two isomers at carbon 8, was dissolved in DMSO at a concentration of 85 μg/20 μl. All compounds were dissolved in ACSF at varying concentrations and were applied by perfusion or drop (Picospritzer). The methyl ester of these products was used to facilitate entry into the cells and because HxA$_3$ is more stable in the methyl ester form than as the free acid [17]. Once in the cell, the methyl ester is hydrolyzed into the free acid by esterases present in the cell cytosol. Preliminary results with the individual isomers of HxA$_3$ show the pattern of action as the data with the mixture of isomers (HxA$_3$).

Biochemical Experiments

Hippocampal slices (78 mg protein) were incubated in 20 ml of ACSF at 34 °C, similar to the conditions used in the electrophysiological experiments. After a 15-min basal period, the bathing solution was removed and immediately frozen, and to the tissue was added fresh medium to which arachidonic acid ($166 \, \mu M$) was added in $10 \, \mu l$ of ethanol. After another 15-min incubation, the medium was removed and immediately frozen, and fresh medium containing BW-755C ($66 \, \mu M$) was added. After a 1-min preincubation with BW-755C, arachidonic acid ($166 \, \mu M$) was added in $10 \, \mu l$ of ethanol. After 15 min, the incubation medium was removed and frozen. The frozen media were thawed, acidified to pH 3, and internal standards were added (50 ng 9-hydroxyoctadecatrienoic acid (9-HOTE) for the HETEs; 40 ng of tetradeuterated TrXA$_3$, the stable hydrolysis product of HxA$_3$). The solution was passed through a C18-SEP PAK cartridge as described previously [22]. The methanol/water/acetic acid (65/35/0.4 v/v) fraction containing the hepoxilins (as trioxilins) from the SEP PAK was taken to dryness, and the residue was converted into the methyl ester t-butyldimethylsilyl ether derivative as previously described [16]. The derivative was analyzed by GCMS using a Hewlett-Packard GCMS-MSD (Model 5970) equipped with a DB1 methyl silicone fused silica capillary column. The methanol/water/acid (90/10/0.4 v/v) fraction containing the HETEs was taken to dryness and analyzed by reversed-phase HPLC on a Spherisorb ODS II (5 μm) using methanol/water/acetic acid (78/22/0.01 v/v) as solvent at a flow rate of 0.5 ml/min. HETEs were detected with a flow-through UV detector.

Results

Arachidonic acid caused an initial membrane hyperpolarization which took from 4 to 15 min (10.5 ± 3.5, n = 23) to develop and lasted for at least 10 min (fig. 1a). This included the perfusion time required for the drug to enter the recording chamber (2–3 min). A similar effect was observed with HxA$_3$ (fig. 1a, table 1). The hyperpolarization was sometimes followed by a depolarization of a few millivolts (not shown) which usually lasted longer than 10 min. These changes in resting membrane potential (RMP) were not associated with any change in the input resistance. During the hyperpolarization, if the RMP repolarized either passively or by intracellular current injection to the predrug control level, spontaneous spiking was noted (data not shown).

The AHP is an intrinsic inhibitory potential generated by a transient rise in intracellular calcium causing increased potassium conductance [18, 19]. Both AA (not shown) and HxA$_3$ (fig. 1b) augmented this potential by increasing its duration and sometimes its amplitude (fig. 1b, table 1). The augmentation of the AHP lasted several minutes following the cessation of the drug-induced hyperpolarization (fig. 1b) and this augmentation was evident in those neurons which also showed a later depolarization.

Synaptic transmission was assessed by measuring the EPSP-IPSP sequence from orthodromic stimulation of the stratum radiatum (fig. 1c, table 1). The amplitude of the EPSPs was not significantly changed with drug perfusion (not shown), whereas the IPSPs were increased by both AA-Me and HxA$_3$. The amplitudes of both the early chloride-dependent

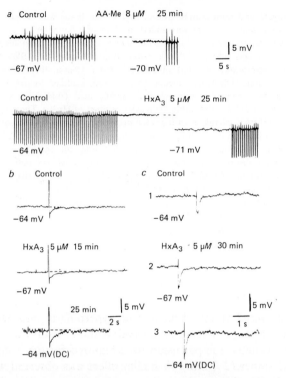

Fig. 1. Effects of AA-Me and HxA$_3$ on *(a)* the RMP, *(b)* the post-spike train AHP, and *(c)* the orthodromic IPSP of hippocampal CA1 neurons. Input resistance as measured by 0.2 nA, 100-ms hyperpolarizing pulses was unchanged. The control RMP is indicated by a dashed line. A 3 *M* KCl electrode was used in figure 1a. The post-spike train AHP *(b)* was caused by 5 spikes. The AHP was increased by HxA$_3$ infusion both in depth and duration. When the membrane was repolarized by DC current injection to the control RMP, the AHP increase was more apparent. Dotted lines show RMP during AHP. Electrode was 3 *M* K acetate. The enhanced IPSP *(c)* due to infusion of HxA$_3$ was made more apparent after repolarization of the RMP during 30 min of drug exposure. Electrode was 3 *M* K acetate.

and the later potassium-dependent phases [20, 21] of the IPSP were augmented. These effects sometimes persisted many minutes after drug washout (fig. 1c). These data show similar inhibitory actions for AA-Me and its 12-LOX metabolite, HxA$_3$. Preliminary data indicated that there were no apparent differences in the actions of the mixture of HxA$_3$ isomers, or the purified individual isomers, HxA$_{3B}$ and HxA$_{3T}$.

Biochemical experiments demonstrated that AA is transformed by hippocampal slices used in the electrophysiological experiments, into a variety of HETEs and HxA$_3$, including 12-HETE. The former was identified by HPLC (fig. 2a) and the latter was identified by GCMS after

Table 1. Responses of hippocampal CA1 neurons (n = 19) to AA and HxA₃ perfusion

Drug	Concentration, μM	RMP hyperpolar- ization	AHP		IPSP increased amplitude
			Prolongation	Increased amplitude	
AA-Me	8–26	6/10[1] (2.8 ± 1.8 mV)[2]	7/10 (167 ± 24%)[3]	6/10 (149 ± 23%)	3/4 (146 ± 46%)
HxA₃-Me	0.5–10	9/9 (2.6 ± 0.9 mV)	9/9 (172 ± 17%)	3/9 (139 ± 15%)	5/6 (161 ± 12%)

RMP = resting membrane potential; AHP = post-spike train after-hyperpolarization; IPSP = inhibitory postsynaptic potential; Me = methyl ester.
[1] Number of neurons responding/total number of neurons tested.
[2] Mean ± SD for neurons exhibiting a response.
[3] % ± SD of controf control values following drug application.

conversion of the product to the methyl ester t-butyldimethylsilyl ether derivative (fig. 2b). Further experiments (not shown) indicated that 12-HPETE is also converted by the intact slices into the trihydroxy derivative of HxA₃, i.e. TrXA₃, indicating not only active synthesis of HxA₃ but also metabolism into TrXA₃, confirming that hepoxilin epoxide hydrolase, previously shown by us to be present in brain homogenates [14], is active in the intact tissue. In the electrophysiological experiments, the concentration of HxA₃ used reflects the amounts formed by homogenates as well as pieces of whole brain and hippocampal slices stimulated by arachidonic acid or various neurotransmitters [17, 22].

HxA₃ has previously been shown to increase the transport of calcium across membranes [13]. This property of HxA₃ may relate to the increase in both the early and later phases of the IPSP, suggesting enhanced neuro-transmitter release, possibly due to altered presynaptic calcium mobiliza-tion. The enhancement of the AHP duration and the later part of the IPSP are presumably related to increased potassium conductance. The effects of hyperpolarization followed by a depolarization with prolongation of the AHP are remarkably similar to those noted with 5-HT application in these neurons [23].

The finding that arachidonic acid has electrophysiological actions on mammalian CA1 neurons is, to our knowledge, new. That this action is mediated via the hepoxilins is also new. Our results provide evidence that these products are formed by the intact and electrophysiologically func-tional mammalian brain. These metabolites may indeed be located within distinct neurons and may represent important messengers coupled to

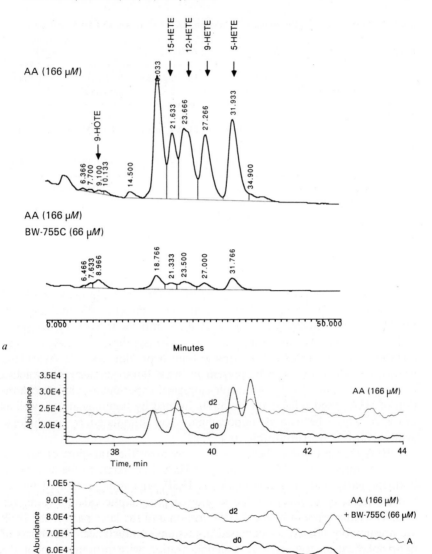

Fig. 2. Analysis of HETEs and HxA₃ released from rat hippocampal slices during incubation with arachidonic acid (133 μM). *a* Reversed-phase HPLC profiles of HETEs in the absence (top) and presence (lower) of BW-755C; 9-HOTE represents an internal standard added prior to sample workup. *b* Selected ion chromatograms (GCMS) of HxA₃ (deuterium internal standard detected at m/z 257 (line A) endogenous compound detected at m/z 255 (line B)) detected as the corresponding methyl ester t-butyldimethylsilyl ether derivative of its trihydroxy metabolite, TrXA₃ in the absence (top), and in the presence (lower) of BW-755C.

specific transmitters. It is possible that these metabolites of arachidonic acid may function as modulators of signal transduction. The mapping of eicosanoid production with the release of specific neurotransmitters from specific neurons may provide key information into the molecular architectural requirements for neurotransmission and membrane excitability.

Acknowledgements

This study was made possible by grants C.R.P.-A. (MRC), to P.L.C. (MRC; OMH) and to E.J.C. (NIH).

References

1   Samuelsson, B.: Identification of a smooth muscle-stimulating factor in bovine brain. Prostaglandins and related factors 25. Biochim. Biophys. Acta *84:* 218–219 (1964).

2   Wolfe, L.S.; Pappius, H.M.; Marion, J.: The biosynthesis of prostaglandins by brain tissue in vitro. Adv. Prostagl. Thromb. Res., *1:* 345–355 (1976).

3   Sautebin, L.; Spragnuolo, C.; Galli, G.: A mass fragmentographic procedure for the simultaneous determination of hydroxy unsaturated fatty acids and prostaglandin F$_{2\alpha}$ in the central nervous system. Prostaglandins *16:* 985–988 (1978).

4   Abdel-Halim, M.S.; Eckstedt, J.; Anggard, E.: Determination of prostaglandin F$_{2\alpha}$, prostaglandin E$_2$, prostaglandin D$_2$ and 6-keto prostaglandin F$_{1\alpha}$ in human cerebrospinal fluid. Prostaglandins *17:* 405–409 (1979).

5   Usui, M.; Asano, T.; Takamura, K.: Identification and quantitative analysis of hydroxyeicosatetraenoic acids in rat brains exposed to regional ischaemia. Stroke *18:* 490–494 (1987).

6   Shohami, E.; Jacobs, T.P.; Hallenbeck, J.M.; Feuerstein, G.: Increased thromboxane A$_2$ and 5-HETE following spinal cord ischaemia in the rabbit. Prostaglandins Leukotrienes Med. *28:* 169–182 (1987).

7   Hulting, A-L.; Lindgren, J.A.; Hokfelt, T.; Eneroth, P.; Werner, S.; Patrono, C.; Samuelsson, B.: Leukotriene C$_4$ as a mediator of luteinizing hormone release from rat anterior pituitary cells. Proc. Natl. Acad. Sci. USA *82:* 3834–3838 (1985).

8   Pellerin, L.; Wolfe, L.S.: Glutamate and norepinephrine induce 12-HETE formation in intact pieces of rat cerebral cortex. Trans. Amer. Soc. Neurochem. *19:* 106 (1988).

9   Piomelli, D.; Shapiro, E.; Feinmark, S.J.; Schwartz, J.H.: Metabolites of arachidonic acid in the nervous system of Aplysia – possible mediators of synaptic modulation. J. Neurosci. *7:* 3675–3686 (1987).

10   Piomelli, D.A.; Volterra, A.; Dale, N.; Sieglebaum, S.A.; Kandel, E.; Schwartz, J.H.; Belardetti, F.: Lipoxygenase metabolites of arachidonic acid as second messengers for presynaptic inhibition of Aplysia sensory neurons. Nature *328:* 38–43 (1987).

11   Pace-Asciak, C.R.: Arachidonic acid epoxides: Demonstration through 18-oxygen studies of an intramolecular transfer of the terminal hydroxyl group of 12(S)-hydroperoxyeicosa-5,8,10,14-tetraenoic acid to form hydroxy epoxides. J. Biol. Chem. *259:* 8332–8337 (1984).

12   Pace-Asciak, C.R.; Martin, J.M.: Hepoxilin, a new family of insulin secretagogues
      formed by intact rat pancreatic islets. Prostaglandins Leukotrienes Med. *16:* 173–180
      (1984).
13   Derewlany, L.O.; Pace-Asciak, C.R.; Radde, I.: Hepoxilin A, hydroxyepoxide metabo-
      lite of arachidonic acid, stimulates transport of ^{45}Ca across the guinea pig visceral yolk
      sac. Can. J. Physiol. Pharmacol. *62:* 1466–1469 (1984).
14   Pace-Asciak, C.R.: Formation and metabolism of hepoxilin A_3 in the rat brain. Biochim.
      Biophys. Res. Commun. *151:* 493–498 (1988).
15   Blaxter, T.J.; Carlen, P.L.; Davies, M.F.; Kujtan, P.W.: Gamma aminobutyric acid
      hyperpolarizes rat hippocampal pyramidal cells through a calcium-dependent potassium
      conductance. J. Physiol. (London) *373:* 181–194 (1986).
16   Pace-Asciak, C.R.; Granstrom, E.; Samuelsson, B.: Arachidonic acid epoxides. Isolation
      and structure of two hydroxy epoxide intermediates in the formation of 8,11,12- and
      10,11,12-trihydroxyeicosatrienoic acids. J. Biol. Chem. *258:* 6835–6840 (1983).
17   Pace-Asciak, C.R.: unpublished observations.
18   Krnjevic, K.; Puil, R.; Werman, R.: EGTA and motoneuronal after-potentials. J.
      Physiol. (London) *275:* 199–223 (1978).
19   Hotson, J.R.; Prince, D.A.: A calcium activated hyperpolarization follows repetitive
      firing in hippocampal neurons. J. Neurophysiol. *43:* 409–419 (1980).
20   Alger, B.E.: Characteristics of a slow hyperpolarizing synaptic potential in rat
      hippocampal pyramidal cells in vitro. J. Neurophysiol., *52:* 892–910 (1984).
21   Newberry, N.R.; Nicoll, R.A.: A bicuculline resistant inhibitory postsynaptic potential
      in rat hippocampal pyramidal cells in vitro. J. Physiol. (London) *348:* 239–254 (1984).
22   Pace-Asciak, C.R.; Asotra, S.; Pellerin, L.; Wolfe, L.S.; Corey, E.J.; Wu, P.; Gurevich,
      N.; Carlen, P.L.: Brain hepoxilins: Formation and biological action. Adv. Prostagl.
      Thromb. Leuk. Res. (1989) in press.
23   Wu, P.H.; Gurevich, N.; Carlen, P.L.: Serotonin-1A receptor activation in hippocampal
      CA1 neurons by 8-hydroxy-2-di-N-propylaminotetralin, 5-methoxytryptamine and 5-
      hydroxytryptamine. Neuroscience Lett. *86:* 72–76 (1988).

Prof. C.R. Pace-Asciak, Research Institute, The Hospital for Sick Children,
555 University Avenue, Toronto, Ont. M5G 1X8 (Canada)

Zor U, Naor Z, Danon A (eds): Leukotrienes and Prostanoids in Health and Disease.
New Trends Lipid Mediators Res. Basel, Karger, 1989, vol 3, pp 183–186

Prostaglandin E$_2$ as a Central Messenger of Fever: Mechanism of Formation

F. Coceani, I. Bishai, N. Hynes, J. Lees, S. Sirko

Research Institute, The Hospital for Sick Children, Toronto, Ont., Canada

Research on the role of eicosanoids in the pathogenesis of fever began with the observation in the early 1970s that E-type prostaglandins are exceedingly potent hyperthermic agents and the almost simultaneous realization that antipyretic drugs exert their effect through interference with arachidonate cyclooxygenase [1]. From that premise and the subsequent demonstration of a more active synthesis of prostaglandins in the brain of the febrile animal [1], a scheme evolved implicating prostaglandin E$_2$ (PGE$_2$) as an intermediary in the central action of pyrogens. Though widely accepted at the outset, through the years this postulate met with increasing criticism, as several data accrued seemingly limiting the importance of PGE$_2$ in the fever process or excluding a prostaglandin involvement altogether [2]. Difficulties with the PGE$_2$ mediator theory were compounded by uncertainties on the site of action of blood-borne pyrogens in brain and the lack of information on changes in prostaglandin synthesis occurring in the hypothalamus during fever.

Our aim has been to verify this proposed function for PGE$_2$ and provide at the same time a better insight into the sequence of events leading to fever. To that end, we adopted a comprehensive approach combining diverse methodologies and lines of investigation.

PGE$_2$ as a Central Mediator of Fever

As a first step, appropriate procedures were set up, and validated, for the in vivo measurement of eicosanoids both in cerebrospinal fluid (CSF) and locally in the hypothalamus [3–5]. Through them, it was proved that PGE$_2$ synthesis is quite low in the absence of fever; in fact, values turned out to be lower than those reported in some previous studies utilizing unspecific radioimmunoassay procedures or a biological assay [3]. This finding removed a major obstacle from the evaluation of earlier data – indeed, it demonstrated the fallacy of certain data – and provided a framework for the analysis of PGE_2 formation during fever [3–5]. Quite appropriately, it was found that PGE_2 elevation precedes the onset of the fever, correlates with the magnitude of the febrile response, and persists for as long as body temperature exceeds the normal range. In addition, no increase in PGE_2 synthesis followed administration of pyrogens at a dose subthreshold for the fever, while animals that had become tolerant to endotoxin exhibited a modest increase. Not only did the pattern of PGE_2 formation conform to a causal role in the onset and progression of the fever, but also the elevation itself was selective and did not extend to other eicosanoids. Thromboxane A_2 (TXA_2), a presumptive alternative mediator of fever [6], was not affected by pyrogen treatment [3–5] and, likewise, negative results were obtained with the peptidoleukotrienes [Hynes et al., unpubl.].[1] Significantly, both such compounds also have a variable effect, or no effect at all, on central thermoregulatory pathways [4, 7, 8].

In brief, the above data provide convincing evidence in support of a mediator role for PGE_2 in the genesis of fever. While documenting this point, our studies also showed that the stimulation of PGE_2 synthesis induced by pyrogens is not confined to the rostral hypothalamus [5]. This unexpected finding has implications (see below) for the definition of the site and mode of action of blood-borne pyrogens in brain.

Mode of Action of Pyrogens in Brain

Even though accepting the concept of an effector role for PGE_2 in the genesis of fever, it remains to be ascertained how the appearance of pyrogens in the circulation is translated into a rise in PGE_2 levels within the brain. Passage of pyrogens across the blood-brain barrier would well explain this response; however, no such transfer could be demonstrated in several studies [1], including a recent study from our laboratory [10]. In addition, any pyrogen seeping into the substance of the brain should have not been selective in its action, but rather should have stimulated the synthesis of both PGE_2 and TXA_2 [4, 5]. Alternatively, PGE_2 could originate from outside the brain and, owing to its peculiar physicochemical properties, be transferred in sufficient amount across the blood-brain

---

[1] Peptidoleukotrienes were measured only in CSF and their concentration was expressed as the total product after enzymatic conversion to LTE_4 [9]. Basal levels varied between 100 and 600 pg/ml (mean 302 pg/ml) and showed no obvious change following intravenous or intracerebroventricular administration of pyrogens.

barrier. Several facts, however, argue against this occurrence and, foremost among them, is the demonstration that intracerebroventricularly injected cyclooxygenase inhibitors interfere with the fever to systemic pyrogens [5]. A possibility considered recently by us is that, upon exposure to blood-borne pyrogens, the cerebral microvasculature may release interleukin-1 (IL-1), PGE$_2$, or both, into the brain parenchyma and may accordingly function as a transducing element in the fever sequence. As it turned out, however, the vasculature does not lend itself to this role [10, 11] and, by exclusion, one must therefore assume that circulating pyrogens act at some place in brain lacking a barrier. Theoretical considerations and findings from other laboratories [1] identify this target site with the circumventricular organs. Among them, the organum vasculosum laminae terminalis (OVLT) and the median eminence are conceivably most important due to their proximity to the thermoregulatory centers and the neural mechanisms subserving the host defense reaction. In this contest, it is significant that the pyrogen-induced activation of the PGE$_2$ mechanism extends beyond the bounds of the anterior hypothalamic/preoptic region (AH/POA) and involves the tuberal region as well [5].

Pathogenesis of Fever: An Updated Scheme

The fever process comprises a series of interdependent events including, in the order, exposure to a noxa (e.g., bacterial endotoxin), formation of a monokine (e.g., IL-1) outside the brain, interaction of the monokine with a specific target in the OVLT (and the median eminence), stimulation of PGE$_2$ synthesis in the AH/POA, and appropriate changes in the activity of the thermoregulatory neurons resulting ultimately in a rise in body temperature. The nature of the stimulus linking the OVLT with AH/POA remains speculative. In our current view, the action of blood-borne IL-1 develops in two steps, namely, initial excitation of neurons projecting from the OVLT to the preoptic region and subsequent acceleration in the local synthesis of PGE$_2$. IL-1 is unlikely to cross the glial layer enclosing as a true barrier the OVLT, for this occurrence should have resulted into the accelerated formation of both PGE$_2$ and TXA$_2$. PGE$_2$, on the other hand, could also be formed within the OVLT (from reticuloendothelial cells?) and act either on OVLT neurons or on neurons of the adjacent neuropile. It must be stressed, however, that the rapidity in the onset of the fever and its consistent time course accord with the operation of a neural mechanism rather than with an uncertain process of diffusion involving IL-1, PGE$_2$, or both.

Conclusion

Our studies prove that PGE_2 formation in brain is stimulated selectively by pyrogens and the activation pattern conforms with a causative role of the compound in the onset and progression of the fever. Blood-borne IL-1, and possibly other monokines as well, are thought to act at discrete sites in the central nervous system, which are identified with the circumventricular organs. This multiplicity of targets is conceivably important in sustaining fever and fever-related events, specifically the sequence of coordinated events forming the host defense against infection.

Acknowledgement

This work was supported by the Medical Research Council of Canada.

References

1   Cooper KE: The neurobiology of fever: thoughts on recent developments. Ann Rev Neurosci 1987;10:297–324.
2   Coceani F, Bishai I, Lees J, et al: Prostaglandin E_2 and fever: a continuing debate. Yale J Biol Med 1986;59:169–174.
3   Coceani F, Bishai I, Dinarello CA, et al: Prostaglandin E_2 and thromboxane B_2 in cerebrospinal fluid of afebrile and febrile cat. Am J Physiol 1983;244:R785–R793.
4   Coceani F, Lees J, Bishai I: Further evidence implicating prostaglandin E_2 in the genesis of pyrogen fever. Am J Physiol 1988;254:R463–R469.
5   Sirko S, Bishai I, Coceani F: Prostaglandin formation in the hypothalamus in vivo: effect of pyrogens. Am J Physiol 1989;256, in press.
6   Laburn H, Mitchell D, Rosendorff C: Effects of prostaglandin antagonism on sodium arachidonate fever in rabbits. J Physiol (Lond) 1977;267:559–570.
7   Gollman HM, Rudy TA: Assessment of the pyrogenic potency of the thromboxane A_2 mimetics, SQ26655 and U46619, and thromboxane B_2 by intrapreoptic injection in the cat. Res Commun Chem Pathol Pharmacol 1985;49:305–308.
8   Mashburn TA Jr, Llanos QJ, Ahokas RA, et al: Thermal and acute-phase protein responses of guinea pigs to intrapreoptic injections of leukotrienes. Brain Res 1986;376:285–291.
9   Hoppe U, Hoppe EM, Peskar BM, et al: Radioimmunoassay for leukotriene E_4. Use for determination of total sulfidopeptide-leukotriene release from rat gastric mucosa. FEBS Lett 1986;208:26–30.
10 Coceani F, Lees J, Dinarello CA: Occurrence of interleukin-1 in cerebrospinal fluid of the conscious cat. Brain Res 1988;446:245–250.
11 Bishai I, Dinarello CA, Coceani F: Prostaglandin formation in feline cerebral microvessels: effect of endotoxin and interleukin-1. Can J Physiol Pharmacol 1987;65:2225–2230.

F. Coceani, MD, Research Institute, The Hospital for Sick Children,
555 University Avenue, Toronto, Ont. M5G 1X8 (Canada)

Metabolism of Leukotrienes and Induction of Anaphylaxis and Shock

Zor U, Naor Z, Danon A (eds): Leukotrienes and Prostanoids in Health and Disease.
New Trends Lipid Mediators Res. Basel, Karger, 1989, vol 3, pp 187–193

Identification and Quantitation of Leukotriene Metabolites in Animal and Man

J. Rokach[a], *P. Tagari*[a], *D. Ethier*[a], *A. Foster*[a], *D. Delorme*[a],
J.L. Malo[b], *A. Cartier*[b], *P. Manning*[c], *P.M. O'Byrne*[c], *Y. Girard*[a]

[a]Merck Frosst Centre for Therapeutic Research, Pointe Claire-Dorval, Que.;
[b]Sacre Coeur Hospital, Montreal, Que., and [c]Department of Medicine,
McMaster University, Hamilton, Ont., Canada

The peptide leukotrienes (LT) C_4, D_4 and E_4 are sequential metabolites of leukotriene A_4, a product of the action of 5-lipoxygenase on arachidonic acid. Elucidation of the subsequent degradation and elimination of these biologically active compounds [1–3] is a prerequisite for understanding their roles in animal models of disease, and in clinical conditions. Accordingly, LT metabolism and excretion was studied in bronchially hyperreactive rats, inbred as an animal model for asthma [4], and in primates as a paradigm for leukotriene elimination in man. Additionally, LTE_4 excretion was estimated in asthmatic patients as an index of leukotriene involvement in this disease.

Results and Discussion

The metabolism and excretion of leukotriene C_4, the precursor of LTD_4 and LTE_4, was studied in male Sprague-Dawley rats by intravenous (IV) administration of $[^3H]$-LTC_4 via the jugular vein (as a model of circulatory elimination), or via tracheal instillation as an index of LT clearance from the lung. Circulatory clearance was rapid [5], and the major route of elimination of tritium was biliary (fig. 1a), comprising $69.0 \pm 4.1\%$ ($n = 6$) within 1 h. Analysis of bile samples obtained by bile duct cannulation after IV $[^3H]LTC_4$ revealed N-acetyl LTE_4 and LTD_4 as major identified metabolites, although polar material, indicating further oxidation, was present (fig. 1b).

After tracheal instillation of $[^3H]$-LTC_4, recovery was smaller and slower, with $[^3H]$-N-acetyl LTE_4 predominating, establishing the sequential formation of LTD_4 and LTE_4, and the latter's N-acetylation, followed by

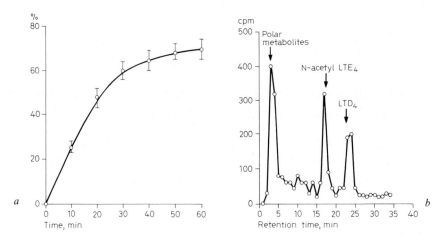

Fig. 1. The biliary recovery of IV [³H]-LTC$_4$ in rats was 69% in 1 h (a). RP-HPLC revealed N-acetyl LTE$_4$, LTD$_4$ and polar material as metabolites (b).

biliary excretion, as a fate of pulmonary peptide LTs. This was additionally confirmed with in vivo generation of leukotrienes by tracheal instillation of antigen in ovalbumin-sensitized bronchially hyperresponsive rats [4], resulting in increased insufflation pressure analogous to an acute asthmatic response. Bile samples obtained from these animals, analyzed by HPLC and RIA, showed elimination of LTC$_4$, D$_4$ and N-acetyl LTE$_4$ immunoreactive material [6]. That this excretion was attenuated by inhibitors of the leukotriene biosynthetic enzyme, 5-lipoxygenase (fig. 2) [6], convincingly demonstrated the ability of metabolic studies to reflect the in vivo involvement of LTs in hyperreactive conditions.

The clinical corollary of this work has been demonstrated by the estimation of urinary LTE$_4$ in asthmatics (as biliary LT elimination is less important in humans [7]), an identified excretory product after LTD$_4$ inhalation in man [8]. Urine samples were obtained before, and between 2–3 h (early phase) and 6–7 h (late phase) after inhaled antigen or appropriate diluent. Reversed-phase HPLC and RIA analysis revealed significant LTE$_4$ excretion 2–3 h after antigen in patients experiencing severe early phase bronchoconstriction (fig. 3a) declining to control values after 6 h (at which time pulmonary function was normal), implying a role for peptido-leukotrienes in this response. In contrast, however, asthmatics showing a progressive impairment of pulmonary function, maximizing 6 h after antigen (isolated late response), showed only a minor increase in urinary LTE$_4$ concentrations (fig. 3b).

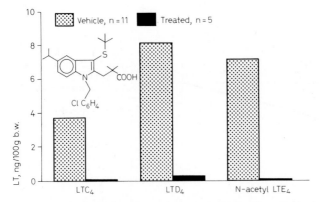

Fig. 2. Oral preadministration of the leukotriene biosynthesis inhibitor L-663,536 (3 mg/kg; ■) completely abolished the biliary excretion of leukotriene metabolites (ng meta-bolite/100 g body weight) elicited by tracheal instillation of antigen in hyperresponsive rodents, as shown by comparison with vehicle-treated animals (⊡).

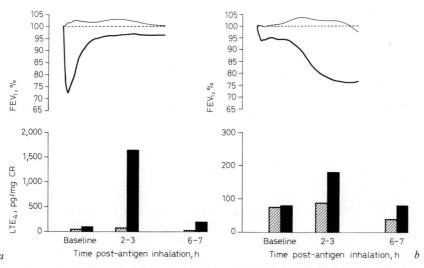

Fig. 3. Five asthmatics showed isolated early bronchoconstriction (depressed FEV_1; bold line, *a*, top) with concomitant increase in urinary LTE_4 concentration (pg LTE_4/mg creatinine; ■, *a*, bottom) after antigen inhalation (0 h). Another 5 suffered prolonged and increasing pulmonary responses (*b*, top) without marked LTE_4 excretion (*b*, bottom; note scale difference). Diluent challenge had no effect on FEV_1 (fine line, *a, b*, top) or LTE_4 concentrations (▨, *a, b*, bottom).

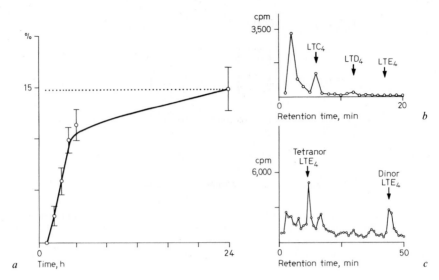

Fig. 4. Urinary [³H] clearance *(a)* in primates was 14.8% 24 h after IV [³H]-LTC₄ (10 µCi/kg). RP-HPLC of urine collected from 0 to 1 h revealed initial excretion of [³H]-LTC₄ and polar material *(b)*. This was resolved (particularly in later samples) to reveal [³H]-18-COOH (dinor) LTE₄ and [³H]-16-COOH (tetranor) LTE₄ as major urinary metabolites *(c)*.

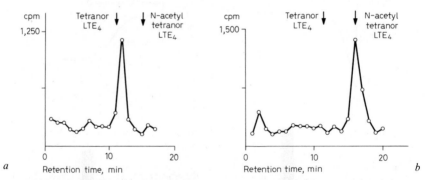

Fig. 5. [³H]-16-COOH (tetranor) LTE₄ was identified as a major urinary [³H]-LTC₄ metabolite in primates by coinjection *(a)*, a correlative shift in retention time after N-acetylation *(b)*, and by oxidative ozonolysis to reveal saturation at the 14,15 position [10].

The occurrence of predominating polar LT metabolites has been noted in the rat (fig. 1b), suggesting that the lack of urinary LTE₄ excretion associated with progressive bronchoconstriction in these asthmatics might be explained by metabolic processes. Resolution of this polar material, obtained from rodent bile after IV [³H]-LTC₄, revealed that N-acetyl LTE₄

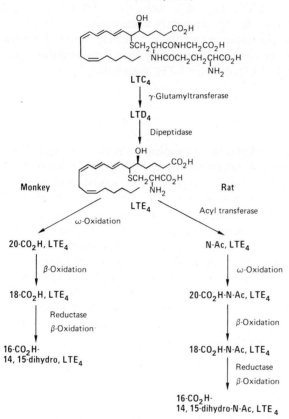

Metabolic pathway

LTC$_4$

γ-Glutamyltransferase

LTD$_4$

Dipeptidase

Monkey   LTE$_4$ Rat

ω-Oxidation   Acyl transferase

20-CO$_2$H, LTE$_4$ N-Ac, LTE$_4$

β-Oxidation ω-Oxidation

18-CO$_2$H, LTE$_4$ 20-CO$_2$H-N-Ac, LTE$_4$

Reductase β-Oxidation
β-Oxidation

16-CO$_2$H-
14, 15-dihydro, LTE$_4$ 18-CO$_2$H-N-Ac, LTE$_4$

Reductase
β-Oxidation

16-CO$_2$H-
14, 15-dihydro-N-Ac, LTE$_4$

Fig. 6. Administration of [14,15-^{3}H]-LTC$_4$ has allowed the elucidation of in vivo leukotriene metabolism in both rodents and primates, with the identification of 16-COOH-LTE$_4$ as a major urinary metabolite in the latter. Excretion of volatile [^{3}H] material (putatively [^{3}H]-H$_2$O) suggests that this product may undergo further beta-oxidation.

underwent omega-oxidation to form 20-COOH-N-Ac LTE$_4$, followed by subsequent cycles of beta-oxidation resulting in the elimination of 18-COOH-N-Ac LTE$_4$ and 16-COOH-14,15-dihydro-N-Ac LTE$_4$ [9, 10]. It was, therefore, of interest to confirm the existence of this metabolic pathway in primates, and establish a suitable urinary (polar) marker metabolite.

Intravenous injection of [^{3}H]-LTC$_4$ in cynomolgus monkeys resulted in significant urinary excretion of [^{3}H] (14.8 $\pm$ 2.1%, n = 4) within 24 h (fig. 4a), consisting largely of polar material (61%). [^{3}H]-LTC$_4$ was observed in urine sampled during the first hour only (fig. 4b). After the first hour, however, $[^3H]$-16-COOH-14,15-dihydro LTE_4, and its precursor beta-metabolite 18-COOH LTE_4 (fig. 4c), were the predominant identified urinary metabolites (see fig. 5), ultimately comprising 2.25 and 0.57% of the original leukotriene dose respectively. Volatile material (putatively $[^3H]$-H_2O) was also in evidence, suggesting that beta-oxidation proceeds beyond the 16-COOH-14,15-dihydro-LTE_4 metabolite.

Conclusion

The definition of leukotriene metabolism in rodents and primates (fig. 6) and the quantitation of metabolite excretion have provided methods for the establishment of leukotriene involvement in animal models of disease, and in clinical settings. The attenuation, by 5-lipoxygenase inhibitors, of biliary LT excretion after antigen challenge in hyperreactive rats [6] evidences a link between biochemical and physiological function in this model of asthma. Similarly, the appearance of high LTE_4 concentrations in urine from certain early-responding asthmatics after antigen inhalation is indicative of a pivotal role of peptide leukotrienes in these patients. Finally, the identification of beta-oxidation metabolites of LTE_4 as important excretory products of leukotrienes, in primates, presages the development of more sensitive estimations of in vivo leukotriene generation in the etiology and progression of inflammatory disease.

References

1   Newcombe DS: Leukotrienes: Regulation of biosynthesis, metabolism and bioactivity. J Clin Pharmacol 1988;28:530–549.
2   Murphy RC, Stene DO: Oxidative metabolism of leukotriene E_4 by rat hepatocytes. Ann NY Acad Sci 1988;524:35–42.
3   Keppler D, Huber M, Hagmann W, et al: Metabolism and analysis of endogenous cysteinyl leukotrienes. Ann NY Acad Sci 1988;524:43–74.
4   Holme G, Piechuta H: The derivation of an inbred line of rats which develop asthma-like symptoms following challenge with aerosolised antigen. Immunology 1981;42:19–24.
5   Foster A, Fitzsimmons B, Rokach J, et al: Metabolism and excretion of peptide leukotrienes in the anaesthetised rat. Biochem Biophys Acta 1987;921:486–493.
6   Foster A, Letts G, Charleson S, et al: The in vivo production of peptide leukotrienes following pulmonary anaphylaxis in the rat. J Immunol 1988;141:3544–3550.
7   Orning L, Kaijser L, Hammarstrom S: In vivo metabolism of leukotriene C_4 in man: urinary excretion of leukotriene E_4. Biochem Biophys Res Commun 1985;130:214–220.
8   Verhagen J, Bel EH, Kijne GM, et al: The excretion of leukotriene E_4 into urine following inhalation of leukotriene D_4 by human individuals. Biochem Biophys Res Commun 1987;148:864–868.

9   Foster A, Fitzsimmons B, Rokach J, et al: Evidence of in vivo omega-oxidation of peptide leukotrienes in the rat: biliary excretion of 20-COOH-N-acetyl LTE$_4$. Biochem Biophys Res Commun 1987;149:1237–1245.
10   Delorme D, Foster A, Girard Y, et al: Synthesis of beta-oxidation products as potential leukotriene metabolites and their detection in bile of anaesthetised rats. Prostaglandins 1988;36:291–303.

J. Rokach, Merck Frosst Centre for Therapeutic Research, P.O. Box 1005, Pointe Claire-Dorval, Que., H9R 4P8 (Canada)

Zor U, Naor Z, Danon A (eds): Leukotrienes and Prostanoids in Health and Disease.
New Trends Lipid Mediators Res. Basel, Karger, 1989, vol 3, pp 194–199

ω-Oxidation of Cysteinyl Leukotrienes in the Rat

Lars Örning[a], *Andrea Keppler*[c], *Tore Midtvedt*[b], *Sven Hammarström*[d]

Departments of [a]Physiological Chemistry and [b]Germ-free Research, Karolinska
Institutet, Stockholm, Sweden; [c]Biochemisches Institut, University of Freiburg, FRG,
and [d]Department of Cell Biology, Division of Medical and Physiological Chemistry,
University of Linköping, Sweden

During the last several years there has been a considerable study on
the metabolism of cysteinyl leukotrienes in vitro and in vivo [reviewed in 1].
This has led to the understanding that cysteinyl leukotrienes undergo
extensive metabolism and rapidly disappear from the general circulation.
After its formation, LTC is rapidly converted to LTD and LTE by
sequential cleavage of glutamate and glycine from the thiol group. LTE
may then be taken up by the liver and the kidney, and, in some species,
N-acetylated to form N-Ac-LTE [2–4]. However, in several investigations,
formation of metabolites more polar than LTC was observed. This material
increased in relative amounts following the administration of LTC, and it
could not be ascribed to the degradation of LTC along the mercapturic
acid pathway. Very recently a specific oxidative metabolism at the ω-end of
LTE_4 and $N-Ac-LTE_4$ by subcellular rat liver fractions [5, 6] and rat
hepatocytes [7] was described. ω-Oxidized metabolites now also have been
demonstrated of LTE_4 in monkey bile and urine [8], of $N-Ac-LTE_4$ in rat
bile [9–11] and feces [10], and β-oxidized metabolites of $N-Ac-LTE_4$ in rat
bile [12]. The present paper describes the characterization of ω-oxidized
metabolites of LTE_4 and $N-Ac-LTE_4$ formed in rat liver, a partial characterization of the involved enzymes, and the identification of these products
in the rat in vivo.

Metabolism of LTE_4 and $N-Ac-LTE_4$ by Subcellular Rat
Liver Fractions [5, 6]

Figure 1 shows a reversed-phase (RP)-HPLC separation of products formed from LTE_4 and $N-Ac-LTE_4$ upon incubation with rat liver
microsomes in the presence of NADPH. Compounds I and II had retention

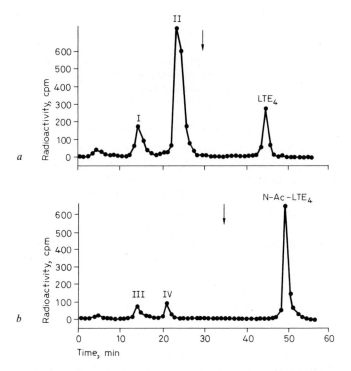

Fig. 1. RP-HPLC of products obtained by incubating rat liver microsomes plus NADPH with *(a)* [³H₈]LTE₄ and *(b)* N-Ac-[³H₈]LTE₄. Incubation conditions *(a)* 18 nmol, 0.9 µCi, *(b)* 8 nmol, 0.4 µCi (37 °C, 30 min). Arrows indicate change of solvent.

times of 0.16 and 0.20 relative to LTE₄, and compounds III and IV of 0.16 and 0.20 relative to N-Ac-LTE₄. The products were isolated using RP-HPLC, and purified from bile acid conjugates by SepPak C₁₈ treatment and Lipidex-DEAP [13]. The structural characterizations were performed using UV-spectroscopy, gas chromatography-electron impact mass spectrometry, fast atom bombardment-mass spectrometry, chemical and enzymatic conversions and comparisons with authentic standards. Table 1 summarizes the indentifications.

Partial Characterization of ω-Oxidizing Enzymes

It was probable that LTE₄ and N-Ac-LTE₄ had been converted to ω-hydroxylated metabolites by an ω-hydroxylase and that the further oxidation to ω-carboxylated products was catalyzed by dehydrogenases. The nature of these enzymes was investigated further.

Table 1. Characterization of compounds I–IV from figure 1

I	ω-COOH-LTE$_4$
II	ω-OH-LTE$_4$
III	N-acetyl-ω-COOH-LTE$_4$
IV	N-acetyl-ω-OH-LTE$_4$

Of substrates investigated, LTE$_4$ was most efficiently transformed by the ω-hydroxylase (1.6 $\pm$ 0.4 nmol/g wet weight of tissue; 37 °C, 10 min). The specific activity with N-Ac-LTE$_4$ as substrate was 16% of that with LTE$_4$. 11-*trans*-LTE$_4$ and N-Ac-11-*trans*-LTE$_4$ were also substrates for the enzyme, whereas LTC$_4$ and LTD$_4$ were not converted.

The ω-hydroxylase reaction required NADPH and molecular oxygen as cofactors. NADH could not substitute for NADPH. At 1 mM concentration, the specific activity using NADH was only 7–8% of that using NADPH.

Time-course experiments demonstrated that while LTE$_4$ decreased, ω-OH-LTE$_4$ accumulated transiently, and the amounts of ω-COOH-LTE$_4$ increased during the entire incubation time. The ω-hydroxylase reaction was dependent on pH, with a maximal activity at pH 7.4, and on concentrations of microsomes, LTE$_4$, and NADPH. Apparent K$_m$ values for LTE$_4$ and NADPH were estimated at 2–3 μM and 0.1 mM, respectively. A 20- to 30-fold excess of arachidonic acid, linoleic acid, PGA$_2$, PGE$_2$, or PGF$_{1\alpha}$, and a 5- to 10-fold excess of LTB$_4$ had no effect on the ω-hydroxylation of LTE$_4$, whereas a 10-fold excess of unlabelled LTE$_4$ inhibited ω-hydroxylation of [^{3}H]LTE$_4$ by more than 70%. Based on these substrate competitions as well as on kinetic data, LTE ω-hydroxylase was distinct from the previously described LTB ω-hydroxylase of rat liver [14].

Subcellular distribution demonstrated the highest specific activity in the microsomal fraction. Kidney and lung microsomes also ω-oxidized LTE$_4$, but with a specific activity of only 1–4% of that with liver microsomes.

The requirements of NADPH and molecular oxygen, the microsomal localization, and the pH optimum all suggested the involvement of a cytochrome P$_{450}$ type of enzyme. This was confirmed by the inhibition of the formation of ω-OH-LTE$_4$ from LTE$_4$ by carbon monoxide, an inhibitor of cytochrome P$_{450}$ type of enzymes (fig. 2).

The oxidative activity responsible for the transformation of ω-OH-LTE$_4$ to ω-COOH-LTE$_4$ was present in all subcellular fractions but was maximal in the cytosol. It required NAD$^+$ as a cofactor, and NADP$^+$ could not substitute. Pyrazole, an inhibitor of some liver alcohol dehydro-

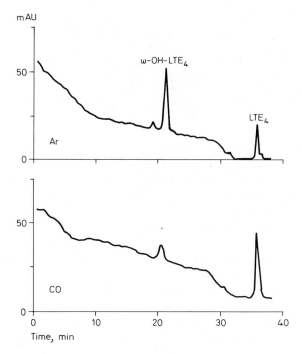

Fig. 2. RP-HPLC of products formed by mixing microsomes (corresponding to 50 mg of tissue), LTE$_4$ (3 nmol), and NADPH (2 μmol) in 200 μl of cold 50 mM potassium phosphate buffer, pH 7.4, with 800 μl of cold argon (Ar)-saturated or carbon monoxide (CO)-saturated buffer. Incubations were performed aerobically, in the dark, for 15 min at 37 $^\circ$C.

genase isoenzymes, and disulfiram, an inhibitor of aldehyde dehydrogenase, at 1 mM concentration inhibited the conversion of ω-OH-LTE$_4$ to ω-COOH-LTE$_4$ by 20 and 85%, respectively, using cytosol corresponding to 100 mg of tissue/ml.

It is concluded that rat liver microsomes contain an LTE ω-hydroxylase which is specific for LTE$_4$ and N-Ac-LTE$_4$ (and their 11-*trans* isomers), and which is probably a cytochrome P$_{450}$ type of enzyme. Furthermore, results suggest that liver alcohol and aldehyde dehydrogenases, present in the cytosol, catalyze the further conversion to ω-carboxylated products.

In vivo Formation of ω-Oxidized Metabolites of LTC$_4$ [10]

[^{3}H$_8$]LTC$_4$ was administered to conventional and germ-free rats, and bile and feces were collected, respectively. Nine biliary metabolites were isolated. Six of these were identified as N-Ac-LTE$_4$ (12.8% of the adminis-

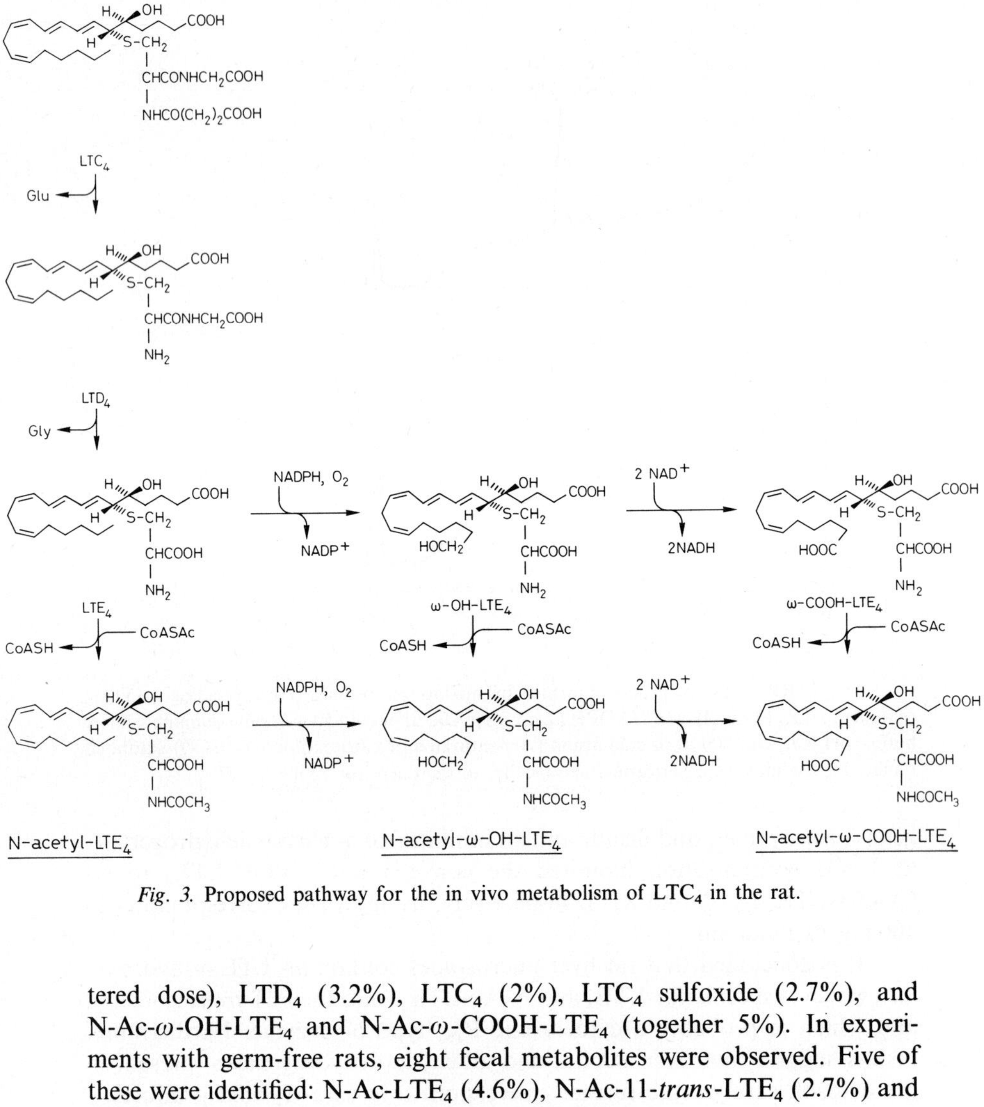

Fig. 3. Proposed pathway for the in vivo metabolism of LTC$_4$ in the rat.

tered dose), LTD$_4$ (3.2%), LTC$_4$ (2%), LTC$_4$ sulfoxide (2.7%), and N-Ac-ω-OH-LTE$_4$ and N-Ac-ω-COOH-LTE$_4$ (together 5%). In experiments with germ-free rats, eight fecal metabolites were observed. Five of these were identified: N-Ac-LTE$_4$ (4.6%), N-Ac-11-*trans*-LTE$_4$ (2.7%) and their sulfoxides, and N-Ac-ω-COOH-LTE$_4$ (3.5%).

The formation of N-Ac-ω-OH-LTE$_4$ and N-Ac-ω-COOH-LTE$_4$ in the rat probably proceeds as outlined in figure 3. Since LTE$_4$ is the preferred substrate for rat liver LTE ω-hydroxylase, and not only LTE$_4$ but also ω-OH-LTE$_4$ and ω-COOH-LTE$_4$ are substrates for microsomal rat liver N-acetyltransferase, it is probable that ω-oxidation of LTE$_4$ precedes the formation N-Ac-ω-OH-LTE$_4$ and N-Ac-ω-COOH-LTE$_4$.

References

1 Hammarström, S.; Örning, L.; Bernström, K.: Metabolism of leukotrienes. Mol. Cell. Biochem. *69:* 7–16 (1985).

2 Bernström, K.; Hammarström, S.: Metabolism of leukotriene E_4 by rat tissues: formation of N-acetyl leukotriene E_4. Arch. Biochem. Biophys. *244:* 486–491 (1986).

3 Örning, L.; Norin, E.; Gustafsson, B., et al.: In vivo metabolism of leukotriene C_4 in germ-free and conventional rats: fecal excretion of N-acetyl leukotriene E_4. J. Biol. Chem. *261:* 766–771 (1986).

4 Hagmann, W.; Denzlinger, C.; Rapp, S., et al.: Identification of the major endogenous leukotriene metabolite in the bile of rats as N-acetyl leukotriene E_4. Prostaglandins *31:* 239–251 (1986).

5 Örning, L.: ω-Hydroxylation of N-acetyl leukotriene E_4 by rat liver microsomes. Biochem. Biophys. Res. Commun. *143:* 337–344 (1987).

6 Örning, L.: ω-Oxidation of cysteine-containing leukotrienes by rat liver microsomes: isolation and characterization of ω-hydroxy and ω-carboxy metabolites of leukotriene E_4 and N-acetyl leukotriene E_4. Eur. J. Biochem. *170:* 77–85 (1987).

7 Stene, D.O.; Murphy, R.C.: Metabolism of leukotriene E_4 in isolated rat hepatocytes: identification of β-oxidation products of sulfidopeptide leukotrienes. J. Biol. Chem. *263:* 2773–2778 (1988).

8 Ball, H.A.; Keppler, D.: ω-Oxidation products of leukotriene E_4 in bile and urine of the monkey. Biochem. Biophys. Res. Commun. *148:* 664–670 (1987).

9 Foster, A.; Fitzsimmons, B.; Rokach, J., et al.: Evidence of in vivo ω-oxidation of peptide leukotrienes in the rat: biliary excretion of 20-COOH N-acetyl LTE_4. Biochem. Biophys. Res. Commun. *148:* 1237–1245 (1987).

10 Örning, L.; Keppler, A.; Midtvedt, T., et al: In vivo formation of ω-oxidized metabolites of leukotriene C_4 in the rat. Prostaglandins *35:* 493–501 (1988).

11 Keppler, D.; Huber, M.; Hagmann, W., et al.: Metabolism and analysis of endogenous cysteinyl leukotrienes. Ann. N.Y. Acad. Sci. *524:* 68–74 (1988).

12 Foster, A.; Delorme, D.; Blacklock, B., et al.: β-Oxidation of peptide leukotrienes in the anesthetized rat. FASEB J. *2:* A410 (1988).

13 Almé, B.; Bremmelgaard, A.; Sjövall, J., et al.: Analysis of metabolic profiles of bile acids in urine using a lipophilic anion exchanger and computerized gas-liquid chromatography-mass spectrometry. J. Lipid Res. *18:* 339–362 (1977).

14 Romano, M.C.; Eckardt, R.D.; Bender, P.E., et al.: Biochemical characterization of hepatic microsomal leukotriene B_4 hydroxylases. J. Biol. Chem. *262:* 1590–1595 (1987).

Lars Örning, PhD, Department of Physiological Chemistry, Karolinska Institutet, S–104 01 Stockholm (Sweden)

Zor U, Naor Z, Danon A (eds): Leukotrienes and Prostanoids in Health and Disease.
New Trends Lipid Mediators Res. Basel, Karger, 1989, vol 3, pp 200–203

Metabolism of 12-Hydroxy-5,8,11,14-Eicosatetraenoic Acid by Porcine Polymorphonuclear Leukocytes

Sandra Wainwright, William S. Powell

Endocrine Laboratory, Royal Victoria Hospital, and Department of Medicine,
McGill University, Montreal, Que., Canada

Leukotriene B_4 (LTB_4) can be metabolized by two major pathways in polymorphonuclear leukocytes (PMNL), the first leading to the formation of omega-oxidation products, and the second to dihydro products. The major metabolites of LTB_4 formed by human PMNL are the corresponding 20-hydroxy and omega-carboxy compounds [1]. Rat PMNL, on the other hand, convert LTB_4 to its 19-hydroxy metabolite, along with a dihydro metabolite in which one of the three conjugated double bonds has been reduced [2]. Rat macrophages, T lymphocytes, mesangial cells, and fibroblasts also convert LTB_4 to a similar reduced product [3]. The reductase pathway is the major pathway of LTB_4 metabolism in porcine PMNL, which possess relatively little omega-oxidation activity [4]. We identified the major metabolites of LTB_4 in these cells as 10,11-dihydro-LTB_4 and 10,11-dihydro-12-oxo-LTB_4 [4]. These products were presumably formed by the actions of a reductase and a dehydrogenase in PMNL. It was important to determine whether other eicosanoids were also metabolized by this pathway and to learn more about its specificity. We therefore investigated the metabolism of a variety of lipoxygenase and cyclooxygenase products by porcine PMNL.

Results and Discussion

Porcine PMNL were incubated with ^{14}C-labeled 12-hydroxy-5,8,10,14-eicosatetraenoic acid (12-HETE) and the products were analyzed by RP-HPLC (fig. 1). The profile of radioactive products formed from this substrate was reminiscent of that from LTB_4 [4], in that in each case two less polar products were formed. As with LTB_4, the UV absorbance spectra of these products indicated that their maximal absorbance had shifted to a

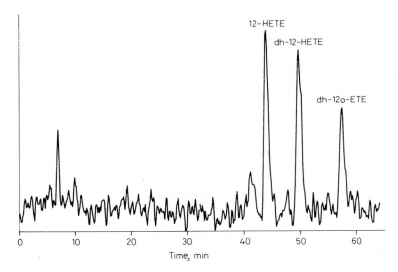

Fig. 1. Reversed-phase high pressure liquid chromatogram of the radioactive products formed after incubation of 12-[1-^{14}C]HETE with porcine leukocytes for 30 min at 37 °C.

lower wavelength, suggesting that one of the conjugated double bonds of 12-HETE had been reduced.

The metabolite of 12-HETE with a retention time of 50 min was further purified by normal phase HPLC and analyzed by gas chromatography-mass spectrometry (GC-MS) after conversion to the trimethylsilyl ether derivative of its methyl ester. The mass spectrum of this compound exhibited intense ions at m/z 408 (M), 393 (M-15), 377 (M-31), 318 (M-90), 297 (C_1 to C_{12}), 213 (C_{12} to C_{20}), 207 (297-90), 175, and 133 (base peak; 207-74) (loss of CH_2=C(OH)–OCH$_3$)). This mass spectrum is quite reminiscent of that of 12-HETE, except that all ions containing the first 12 carbons are observed at m/z values 2 units higher, suggesting that one of the double bonds in this part of the molecule had been reduced. Although it is not possible to determine from the above mass spectrum the location of the reduced double bond, it would appear likely to be the one between carbons 10 and 11, by analogy with the 10,11-dihydro metabolite of LTB$_4$. This would suggest that this metabolite of 12-HETE is 12-hydroxy-5,8,14-eicosatrienoic acid (10,11-dihydro-12-HETE).

The second metabolite of 12-HETE had a longer retention time (57 min) than dihydro-12-HETE, suggesting that it could be a dihydro-oxo product, analogous to the dihydro-oxo metabolite of LTB$_4$. In agreement with this, it reacted with hydroxylamine hydrochloride to give an oxime derivative, which was analyzed by GC-MS after conversion to its

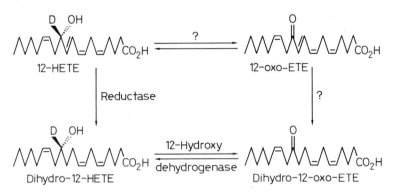

Fig. 2. Scheme for the formation of dihydro and dihydro-oxo metabolites of 12-HETE.

trimethylsilyl ether, ester derivative. The mass spectrum of this derivative had intense ions at m/z 479 (M), 464 (M-15), 436 (C_1 to C_{17}), 422 (C_1 to C_{16}), 348 (C_3 to C_{20}), 280 (C_8 to C_{20}), 266, 210, 170, 147, and 117 (base peak). This mass spectrum indicates that this substance was a dihydro-12-oxo metabolite of 12-HETE, presumably 12-oxo-5,8,14-eicosatrienoic acid (10,11-dihydro-12-oxo-ETE).

Dihydro-12-HETE could have been formed either by the direct reduction of 12-HETE, or via an oxo intermediate (i.e. dihydro-12-oxo-ETE) (fig. 2). To investigate the pathway for the formation of dihydro-12-HETE, 12-[5,6,8,9,11,12,14,15-^{2}H] HETE was incubated with porcine PMNL and the resulting dihydro product analyzed by GC-MS. The mass spectrum of this compound indicated that about 50% of the deuterium in the 12-position had been retained, suggesting that it could be formed both by the direct reduction of 12-HETE and also via a 12-oxo intermediate. We have not yet determined whether 12-oxo-ETE is formed in this reaction, or whether the deuterium at carbon-12 is lost due to reversible formation of dihydro-12-oxo-ETE, analogous to the situation with dihydro-LTB$_4$ [4].

To examine the specificity of the reductase/dehydrogenase pathway in PMNL, we investigated the metabolism of a number of related eicosanoids by porcine PMNL. LTB$_4$ and 12-HETE were about equally good substrates, with approximately 40–50% metabolism to dihydro and dihydro-oxo products. Of the other monohydroxy fatty acids investigated, 13-hydroxy-9,11-octadecadienoic acid (13-HODE) (12% metabolism) was the best substrate. Other related compounds (9-HODE, 5-HETE, 15-HETE, and 12-hydroxy-5,8,10-heptadecatrienoic acid (HHT)) were metabolized much more slowly, whereas prostaglandins E$_2$ and F$_{2\alpha}$ were not metabolized to any detectable products. These results suggest that this pathway is specific for eicosanoids with a 12-hydroxyl group, and is quite distinct from the major metabolic pathway of prostaglandins.

Although we have not examined the biological activities of dihydro-12-HETE, a similar substance (10,11-dihydro-12(R)-HETE) which was isolated after incubation of arachidonic acid with bovine corneal microsomes was reported to have potent proinflammatory properties [5]. It would seem likely that the latter product is identical to the one formed by porcine leukocytes in the present study. Since 12-HETE is the major metabolite of arachidonic acid in porcine PMNL, its conversion to a more biologically active dihydro metabolite could play an important role in inflammatory reactions.

References

1   Powell WS: Properties of leukotriene B_4 20-hydroxylase from polymorphonuclear leukocytes. J Biol Chem 1984;259:3082–3089.
2   Powell WS: Conversion of leukotriene B_4 to dihydro and 19-hydroxy metabolites by rat polymorphonuclear leukocytes. Biochem Biophys Res Commun 1987;145:991–998.
3   Kaever V, Martin M, Fauler J, et al: A novel metabolic pathway for leukotriene B_4 in different cell types: primary reduction of a double bond. Biochim Biophys Acta 1987;922:337–344.
4   Powell WS, Wainwright S, Gravelle F: Metabolism of leukotriene B_4 and related substances by polymorphonuclear leukocytes. Adv Prostaglandin Thromboxane Leukotriene Res 1989; in press.
5   Murphy RC, Falck JR, Lumin S, et al: 12(R)-hydroxyeicosatrienoic acid: a vasodilator cytochrome P-450-dependent arachidonate metabolite from the bovine corneal epithelium. J Biol Chem 1988;263:17197–17202.

William S. Powell, PhD, Endocrine Laboratory, Royal Victoria Hospital, 687 Pine Avenue West, Montreal, Que. H3A 1A1 (Canada)

Zor U, Naor Z, Danon A (eds): Leukotrienes and Prostanoids in Health and Disease.
New Trends Lipid Mediators Res. Basel, Karger, 1989, vol 3, pp 204–209

Generation of Endogenous Cysteinyl Leukotrienes during Anaphylactic Shock in the Guinea Pig

Andrea Keppler[a], *Albrecht Guhlmann*[b], *Stefanie Kästner*[b],
Dietrich Keppler[b]

[a]Biochemisches Institut der Universität Freiburg, Freiburg, and
[b]Deutsches Krebsforschungszentrum, Abteilung Tumorbiochemie, Heidelberg, FRG

Introduction

Since the work of Brocklehurst [1] with sensitized guinea pigs, SRS-A has been considered a major component in the development of allergy, asthma, and anaphylaxis. The life-threatening condition of an anaphylactic shock is characterized by airway obstruction and cardiovascular collapse [2, 3]. It is well documented that exogenously administered LTC_4 or LTD_4 can produce all major consequences of acute systemic anaphylaxis, such as bronchoconstriction, plasma extravasation, mucus secretion, and cardiac failure [4]. To ascertain the potential role of cysteinyl leukotrienes in anaphylaxis it was necessary to demonstrate that production of cysteinyl leukotrienes occurs in vivo in amounts sufficient to provoke an anaphylactic reaction [5].

Results and Discussion

Elimination and Metabolism of Labeled LTC_4 in the Guinea Pig

Following intravenous injection of $[^3H_2]LTC_4$ in the guinea pig, radioactivity was eliminated from the blood circulation with an initial half-life of about 40 s. The predominant hepatobiliary elimination and the biliary metabolite pattern were analyzed after intravenous *infusion* of $[^3H_2]LTC_4$ over a 15-min period. The infusion was performed to mimic the time course of elimination of endogenous LTC_4 production during anaphylaxis. Within 30 min, 65% of the infused radioactivity were recovered in bile, and 38% of the administered dose corresponded to biliary LTD_4 (fig. 1). Further metabolism at later times led to increased proportions of first LTE_4

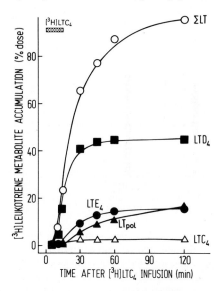

Fig. 1. Distribution of [³H]leukotriene metabolites in bile. After intravenous infusion of [14,15-³H₂]LTC₄ over a 15-min period, bile was sampled continuously. Bile fractions were analyzed by RP-HPLC with continuous detection of radioactivity in cysteinyl leukotriene metabolites. The amounts of the metabolites accumulated in bile are expressed as percentages of the infused dose. LT$_{pol}$ represents the polar metabolites; Σ LT represents the amount of total leukotriene radioactivity excreted into bile.

and then polar metabolites. Infusions were carried out in sensitized animals exposed to antigen challenge as well as in sensitized controls. Between these two groups no major differences were observed, both with respect to the excretion rate in bile and to the leukotriene metabolite pattern in bile.

Endogenous Generation of LTD₄ during Anaphylaxis

Based on the tracer studies, a model was developed which enabled measurements of systemic LTC₄ production during anaphylaxis [5]: Guinea pigs were sensitized to ovalbumin (3 μg/kg with Al(OH)₃ at 300 mg/kg, i.p.) 3 weeks before challenge to produce IgE and IgG antibodies [6]. Cutaneous testing of the sensitization was performed in each animal by examination of the local extravasation of intravenously injected Evans blue after intradermal injection of ovalbumin (1 μg). The actual experiments were performed in anesthetized animals equipped with a bile duct cannula and a jugular vein catheter. A 3.5-hour time interval was kept between surgery and antigen challenge to avoid trauma-induced interference of the leukotriene measurements [7]. The animals received the histamine blocker pyrilamine maleate (1 mg/kg, i.v.) 5 min prior to the challenge.

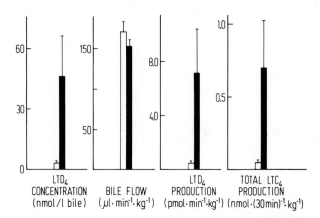

Fig. 2. Effect of antigen challenge on biliary LTD_4 concentration, bile flow, biliary LTD_4 production rate, and total LTC_4 production during 30 min. Open bars represent prechallenge values, closed bars indicate postchallenge values. Mean values $\pm$ SEM from 5 animals are given. Prechallenge fractions collected 30 to 0 min before challenge and post-challenge fractions collected 0 to 15 min after challenge were analyzed by RP-HPLC and subsequent radioimmunoassay. Calculations for total LTC_4 production during a 30-min period are based on the total recovery of LTD_4 in our analyses (77%) and on the finding that 38% of the infused [^{3}H]LTC_4 are recovered from bile as [^{3}H]LTD_4 during 30 min (fig. 1).

Bile samples, collected before and after challenge, were analyzed by the combined use of reversed-phase high-performance liquid chromatography (RP-HPLC) and subsequent radioimmunoassay [5, 8]. In accordance with the tracer studies, LTD_4 was the major endogenous immunoreactive LTC_4 metabolite in anaphylactic guinea pig bile. Antigen challenge was followed by a 15-fold increase of the biliary LTD_4 concentration which rose from 3.1 ± 1.4 nmol/l bile (SEM, n = 5) to 46.2 ± 20.0 nmol/l (SEM, n = 5) (fig. 2). Only a minor decrease in bile flow was observed in this series, and the biliary LTD_4 production rate increased 14-fold.

The total amount of LTC_4 produced during anaphylaxis within a 30-min period was calculated to be 0.7 ± 0.3 nmol/kg (SEM, n = 5) (fig. 2). This amount should be sufficient to induce cardiovascular and hemo-dynamic effects leading to severe shock reactions in guinea pigs. Intravenous administration of 1 nmol of LTC_4 or LTD_4 per kilogram evoked widespread plasma extravasation in the guinea pig [9] and 2 nmol of LTD_4 per kilogram intravenously can be lethal in spontaneously breathing guinea pigs [S.-E. Dahlén, pers. commun.].

Table 1. Influence of dexamethasone on the LTD_4 production rate $(pmol \cdot min^{-1} \cdot kg^{-1})$ during anaphylactic shock

Time interval min	Control	Dexamethasone[1]
−30–0	0.5 ± 0.2	0.6 ± 0.2
0–15	7.2 ± 3.3	6.4 ± 2.9
15–30	5.9 ± 2.2	3.3 ± 1.3
30–90	1.9 ± 0.8	1.4 ± 0.5

Mean values ± SEM, 5 guinea pigs in each group.
[1] 10 mg/kg i.v., 3.5 h before challenge.

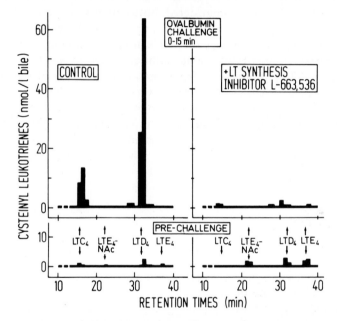

Fig. 3. Effect of the leukotriene biosynthesis inhibitor L-663,536 on antigen-induced cysteinyl leukotriene generation in vivo in the guinea pig. Biliary cysteinyl leukotrienes were separated by RP-HPLC and the fractions were analyzed by radioimmunoassay based on the individual cross-reactivities of the metabolites [8]. Nonpeak areas were calculated as LTE_4 immunoreactivity. Arrows indicate retention times of internal [^3H]leukotriene standards added to each sample in small amounts not interfering with the radioimmunoassay. *Lower panels:* Immunochromatograms of the bile samples collected 30 to 0 min before antigen challenge. *Upper panel, left:* Bile sample of a sensitized animal collected 0 to 15 min after ovalbumin challenge. *Upper panel, right:* Suppression of antigen-induced rise in cysteinyl leukotriene concentration in the postchallenge sample of an animal pretreated with the inhibitor L-663,536 [10, 11] in a dose of 10 mg/kg, i.v., 15 min before challenge. Similar results were obtained with 4 animals in each group.

Influence of Leukotriene Biosynthesis Inhibitors on LTC₄ Formation during Anaphylaxis in the Guinea Pig

After pretreatment of the animals with dexamethasone (10 mg/kg, i.v., 3.5 h before antigen challenge) no significant decrease of the antigen-induced cysteinyl leukotriene production was observed under the conditions described (table 1). Moreover, administration of dexamethasone at a higher dose (40 mg/kg, i.p., 16 h before challenge) did not suppress the ana-phylaxis-induced LTC_4 production. In the guinea pig the target cells of dexamethasone action in vivo may be different from the cell types which are responsible for the IgE-mediated leukotriene production during anaphylac-tic shock in vivo.

The potent inhibitor of 5-lipoxygenase translocation and leukotriene synthesis, L-663,536 [10, 11], administered intravenously 15 min before challenge, completely blocked the rise in LTC_4 production after ovalbumin challenge (fig. 3). This inhibition was reversible after several hours: rechal-lenge of the animals 3 h after intravenous inhibitor treatment was followed by an increase in LTC_4 production. Symptoms usually observed during guinea pig anaphylaxis, such as acute respiratory distress, cyanosis, mic-turation, and sudden death were abolished by pretreatment with L-663,536. This potent inhibitor of leukotriene formation in vivo may be of therapeu-tic value also in human anaphylaxis and asthma as well as in other inflammatory disorders where an excess of 5-lipoxygenase products is generated.

Acknowledgements

We are indebted to Dr. J. Rokach, and Merck Frosst Canada for providing L-663,536, and to Dr. M. Huber for constructive discussion and support. This work was supported in part by grants from the Deutsche Forschungsgemeinschaft through SFB 154, Freiburg.

References

1   Brocklehurst WE: The release of histamine and formation of a slow-reacting substance (SRS-A) during anaphylactic shock. J Physiol 1960;151:416–435.
2   Austen KF: The anaphylactic syndrome; in Samter M (ed): Immunological Diseases. Boston, Little, Brown, 1978, pp 885–899.
3   Smith PL, Kagey-Sobotka A, Bleeker ER, et al: Physiological manifestations of human anaphylaxis. J Clin Invest 1980;66:1072–1080.
4   Feuerstein G: Autonomic pharmacology of leukotrienes. J Auton Pharmacol 1985;5:149–168.
5   Keppler A, Örning L, Bernström K, et al: Endogenous leukotriene D₄ formation during anaphylactic shock in the guinea pig. Proc Natl Acad Sci USA 1987;84:5903–5907.

6   Andersson P: Effects of inhibitors of anaphylactic mediators in two models of bronchial anaphylaxis in anaesthetized guinea pigs. Br J Pharmacol 1982;77:301–307.

7   Denzlinger C, Rapp S, Hagmann W, et al: Leukotrienes as mediators in tissue trauma. Science 1985;230:330–332.

8   Denzlinger C, Guhlmann A, Scheuber PH, et al: Metabolism and analysis of cysteinyl leukotrienes in the monkey. J Biol Chem 1986;261:15601–15606.

9   Hua XY, Dahlén SE, Lundberg JM, et al: Leukotrienes C_4, D_4 and E_4 cause widespread and extensive plasma extravasation in the guinea pig. Naunyn Schmiedebergs Arch Pharmacol 1985;330:136–141.

10   Gillard JW, Girard Y, Morton HE, et al: The discovery and optimization of new classes of thromboxane antagonists and leukotriene biosynthesis inhibitors; in Zor U, Naor Z, Danon A (eds): Leukotrienes and prostanoids in health and disease. New Trends Lipid Mediators Res. Basel, Karger, 1989, vol 3, pp 46–49.

11   Gillard J, Ford-Hutchinson AW, Chan C, et al: L-663,536 (3-[1-(4-chlorobenzyl)-3-t-butyl-thio-5-isopropylindol-2-yl]-2,2-dimethylpropanoic acid), a novel, orally active leukotriene biosynthesis inhibitor. Canad J Physiol Pharmacol 1989; in press.

Dr. Andrea Keppler, Biochemisches Institut, Universität Freiburg,
D–7800 Freiburg (FRG)

Zor U, Naor Z, Danon A (eds): Leukotrienes and Prostanoids in Health and Disease.
New Trends Lipid Mediators Res. Basel, Karger, 1989, vol 3, pp 210–215

Effects of Metamizol and Its Metabolites on Eicosanoid Release from Anaphylactic Guinea Pig Lungs and Rat Gastric Mucosa

M. Trautmann, B.M. Peskar, S. Pritze, W. Luck, U. Hoppe, B.A. Peskar

Departments of Experimental Clinical Medicine and Pharmacology, Ruhr University, Bochum, FRG

Introduction

We have described previously [1] that high concentrations of the biologically active metamizol metabolites 4-monomethylaminoantipyrine (MAA) and 4-aminoantipyrine (AA) inhibit release of cyclooxygenase products of arachidonate metabolism from isolated perfused anaphylactic guinea pig hearts, while release of cysteinyl-leukotrienes (LT) was simultaneously increased. Similarly, only high concentrations of MAA affect eicosanoid release from mouse macrophages [2]. From these results it seems that the effects of these compounds on the eicosanoid system may correlate more closely to the anti-inflammatory activity of nonacidic pyrazolones than to their analgesic and antipyretic activity, which is observed at lower doses [2]. These latter effects have been explained by their more pronounced inhibition of prostaglandin (PG) synthesis in the central nervous system as compared to peripheral tissues [3]. Alternatively, PG-independent peripheral effects of metamizol may mediate analgesia [4]. We have now investigated the effects of MAA and AA as well as of 4-formylaminoantipyrine (FAA) and 4-acetylaminoantipyrine (AAA) on eicosanoid release from the isolated perfused anaphylactic guinea pig lung. These data were compared with the effects of this class of compounds on gastric mucosal integrity and ex vivo release of eicosanoids from rat gastric mucosa after oral instillation of ethanol. We had demonstrated previously [5] that this treatment stimulates release of LTC_4 without affecting release of PG and thromboxane (TX) B_2. Finally, we have tried to correlate the effects of metamizol on gastric mucosa with the tissue levels of its active metabolites using a newly developed radioimmunoassay.

Methods

Isolated lungs from ovalbumin-sensitized guinea pigs were perfused and challenged with antigen as described previously [6]. MAA, AA, FAA and AAA were infused at a rate of 2.8×10^{-6} mol/min resulting in final perfusate concentrations of $> 10^{-4} M$. Drug infusions started 4 min before antigen injection and were continued throughout the experiment. Eicosanoids in the perfusates were determined radioimmunologically as described previously [6–9]. In other experiments, male Wistar rats (170–200 g) were fasted for 24 h with free access to water. They were then treated orally with various doses of metamizol or its metabolites dissolved in distilled water. Controls received the same volume (2.5 ml/kg) of distilled water. Thirty minutes after pretreatment the rats received 1.5 ml absolute ethanol by oral instillation and were killed 5 min later. The stomach was opened, and the severity of gastric damage was evaluated as described previously [5]. Fragments of gastric corpus mucosa were then excised and incubated as described elsewhere [5]. Release of LTC_4 and 6-keto-$PGF_{1\alpha}$ into the incubation media was determined radioimmunologically [10].

Antibodies binding MAA, AA, FAA and AAA with about equal avidity were obtained from rabbits immunized with an AA-bovine serum albumin (BSA) conjugate. The conjugate was prepared by incubation of 10 mg AA, 10 mg BSA, both dissolved together in 2 ml distilled water, and 1 ml glutaraldehyde (0.021 M) for 1 h at room temperature followed by exhaustic dialysis. Using ^{14}C-aminopyrine (Amersham, spec. act. 118 mCi/mmol) as tracer the linear range of the standard curves for the metamizol metabolites was between 10 and 300 ng.

Means $\pm$ SEM were calculated. Statistical analysis was performed using Student's t test.

Results

The effects of the metamizol metabolites on eicosanoid release from isolated perfused anaphylactic guinea pig lungs are illustrated in figure 1. While FAA and AAA did not affect release of immunoreactive LTC_4 or of cyclooxygenase products of arachidonate metabolism, MAA inhibited release of PGD_2, TXB_2, 6-keto-$PGF_{1\alpha}$ and 15-keto-13,14-dihydro-PGE_2 and simultaneously increased release of immunoreactive LTC_4. The effect of MAA on 15-keto-13,14-dihydro-$PGF_{2\alpha}$ did not reach the level of significance (p > 0.05). AA at the infusion rate used (2.8×10^{-6} mol/min) inhibited release of PGD_2 and TXB_2 significantly, but the effect on release of immunoreactive LTC_4, 6-keto-$PGF_{1\alpha}$ and the 15-keto-13,14-dihydro metabolites of $PGF_{2\alpha}$ and PGE_2, although qualitatively similar to that of MAA, was statistically not significant.

The effects of metamizol and its metabolites on ethanol-induced gastric mucosal damage and ex vivo mucosal release of LTC_4 and 6-keto-$PGF_{1\alpha}$ are shown in figure 2. While metamizol, MAA and AA protect the gastric mucosa in a dose-dependent manner, FAA and AAA (both tested at 100 mg/kg) had no significant effect. None of the drugs affected release of 6-keto-$PGF_{1\alpha}$ from the rat gastric mucosa. Release of LTC_4 was significantly (p < 0.05) inhibited by 300 mg/kg MAA only. The protective effect of

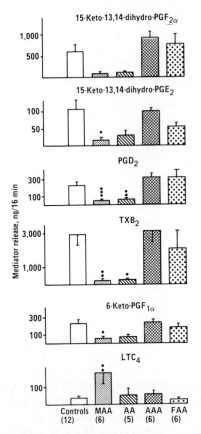

Fig. 1. Release of immunoreactive eicosanoids from isolated perfused anaphylactic guinea pig lungs and the effect of four metamizol metabolites. Perfusates were collected for a total of 16 min after challenge and drugs were infused from 4 min before antigen injection until the end of the experiments. Number of experiments shown in parentheses. ●●●$p < 0.01$; ●●$p < 0.025$; ●$p < 0.05$.

metamizol administered 30 min prior to ethanol correlated with the tissue concentrations of immunoreactive metamizol metabolites determined in the tissue incubation media. These concentrations were $10.4 \pm 1.7\ \mu g/g$ wet weight (n = 6) for a dose of 10 mg/kg, $37.0 \pm 4.8\ \mu g/g$ (n = 6) for 30 mg/kg and $112.2 \pm 21.2\ \mu g/g$ (n = 6) for 100 mg/kg. For validation of these data plasma levels of immunoreactive metamizol metabolites were also determined. Thirty minutes after oral administraton of 30 mg/kg metamizol, plasma concentrations were $17.9 \pm 6.8\ \mu g/ml$ (n = 6). These data are in satisfactory agreement with results obtained by Christ et al. [11] using ^{14}C-labelled metamizol.

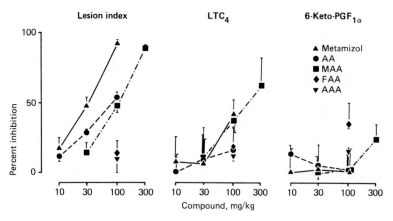

Fig. 2. Effect of metamizol and four of its metabolites on ethanol-induced rat gastric injury (determined as lesion index) and on ex vivo release of immunoreactive LTC$_4$ and 6-keto-PGF$_{1\alpha}$ from gastric mucosa incubated ex vivo.

Discussion

The results demonstrate organ-specific effects of the metamizol metabolites MAA and AA as to eicosanoid release. While high concentrations of these compounds inhibit formation of cyclooxygenase products of arachidonate metabolism in anaphylactic guinea pig hearts [1] as well as lungs, they have no effect on release of 6-keto-PGF$_{1\alpha}$ from rat gastric mucosa incubated in vitro after drug administration in vivo. On the other hand, release of cysteinyl-LT from isolated perfused anaphylactic hearts [1] and lung is increased by MAA, while with a very high dose of MAA (300 mg/kg) inhibition of LTC$_4$ release was observed in rat gastric mucosa challenged with ethanol, a known stimulus for gastric LT synthesis [5]. The organ specificity of the drug effect is obviously not due to lower drug concentrations in the rat gastric mucosa after in vivo treatment as compared to the concentrations used in the perfusion experiments. As measured by the radioimmunoassay for the metamizol metabolites, concentrations as high as about 3×10^{-4} M do occur in rat gastric mucosa 30 min after oral administration of 100 mg/kg metamizol. Metamizol is not stable in aqueous solutions, but rapidly hydrolyzed nonenzymatically [12]. The hydrolysis product MAA is further metabolized more slowly, and from kinetic studies [12] it can be assumed that 30 min after drug administration most of the immunoreactive material is MAA.

We had demonstrated previously that a number of drugs protect the rat gastric mucosa against ethanol-induced injury and inhibit LTC$_4$ release in a parallel manner [13]. Interestingly, metamizol and its active metabolites are protective, but have only marginal effects on LTC$_4$ release. The metamizol-induced gastroprotection seems to be independent of endogenous PG, since metamizol in high doses (>100 mg/kg) is also protective against indomethacin-induced gastric mucosal injury (data not shown).

While inhibition of fatty acid cyclooxygenase by the active metamizol metabolites might be related to their anti-inflammatory activity [1, 2], the possible relation of the metamizol-induced preservation of gastric mucosal integrity to therapeutically relevant drug effects is not clear. It has, indeed, been suggested that the analgesic effect of metamizol is unrelated to the PG system and may be rather mediated by blocking increased calcium fluxes into nociceptors [4]. In this context it is of interest that Tarnawski et al. [14, 15] have demonstrated that increased calcium influx is also a key factor in ethanol- and indomethacin-induced gastric mucosal injury. Thus, although the protective effect of metamizol and its active metabolites is observed with rather high doses, the possibility that the mechanisms of actions underlying the metamizol-induced gastroprotection and analgesia are similar deserves further investigation.

References

1   Coersmeier, C, Wittenberg HR, Aehringhaus U, et al: Effect of anti-inflammatory and analgesic pyrazoles on arachidonic acid metabolism in isolated heart and gastric mucosa preparations. Agents Actions 1986;19(suppl):137–154.

2   Lanz R, Peskar BA, Brune K: The effects of acidic and nonacidic pyrazoles on arachidonic acid metabolism in mouse peritoneal macrophages. Agents Actions 1986;19(suppl):125–135.

3   Dembińska-Kieć A, Zmuda A, Krupinska I: Inhibition of prostaglandin synthetase by aspirin-like drugs in different microsomal preparations. Adv Prostaglandin Thromboxane Res 1976;1:99–103.

4   Ferreira SH: Prostaglandins, pain and inflammation. Agents Actions 1986;19(suppl): 91–97.

5   Peskar BM, Lange K, Hoppe U, et al: Ethanol stimulates formation of leukotriene C$_4$ in rat gastric mucosa. Prostaglandins 1986;31:283–293.

6   Liebig R, Bernauer W, Peskar BA: Release of prostaglandins, a prostaglandin metabolite, slow-reacting substance and histamine from anaphylactic lungs, and its modification by catecholamines. Naunyn Schmiedebergs Arch Pharmacol 1974;284:279–293.

7   Peskar BM, Günter B, Steffens C, et al: Antibodies against dehydration products of 15-keto-13,14-dihydro-prostaglandin E$_2$. FEBS Lett 1980;115:123–126.

8   Lange K, Peskar BA: Radioimmunological determination of prostaglandin D$_2$ after conversion to a stable degradation product. Agents Actions 1984;15:87–89.

9   Juan H, Peskar BA, Simmet T: Effect of exogenous 5,8,11,14,17-eicosapentaenoic acid on cardiac anaphylaxis. Br J Pharmacol 1987;90:315–325.

10   Peskar BM, Hoppe U, Lange K, et al: Effects of non-steroidal anti-inflammatory drugs on rat gastric mucosal leukotriene C_4 and prostanoid release: relation to ethanol-induced injury. Br J Pharmacol 1988;93:937–943.

11   Christ O, Kellner HM, Ross G, et al: Biopharmazeutische Untersuchungen nach Gabe von Metamizol-^{14}C an Ratte, Hund und Mensch. Arzneimittelforschung 1973;23:1760–1767.

12   Weiss R, Brauer J, Goertz U, et al: Vergleichende Untersuchungen zur Frage der Absorption und Metabolisierung des Pyrazolderivates Metamizol nach oraler und intramuskulärer Gabe beim Menschen. Arzneimittelforschung 1974;24:345–348.

13   Lange K, Peskar BA, Peskar BM: Stimulation of rat gastric mucosal leukotriene C_4 formation by ethanol and effect of gastric protective drugs. Adv Prostaglandin Thromboxane Leukotriene Res 1987;17:299–302.

14   Tarnawski A, Piastucki I, Hollander D, et al: Ionophore A23187-induced influx of extracellular calcium potentiates indomethacin injury of isolated gastric gland cells and abolishes prostaglandin cytoprotection (abstract). Gastroenterology 1987;92:1666.

15   Tarnawski A, Piastucki I, Hollander D, et al: Influx of extracellular calcium – a key factor in ethanol injury of isolated gastric gland cells (abstract). Gastroenterology 1987;92:1666.

Dr. M. Trautmann, Abteilung für Pharmakologie und Toxikologie, Ruhr-Universität, Bochum, Universitätsstrasse 150, D-4630 Bochum (FRG)

Zor U, Naor Z, Danon A (eds): Leukotrienes and Prostanoids in Health and Disease.
New Trends Lipid Mediators Res. Basel, Karger, 1989, vol 3, pp 216–220

Pulmonary and Circulatory Effects of the PAF Antagonists WEB 2086 and WEB 2170 in Anaphylactic and Septic Shock

H. Heuer[a], *J. Casals-Stenzel*[a], *K.H. Weber*[a], *L.G. Letts*[b]

[a]Boehringer Ingelheim KG, Ingelheim, FRG, and
[b]Boehringer Ingelheim Pharmaceuticals Inc., Ridgefield, Conn., USA

Introduction

Platelet-activating factor (PAF, PAF-acether) is a potent, naturally occurring phospholipid that has been implicated in several pathophysiological responses in man [1]. Since its discovery in the early 1970s, it has been shown to be produced by numerous human cell types, including platelets, neutrophils, macrophages, lymphocytes, and endothelial cells [for review, see 1]. Exogenous PAF exhibits extremely potent actions both in vitro and in vivo. These include aggregation and activation of secretory events in platelets and neutrophils, neutrophil and macrophage chemotaxis, bronchonconstriction, systemic hypotension, enhanced fluid and solute exchange, as well as release of a number of mediators. Secondary mediators released following cellular exposure to PAF include thromboxane A_2 and prostaglandins, leukotrienes, lysosomal enzymes, superoxide, IL-1 and TNF [1].

Whether a PAF antagonist is developed as an important therapy for human diseases such as asthma or septic shock is largely dependent on the functional and biochemical effectiveness of potent, selective antagonists in relevant animal models. In these studies we recount that two structurally similar PAF receptor antagonists, WEB 2086 and WEB 2170 [2, 3], exhibit beneficial activities in anaphylactic and endotoxin-induced shock.

Materials and Methods

The data and methodology have been published earlier: intravascular platelet aggregation in vivo [2]; platelet aggregation in vitro [2]; platelet binding [9, 10]; guinea pig pulmonary anaphylaxis (passive and actively sensitized) [4]; *Escherichia coli* endotoxic shock in rodents [5].

Results

The actions of WEB 2086 and WEB 2170 in vitro are summarized in table 1. The data show that both antagonists are potent, selective antagonists of the effects of synthetic PAF.

The potent actions of the two hetrazepine antagonists against the effects of intravenous PAF in vivo are also shown (table 2). These results clearly show the WEB 2086 and WEB 2170 are potent antagonists of the effects of infused PAF on platelet accumulation in the lung, bronchoconstriction and systemic blood pressure. It should be noted that neither of the antagonists possess sedative actions in conscious animals [6].

In actively sensitized guinea pigs, WEB 2086 administered perorally alone at doses of 0.05–0.5 mg/kg 1 h before the intravenous antigen challenge (ovalbumin) did not inhibit the pulmonary or systemic anaphylactic reaction. When administered in animals pretreated with a low dose of the antihistamine mepyramine (5 µg/kg, i.v., 10 min prior), however, WEB 2086 effectively inhibited the bronchoconstriction but had less of an effect on the shock-induced hypotension. A similar profile was obtained using WEB 2170 at doses of 0.1–1.0 mg/kg, p.o. [9].

In passively sensitized guinea pigs that were pretreated with mepyramine, peroral WEB 2086 at the above doses and pretreatment times protected against both the pulmonary and systemic effects of antigen challenge.

In anesthetized rats the intravenous injection of *E. coli* endotoxin (15 mg/kg, type 0111:B4, Sigma) or PAF (30 µg/kg) caused a sharp drop in the systemic blood pressure that peaked after 5–10 min. This blood pressure response to both agents was dose dependently inhibited by pretreatment with either WEB 2086 (1–10 mg/kg, p.o.; 0.1–5 mg/kg, i.v.) or WEB 2170 (0.5–10 mg/kg, p.o.; 0.01–1 mg/kg, i.v.) [10].

In conscious rats, WEB 2086 protected against the lethal dose of endotoxin (7.5 mg/kg, i.v., 24 h). The results are summarized in table 3. The mean effective doses that prevented lethality were 6.3 (1.6–49.0) mg/kg, p.o. and 2.17 (0.89–4.04) mg/kg, i.v.

Discussion

The pulmonary and systemic response to intravenous PAF in the anesthetized guinea pig is similar to that seen when antigen is administered to a sensitized animal. The response is characterized by a rapid and marked pulmonary hypertension and bronchoconstriction accompanied by systemic hypotension, thrombocytopenia, neutropenia, increased vascular permeability and death. In the rat, the blood pressure response to intravenous

Table 1. Effects of WEB 2086 and WEB 2170 in vitro

Cell	Assay	Agonist	IC_{50}, μM	
			WEB 2086	WEB 2170
Platelet	ligand binding	^{3}H-PAF	15	16
	aggregation	PAF	0.17	0.31
		5-HT	156	>150
		AA	190	>200
		collagen	341	>250
		ADP	>1,000	>1,000
Neutrophil	aggregation	PAF	0.37	0.83
		ConA	>1,000	>1,000
		LTB_4	>1,000	>1,000
		FMLP	>1,000	>1,000

See references 2, 4 and 5, and data on file, Boehringer Ingelheim.

Table 2. Actions of WEB 2086 and WEB 2170 against intravenous PAF in vivo

Animal	Parameter	ED_{50}, mg/kg i.v.	
		WEB 2086	WEB 2170
Guinea pig	bronchoconstriction	0.018	0.008
	blood pressure	0.016	0.006
Rat	blood pressure	0.052	0.016

See references 2 and 5 for details.

Table 3. Protective effects of WEB 2086 against lethal dose of endotoxin in conscious rats

Dose, mg/kg	Route	n	Survival, %
0	p.o.	20	0
	i.v.	18	0
1.0	p.o.	12	27
	i.v.	12	25
5.0	p.o.	12	48
	i.v.	12	63
10.0	p.o.	24	58
	i.v.	12	100

PAF is near identical to that obtained with intravenous *E. coli* endotoxin. In both animal models, the monitored variables are inhibited in a dose-related fashion by pretreatment with the potent and selective PAF antagonists WEB 2086 and WEB 2170. In some cases, administration of the PAF antagonists after the induction of shock results in the reversal of PAF effects [7].

The data show that WEB 2086 and WEB 2170 effectively antagonize the severe abnormality of respiratory function that occurs with an acute pulmonary anaphylactic reaction. In the guinea pig this is best illustrated in actively or passively sensitized animals pretreated with an antihistamine since histamine is the predominant mediator released under these experimental conditions.

In the rat, both WEB 2086 and WEB 2170 prolong survival after *E. coli* shock. The mechanism(s) by which these compounds act, however, has not been addressed. It is tempting to speculate that these compounds act, in part, by improving pulmonary function as suggested to occur in sheep by Purvis et al. [8]. Further studies, including close examination of the compounds on pulmonary function in different species, will be necessary.

The role that PAF may play in the pathophysiology of human asthma or septic shock remains unclear. Both anaphylactic and endotoxic shocks are complex events that are not fully understood. Moreover, the different reactivity of animal species to PAF, including methodological differences between investigators leading to diversities such as the cell types initially exposed, the variables monitored, secondary mediator release, tissue metabolism/inactivation, etc., obscures the true pathophysiological importance of PAF in relation to human disease. As with other purported mediators, clinical significance is best determined by controlled clinical trials using potent, selective PAF antagonists. In the case of both WEB 2086 and WEB 2170, the hypothesis that PAF is a major mediator of some aspects of shock can be answered. The data provided support earlier indications that PAF antagonists may protect against respiratory dysfunction in allergic pulmonary disorders as well as prevent some of the inflammatory events during sepsis.

References

1   Braquet P, Touqui L, Shen TY, et al: Perspectives in platelet-activating factor research. Pharmacol Rev 1987;39:98–145.
2   Casals-Stenzel J, Muacevic G, Weber K-H: Pharmacological actions of WEB 2086, a new specific antagonist of platelet-activating factor. J Pharmacol Exp Ther 1987;241:974–981.
3   Heuer H, Casals-Stenzel J, Muacevic G, et al: Activity of the new and specific PAF antagonists WEB 2170 and STY 2108 on PAF-induced bronchoconstriction and intrathoracic accumulation of platelets in the guinea pig. Prostaglandins 1988;35:798.

4   Casals-Stenzel J: Effects of WEB 2086, a novel antagonist of platelet-activating factor, in active and passive anaphylaxis. Immunopharmacology 1987;13:117–124.

5   Casals-Stenzel J: Protective effect of WEB 2086, a novel antagonist of platelet-activating factor, in endotoxin shock. Eur J Pharmacol 1987;135:117–122.

6   Casals-Stenzel J, Weber KH: Triazolodiazepines: dissociation of their PAF (platelet-activating factor) antagonistic and CNS activity. Br J Pharmacol 1987;90:139–146.

7   Pretolani M, Lefort J, Malanchere E, et al: Interference by the novel PAF-acether antagonist WEB 2086 with the bronchopulmonary responses to PAF-acether and to active and passive anaphylactic shock in guinea pigs. Eur J Pharmacol 1987;140:311–371.

8   Purvis AW, Christman BW, McPherson CD, et al: WEB 2086, a platelet-activating factor receptor antagonist attenuates the response to endotoxin in wake sheep. Am Rev Resp Dis, in press.

9   Heuer H, Birke F, Brandt K, et al: Biological characterisation of the enantiomeric hetrazepines of the PAF-antigonist WEB 2170. Prostaglandins 1988;35;850.

10   Birke FW, Weber, KH: Specific and high affinity binding of WEB 2086 to human platelet PAF receptors. Clin Exp Pharmacol Physiol 1988(suppl 13):2.

L.G. Letts, PhD, Boehringer Ingelheim Pharmaceuticals Inc., 90 East Ridge, PO Box 368, Ridgefield, CT 06877 (USA)

Zor U, Naor Z, Danon A (eds): Leukotrienes and Prostanoids in Health and Disease.
New Trends Lipid Mediators Res. Basel, Karger, 1989, vol 3, pp 221–224

The Adrenal-Pituitary Axis, Acute Phase Response and Prostanoid Synthesis

Abraham Danon[a], *George Prajgrod*[a], *Shulamit Zimlichman*[a,b], *Ruth Shainkin-Kestenbaum*[b]

Departments of [a]Clinical Pharmacology and [b]Biochemistry, Ben-Gurion University and Soroka Medical Center, Beer-Sheva, Israel

A decade ago we pioneered the observation that corticosteroids (CS) inhibit prostaglandin (PG) synthesis by a mechanism involving new RNA and protein synthesis [1]. This observation has been confirmed in many cells and tissues and led to the characterization of the phospholipase A_2 (PLA_2)-inhibiting protein lipocortin [2]. More recently, an increasing number of authors have been unable to show this effect of CS in the in vivo situation. Thus, in vivo CS are reported to either decrease [3], increase [4] or not alter [5] the rate of eicosanoid synthesis. The present experiments were designed to shed light on this in vitro-in vivo discrepancy. The working hypothesis has been that multiple factors operate in vivo, notably neuroendocrine mechanisms on the one hand and immune mediators on the other hand, and the complex interactions therein [6] may mask the anticipated action of CS via lipocortin.

Male Charles River rats were used throughout. We experimented with two assay systems, namely the rat renal papilla (RRP) and rat aortic rings (RAR). PGE_2 and 6-keto-$PGF_{1\alpha}$ in the media were measured by radio-immunoassay (RIA), after in vitro incubations of RRP and RAR, respectively.

Effect of Dexamethasone

Dexamethasone (Dex) added in vitro produced typical inhibition of PGE_2 synthesis in RRP, probably mediated by lipocortin. However, we were unable to detect an effect of CS on RAR prostacyclin (PGI_2) synthesis (measured as 6-keto-$PGF_{1\alpha}$) in short-term incubations up to 5.5 h in DMEM, or in organ culture of 26 h.

Ex vivo experiments also revealed different responses of the two assay systems. Acute treatment with Dex, 5 mg/kg i.p. 2 h prior to sacrifice, was without effect on RRP PGE_2 synthesis, while augmenting PGI_2 by 81% in RAR. Chronic Dex treatment, 5 mg/kg/day for 7 days, produced inhibition of RRP PGE_2, but not of RAR PGI_2.

Contribution of Endogenous CS

To study the contribution of endogenous CS on prostanoid synthesis, adrenalectomy (ADX) was performed and prostanoid synthesis was assessed 14 days later. ADX produced a 77% decrease in PGI_2 synthesis by RAR ex vivo, while the RRP increased its PGE_2 synthesis more than twofold. Replacement of Dex for the last 7 days (0.2 mg/kg/day, i.p.) restored RAR PGI_2 synthesis. Because ACTH levels increase after ADX, we tested the effect of ACTH on prostanoid synthesis. ACTH was injected (5 μg/kg, i.p.), and prostanoid synthesis was assessed ex vivo 2 h later. Both RRP PGE_2 synthesis and RAR PGI_2 increased following ACTH, by 64 and 183% respectively. However, in in vitro incubations with ACTH, prostanoid production was not stimulated. On the contrary, a 36% decrease in RAR PGI_2 synthesis was observed. Therefore, the ex vivo stimulation by ACTH is probably indirect.

Effects of IL-1 and Lipopolysaccharide

Tissues were incubated with 10 U/ml human recombinant IL-1 (rh-IL-1α). RRP PGE_2 was stimulated after a lag time of 90 min. In similar incubations, RAR did not respond to IL-1, but in organ culture PGI_2 production was enhanced.

Lipopolysaccharide (LPS) injection (250 μg/rat, i.p.) to normal rats resulted in marked reductions in RAR PGI_2 synthesis, that became evident after 1 h and lasted for at least 24 h, returning to baseline by 96 h. Similarly, RRP PGE_2 decreased 24 h after LPS. On the other hand, prostanoid production by tissues from ADX rats was stimulated by LPS, compared to baseline ADX levels.

Effect of Serum Amyloid A on Human Platelet Aggregation

Finally we investigated the effect of the acute phase protein serum amyloid A (SAA) on human platelet aggregation. SAA levels increase many hundredfold in patients during the acute phase response, via the mediation of cytokines, particularly IL-1, IL-6 and TNF [7], but their biological function remains uncertain. We used thrombin-induced gel-filtered human platelets. SAA was purified from the serum of patients after musculoskeletal trauma. SAA (25–100 μg/ml) produced marked inhibition of thrombin-induced platelet aggregation. There was a concomitant decrease in thromboxane B_2 synthesis (measured by RIA), as well as in serotonin release and in the rise in cytosolic calcium (by the Quin-2 method). Therefore, SAA, which is an HDL apoprotein, may play a role in attenuating platelet aggregation and thromboxane synthesis, and may mediate the well-established protective effect of HDL in cardiovascular disease [8].

Conclusions

The data indicate the CS and IL-1, whether endogenous or exogenous, exert different effects on prostanoid synthesis in each of the two tissues that were studied. The RRP responds in vitro and in the chronic ex vivo model with the well-recognized inhibition of prostanoid synthesis. However, this effect is masked in the acute ex vivo experiment. On the other hand, the RAR probably lack the lipocortin mechanism and, conversely, CS may exhibit a prostanoid stimulatory activity in this tissue. Moreover, the overall in vivo effects involve interactions of multiple factors, including the hypothalamus-pituitary-adrenal axis on the one hand, and the immune system and acute phase reactants on the other hand. Thus, cytokines may affect prostanoid synthesis either directly [9, 10] or indirectly through the acute phase proteins [11], which could in turn feed back and attenuate the effects of the cytokines [12]. IL-1 also activates the pituitary-adrenal system [13, 14]. CS, in turn, may suppress the immune response by reducing IL-1 production [15].

References

1   Danon A, Assouline G: Inhibition of prostaglandin biosynthesis by corticosteroids requires RNA and protein synthesis. Nature 1978;273:552–554.
2   Flower RJ: The lipocortins and their role in controlling defense reactions. Adv Prostaglandin Thromboxane Leukotriene Res 1985;15:201–203.
3   Van de Velde VJS, Herman AG, Bult H: Effects of dexamethasone on prostacyclin biosynthesis in rabbit endothelial cells. Prostaglandins 1986;32:169–178.
4   Erman A, Hassid A, Baer PG, et al: Treatment with dexamethasone increases glomerular prostaglandin synthesis in rats. J. Pharmacol exp Ther 1986;239:296–301.

5 Tsai MY: Glucocorticoid and prostaglandins: lack of an inhibitory effect by dexamethasone on the synthesis of 6-keto-prostaglandin $F_{1\alpha}$ in rat lung. Prostaglandins Leukotrienes Med 1987;28:119–125.

6 Goetzl EJ, Sreedharan SP, Harlbnen S: Pathogenetic roles of neuroimmunologic mediators. Immunol Allergy Clin North Am 1988;8:183–199.

7 Kampschmidt RF: The numerous postulated biological manifestations of interleukin-1. J Leukocyte Biol 1984;36:341–355.

8 Miller GJ, Miller NE: Plasma high density lipoprotein concentration and development of ischemic heart disease. Lancet 1975;i:16–19.

9 Rossi V, Breviario F, Ghezzi P, et al: Prostacyclin synthesis induced in vascular cells by interleukin-1. Science 1985;229:174–176.

10 Xiao D, Levine L: Stimulation of arachidonic acid metabolism: differences in potencies of recombinant human interleukin 1α and interleukin 1β on two cell types. Prostaglandins 1986;32:709–718.

11 Dinarello CA: Interleukin-1 and the pathogenesis of the acute-phase response. N Engl J Med 1984;311:1413–1418.

12 Benson MD, Aldo-Benson MA: Effect of purified protein SAA on immune response in vitro: Mechanisms of suppression. J Immunol 1979;122:2077–2082.

13 Bernton EW, Beach JE, Holaday JW, et al: Release of multiple hormones by a direct action of interleukin-1 on pituitary cells, Science 1987;238:519–521.

14 Roh MS, Drazenovich KA, Barbose JJ, et al: Direct stimulation of the adrenal cortex by interleukin-1. Surgery 1987;102:140–146.

15 Besedovsky H, Del Rey A, Sorkin E, et al: Immunoregulatory feedback between interleukin-1 and glucocorticoid hormones. Science 1986;233:652–654.

Abraham Danon, MD, PhD, Department of Clinical Pharmacology, Ben-Gurion University and Soroka Medical Center, P.O. Box 653, Beer-Sheva 84105 (Israel)

Zor U, Naor Z, Danon A (eds): Leukotrienes and Prostanoids in Health and Disease.
New Trends Lipid Mediators Res. Basel, Karger, 1989, vol 3, pp 225–230

Cysteinyl Leukotrienes in Experimental Ulcers in Rats

B.M. Peskar

Department of Experimental Clinical Medicine, Ruhr-University, Bochum, FRG

Gastric Mucosal Biosynthesis and Vascular Effects of Cysteinyl Leukotrienes

Microcirculatory changes are a prominent feature of ethanol-induced gastric mucosal damage. In the rat stomach perfused in situ via the vasculature infusions of the leukotrienes (LT) LTC_4 and LTD_4 cause a rapid decrease in vascular flow (fig. 1) indicating potent gastric vasoconstrictor actions of exogenous cysteinyl-LT. Furthermore, LTC_4, but not LTD_4, elicited flow disturbances in the rat gastric submucosal microcirculation resembling those observed after ethanol [1] and both LTC_4 and LTD_4, although not ulcerogenic themselves, enhanced the noxious effects of various ulcerogens including ethanol [2, 3]. We have shown previously that intragastric instillation of ethanol causes a marked and dose-dependent increase in the release of LTC_4 from gastric mucosal fragments incubated ex vivo [4]. In contrast, neither mucosal formation of prostaglandins (PG) nor of thromboxane was significantly increased after ethanol exposure [4].

Effect of Protective Drugs on Gastric Mucosal Leukotriene Formation

A great number of protective compounds inhibit rat gastric mucosal LT formation. Thus, the nonselective lipoxygenase inhibitor nordihydroguaiaretic acid (NDGA) [4] and the dual lipoxygenase and cyclooxygenase inhibitor BW755C [5] significantly reduced rat gastric mucosal LTC_4 formation and simultaneously protected against the damage caused by ethanol. Sulfhydryl-containing agents such as cysteamine, *L*-cysteine, N-acetylcysteine and dimercaprol, sulfhydryl-blocking compounds such as N-ethylmaleimide, diethylmaleate and iodacetamide as well as metals have been reported to protect the rat gastric mucosa against ethanol-induced

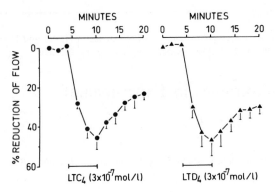

Fig. 1. Effect of LTC$_4$ and LTD$_4$ on vascular flow in the isolated perfused rat stomach. LT were infused for 6 min (n = 6 each). Basal flow before LT infusion was 3.3 ± 0.3 ml/min.

Table 1. Effect of dimercaprol (DMC) on rat gastric mucosal damage and eicosanoid release after ethanol challenge

	Lesion index	LTC$_4$	PGE$_2$
Controls	45 ± 7	421 ± 15	780 ± 69
DMC	$3 \pm 1^*$	$26 \pm 5^*$	$40 \pm 10^*$

Groups of 6 rats were treated orally with DMC (30 μl/kg) or water 30 min prior to intragastric ethanol (1.5 ml, 100%) and were killed 5 min later. Lesion production and mucosal eicosanoid release (ng/g/10 min) was determined as described [4, 9]. *p < 0.001 compared to controls (Student's t test).

damage [6–9], although the mechanism of this protection has not been elucidated. Oral treatment with these compounds dose-dependently prevented the stimulatory action of ethanol on gastric mucosal LTC$_4$ formation with only minor differences in the ID$_{50}$ values for protection and inhibition of LTC$_4$ release [9, 10]. None of the agents increased mucosal PG formation, indicating that inhibition of gastric 5-lipoxygenase does not shift arachidonate metabolism to the cyclooxygenase pathway. Most of the sulfhydryl-modulating agents and metals even inhibited rat gastric PG formation [9, 10]. Thus, for example, dimercaprol in doses that completely prevented the development of ethanol-induced gastric lesions virtually abolished mucosal PG release (table 1) indicating that the protective effect of sulfhydryl-modulating agents is not mediated by endogenous PG. Furthermore, these data demonstrate that drug-induced gastroprotection can occur although mucosal PG formation is severely suppressed.

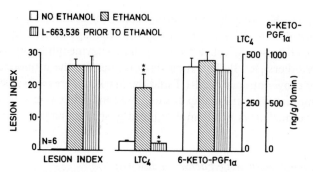

Fig. 2. Rat gastric lesion production and eiconsanoid release under basal conditions and after ethanol challenge and effect of the LT biosynthesis inhibitor L-663,536. Intragastric instillation of 1.5 ml ethanol (50%) caused severe mucosal damage and increased the release of gastric mucosal LTC_4, but not 6-keto-PGF_{1a}. Treatment with L-663,536 (5 mg/kp, i.p.) 30 min prior to ethanol completely blocked the stimulatory action on LTC_4 formation, but did not protect against mucosal damage caused by ethanol. **$p < 0.01$ compared to rats not receiving ethanol; *$p < 0.01$ compared to rats treated with ethanol only. For details of methods, see Peskar et al. [4].

Contrary to most nonsteroidal anti-inflammatory drugs (NSAID), sodium salicylate is not ulcerogenic in rats, but even protects against gastric damage caused by ethanol [for references, see 11]. Although sodium salicylate does not inhibit gastric PG and LTC_4 formation in rats not challenged with ethanol, the drug dose-dependently prevents the stimulatory action of the irritant on mucosal LTC_4 formation [11]. The inhibitory effect on LTC_4 formation (ID_{50} 40 mg/kg) parallels the protective activity (ID_{50} 12 mg/kg) of the drug. The lack of inhibitory action of sodium salicylate on gastric LT formation under basal conditions suggests that the mechanism underlying the blockade of ethanol-stimulated LT release does not involve direct enzyme inhibition.

Certain drugs, which are highly effective in protecting against gastric mucosal injury, do not prevent the stimulatory action of ethanol on mucosal LTC_4 formation. These drugs include Al-containing antacids [12], sucralfate [12], PGE_2 [10] and various synthetic PG analogs [unpubl. results]. These findings indicate that the inhibition of gastric mucosal LTC_4 formation by certain protective agents is not the result of tissue preservation, but is an event independent of the damage to the mucosa.

The close interrelationship between inhibition of gastric mucosal damage and LTC_4 formation observed with numerous gastroprotective drugs has suggested that cysteinyl-LT may play a mediator role in ethanol-induced gastric damage. Recently, selective inhibitors of LT biosynthesis have become available. As shown in figure 2, treatment of rats with the LT biosynthesis inhibitor L-663,536 [J.W. Gillard et al., see pp. 46–49] in a dose that completely blocked the stimulatory action of ethanol on gastric mucosal LTC_4 formation did not prevent gastric damage caused by the irritant. Similar results using selective 5-lipoxygenase enzyme inhibitors have been reported by others [5, 13]. This indicates that despite the pronounced biologic effects of exogenous cysteinyl-LT and the marked increase in gastric mucosal LT formation after ethanol exposure, cysteinyl-LT may not be the major or exclusive mediators of ethanol-induced gastric damage.

Effect of Platelet-Activating Factor on Rat Gastric Cysteinyl Leukotriene Formation

Platelet-activating factor (PAF) is a potent ulcerogen in the rat stomach [14]. We have studied the actions of PAF on gastric flow and cysteinyl-LT release in the isolated, vascularly perfused rat stomach [15]. It was found that bolus injection of PAF (3–50 ng) dose-dependently reduced vascular flow and simultaneously increased the release of LTC_4, LTD_4 and LTE_4 into the perfusate. Infusions of NDGA in concentrations that significantly inhibited the PAF-induced release of cysteinyl-LT did not, however, affect the flow reduction caused by PAF [15]. Likewise, the cysteinyl-LT receptor antagonist FPL 55712 did not counteract the PAF-induced flow reduction, although the flow reduction caused by exogenous LTC_4 was fully blocked [15]. Both flow reduction and increase in cysteinyl-LT release by PAF were dose-dependently attenuated by the PAF receptor antagonist BN 52021 [15]. These results demonstrate that PAF and LTC_4 induce flow reductions in the rat gastric vascular bed by activating different receptors and that endogenous cysteinyl-LT released by PAF do not contribute significantly to the PAF effect on gastric vascular flow.

Leukotrienes in Gastric Mucosal Damage Induced by Nonsteroidal Anti-Inflammatory Drugs

With respect to the pro-ulcerogenic properties of cysteinyl-LT, it was hypothesized that increased LT formation in addition to inhibition of PG generation may be involved in the gastric irritancy caused by NSAID. However, in rats, oral treatment with indomethacin or aspirin inhibited release of both cyclooxygenase products and LTC_4 from gastric mucosal fragments incubated ex vivo [11]. As the inhibitory action of NSAID on gastric mucosal PG formation is more pronounced than that on LTC_4 release, treatment with NSAID results in a shift in the balance between

Table 2. Effect of L-663, 536 on gastric lesion production and eicosanoid release in in-domethacin (IND)-treated rats

	Lesion index	LTC$_4$	6-Keto-PGF$_{1a}$
Controls	1 ± 1	37 ± 4	943 ± 75
IND	23 ± 4**	24 ± 3*	108 ± 9**
L-663,536 + IND	17 ± 1**	4 ± 1**	122 ± 35**

Groups of 6 rats were treated with L-663,536 (15 mg/kg, p.o.) or vehicle 30 min prior to IND (20 mg/kg, p.o.) and were killed 5 h after IND. Lesion production and mucosal eicosanoid release (ng/g/10 min) were determined as described [4, 11]. **$p < 0.001$, *$p < 0.05$ compared to controls not receiving IND.

protective PG and potentially ulcerogenic LTC$_4$. However, pretreatment of rats with the LT biosynthesis inhibitor L-663,536 did not significantly reduce the number or severity of lesions caused by indomethacin (table 2).

Concluding Comments

Cysteinyl-LT cause marked flow reductions in the rat gastric vascular bed and are synthesized in increased amounts after certain ulcerogens such as ethanol or PAF. There are numerous examples of a close interrelation-ship between drug-induced gastroprotection and inhibition of mucosal cysteinyl-LT formation. This interrelationship is, however, not complete since selective LT biosynthesis inhibitors are not gastroprotective. This indicates that cysteinyl-LT are not major mediators of gastric mucosal damage caused by ethanol or PAF. Oral NSAID in doses that produce severe gastric lesions inhibit mucosal formation of PG and, to a lesser extent, LTC$_4$. Although this shifts the balance from protective PG to potentially ulcerogenic LT, a LT biosynthesis inhibitor did not protect against indomethacin-induced gastric lesions. The close interrelationship between gastroprotection and inhibition of LT formation found with a number of drugs may result from effects on a target which is relevant to both mucosal damage and stimulation of 5-lipoxygenase. The nature of this common target remains to be elucidated.

References

1   Whittle BJR, Oren-Wolman N, Guth PH: Gastric vasoconstrictor actions of leukotriene C$_4$, PGF$_{2a}$, and thromboxane mimetic U-46619 on rat submucosal microcirculation in vivo. Am J Physiol 1985;248:580–586.

2   Pihan G, Rogers C, Szabo S: Vascular injury in acute gastric mucosal damage. Mediatory role of leukotrienes. Dig Dis Sci 1988;33:625–632.

3   Konturek SJ, Brzozowski T, Drozdowicz D, et al: Role of leukotrienes in acute gastric lesions induced by ethanol, taurocholate, aspirin, platelet-activating factor and stress in rats. Dis Dis Sci 1988;33:806–813.

4   Peskar BM, Lange K, Hoppe U, et al: Ethanol stimulates formation of leukotriene C_4 in rat gastric mucosa. Prostaglandins 1986;31:283–293.

5   Wallace JL, Beck PL, Morris GP: Is there a role for leukotrienes as mediators of ethanol-induced gastric mucosal damage? Am J Physiol 1988;245:G117–123.

6   Szabo S, Trier JS, Frankel PW: Sulfhydryl compounds may mediate gastric cytoprotection. Science 1981;214:200–202.

7   Robert A, Eberle D, Kaplowitz N: Role of glutathione in gastric mucosal cytoprotection. Am J Physiol 1984;247:G296–304.

8   Dupuy D, Szabo S: Protection by metals against ethanol-induced gastric mucosal injury in the rat. Comparative biochemical and pharmacologic studies implicate protein sulfhydryls. Gastroenterology 1986;91:966–974.

9   Peskar BM, Lange K: Role of leukotriene C_4 in gastric protection by sulfhydryl-modulating agents and metals in the rat. Gastroenterology 1987;92:1573.

10 Lange K, Peskar BA, Peskar BM: Stimulation of rat gastric mucosal leukotriene C_4 formation by ethanol and effect of gastric protective drugs; in Samuelsson B, Paoletti R, Ramwell PW (eds): Advances in Prostaglandin, Thromboxane, and Leukotriene Research. New York, Raven Press, 1987, vol 17, pp 299–302.

11 Peskar BM, Hoppe U, Lange K, et al: Effects of non-steroidal anti-inflammatory drugs on rat gastric mucosal leukotriene C_4 and prostanoid release: relation to ethanol-induced injury. Br J Pharmacol 1988;93:937–943.

12 Lange K, Peskar BA, Peskar BM: Inhibition of leukotriene formation as a mode of action of gastroprotective drugs; in Sinzinger H, Schrör K (eds): Prostaglandins in Clinical Research. New York, Liss, 1987, vol 242, pp 283–288.

13 Boughton-Smith NK, Whittle BJR: Failure of the inhibition of rat gastric mucosal 5-lipoxygenase by novel acetohydroxamic acids to prevent ethanol-induced damage. Br J Pharmacol 1988;95:155–162.

14 Rosam AC, Wallace JL, Whittle BJR: Potent ulcerogenic actions of platelet-activating factor on the stomach. Nature 1986;319:54–56.

15 Dembinska-Kiec A, Peskar BA, Müller MK, et al: The effect of platelet-activating factor on flow rate and eicosanoid release in the isolated perfused rat gastric vascular bed. Prostaglandins 1989;37:69–91.

Dr. Brigitta M. Peskar, Abteilung für Experimentelle Klinische Medizin, Ruhr-Universität Bochum, Im Lottental, D–4630 Bochum (FRG)

Cardiovascular

Zor U, Naor Z, Danon A (eds): Leukotrienes and Prostanoids in Health and Disease.
New Trends Lipid Mediators Res. Basel, Karger, 1989, vol 3, pp 231–235

The Contribution of Chemical Mediators to the Anti-Thrombotic Properties of the Endothelial Cell

Regina Botting, John R. Vane

The William Harvey Research Institute, St. Bartholomew's Hospital Medical College, London, UK

Endothelial cells (EC) envelop the bloodstream in a monolayer and help to maintain the patency of blood vessels and the fluidity of blood by the formation and/or secretion of an extraordinary array of chemicals. The production of these substances is modulated by interactions between the EC and white blood cells, platelets or constituents of plasma. The EC can be activated by amines, peptides, proteins, nucleotides and arachidonic acid (AA) and its metabolites as well as by physical changes (pulse pressure). This activation of EC is often mediated by specific receptors leading to, for example, the generation of prostacyclin and endothelium-derived relaxing factor (EDRF) which in turn cause vasodilatation and inhibition of platelet aggregation. 13-Hydroxyoctadecadienoic acid (13-HODE) formation may also contribute to the anti-adhesive properties of the EC. The mechanisms by which EC receive these chemical messages and translate them into such different chemical signals is reviewed, including the importance of receptor-mediated activation of phospholipase C. An important conclusion is that the generation by the EC of prostacyclin and EDRF is coupled, lending a new dimension to the synergism between the activities of the two substances.

Prostacyclin

The discoveries of thromboxane A_2 (made by platelets) and of prostacyclin (made by the vessel wall) have led to many important new concepts in vascular pathophysiology [1].

Prostacyclin is a dienoic bicyclic eicosanoid which derives from the membrane-bound fatty acid, AA. The chemical instability of prostacyclin at physiological pH ($t\frac{1}{2} \sim 3$ min) is a result of the clevage of its furan ring with subsequent formation of the prostaglandin (PG) 6-keto-PGF$_{1\alpha}$. The biosynthesis of prostacyclin by EC is catalysed by a haeme-containing oxygenase (cyclo-oxygenase) which is inhibited by aspirin and aspirin-like drugs.

Several drugs, as well as endogenous mediators, stimulate the generation of prostacyclin by cultured EC or smooth muscle cells. Endogenous prostacyclin generators include bradykinin (BK), choline esters, AA, PGH$_2$, thrombin, trypsin, platelet-derived growth factor, epidermal growth factor, interleukin-1 and adenine nucleotides.

Glucocorticosteroids and lipocortin [2], cyclo-oxygenase inhibitors, low density lipoproteins loaded with lipid hydroperoxides [3] and, unexpectedly, vitamin K$_1$, all inhibit the biosynthesis of prostacyclin in EC.

The platelet-suppressant and vasodilator actions of prostacyclin are mediated by stimulation of adenylate cyclase. Its fibrinolytic and 'cytoprotective' properties, though, do not seem to be mediated by cAMP and yet their impact on the therapeutic prospects of prostacyclin and its analogues is noteworthy.

In man, intravenous infusion of prostacyclin for several hours on each of 4 days promotes the healing of ischaemic skin ulcers [4]. Prostacyclin also protects against post-ischaemic reperfusion damage to animal brains and hearts [5]. Higenbottam [6] showed that continuous infusions of prostacyclin for up to 2 years allowed patients with severe pulmonary hypertension to live independently, whilst awaiting a heart-lung transplant.

13-HODE

In 1985, Buchanan et al. [7] announced that monolayers of human cultured EC synthesized a cytosol-associated lipoxygenase metabolite which they called LOX. Subsequently [8], LOX was identified by HPLC and gas chromatography/mass spectrography as 13-HODE, a product of linoleic acid (18:2 n − 6). Adhesion of platelets to EC was reduced if the EC were pre-incubated with 13-HODE [9]. In rabbit EC, 13-HODE concentrations increase with time. This accumulation is inhibited by superoxide dismutase (SOD), suggesting that in these cells 13-HODE is a product of autolysis [10].

Endothelium-Derived Relaxing Factor

EDRF is a labile vasodilator with a half-life counted in seconds. Furchgott and Zawadzki in 1980 showed that the presence of the endothelial lining is obligatory for the relaxant action of acetylcholine (ACh) on aortic and arterial strips or rings isolated from rabbits. Some other substances which release EDRF from EC include BK, angiotensin, histamine, noradrenaline, 5-hydroxytryptamine, calcium ionophore A23187, adenine nucleotides, thrombin, AA and leukotrienes [11]. Pulsatile pressure [12], visible light and electrical field stimulation also release and more EDRF is formed in arteries than in veins [13].

Vascular smooth muscle is the obvious target for the biological action of EDRF, but it also potently inhibits platelet aggregation and adhesion [14, 15]. EDRF stimulates guanylate cyclase resulting in an increase in intracellular cGMP.

The chemical structure of EDRF has been intensively investigated. EDRF released from cultured porcine EC and from EC of bovine pulmonary artery and vein has been identified as NO [15, 16]. NO is formed in EC from either L-arginine [17] or arginine derivatives by an enzymatic pathway, possibly a peptidyl arginine deiminase and an oxygenase [18]. Both EDRF and NO showed similar anti-aggregatory and anti-adhesive activity for platelets and there was synergism between the anti-aggregatory effect of prostacyclin and subthreshold concentrations of EDRF or NO.

The Release of Prostacyclin and EDRF Is Coupled

Substances which generate prostacyclin also largely stimulate the release of EDRF and we have proposed that the receptor-mediated release of prostacyclin and EDRF is coupled [19].

Our experiments suggest that release of EDRF as well as prostacyclin is due to activation of a phospholipase C (PLC)-mediated mechanism of transduction. Activation of a PLC by BK or ADP provides a common transduction mechanism for the generation of both prostacyclin and EDRF in EC. This is mediated by increased inositol phospholipid metabolism and activation of a protein kinase C.

The DAG kinase inhibitor, R59022, had two effects: increased basal release of both EDRF and prostacyclin, and thereafter, a substantially reduced release induced by BK and ADP. This enhancement in basal release of both substances reinforces our view on the coupling of EDRF and prostacyclin release. The reduction of receptor-stimulated release would follow from increased protein kinase C activity.

How is the release of EDRF coupled to that of prostacyclin? Clearly, ligand/receptor interactions are a common step and an increase in cytosolic Ca^{++} is needed for release of both EDRF and prostacyclin, so this may also be a common trigger. Interestingly, endothelin, the potent vasoconstrictor peptide made by EC, strongly releases prostacyclin and EDRF, thus limiting its own action in the circulation [20].

Our conclusion that activation of the same receptors ultimately leads to release of both EDRF and prostacyclin suggests that these substances act in concord as a common mechanism of defence for the EC. The synergism already shown between EDRF (NO) and prostacyclin in preventing platelet activation then takes on a new dimension, especially if there is also concerted and synergistic action against interaction of other circulating cells with the EC. Indeed, neutrophils and mononuclear cells reinforce this endothelial function by elaborating a vasorelaxant and anti-aggregatory factor indistinguishable from EDRF [21–23].

Acknowledgement

The William Harvey Research Institute is supported by a grant from Glaxo Group Research Ltd.

References

1   Moncada S, Vane JR: Pharmacology and endogenous roles of prostaglandin endoperoxides, thromboxane A_2 and prostacyclin. Pharmacol Rev 1979;30:293–331.
2   Flower RJ: Lipocortin and the mechanism of action of the glucocorticoids. Br J Pharmacol 1988;94:987–1015.
3   Szczeklik A, Gryglewski RJ: Low density lipoproteins (LDL) are carriers for lipid peroxides and invalidate prostacyclin (PGI_2) biosynthesis in arteries. Artery 1970;7:489–491.
4   Niżankowski R, Królikowski W, Bielatowicz J, et al: Prostacyclin for ischaemic ulcers in peripheral arterial disease: a random assignment, placebo-controlled study; in Gryglewski RJ, Szczeklik A, McGiff JC (eds): Prostacyclin – Clinical Trials. New York, Raven Press, 1985, pp 15–22.
5   Simpson PJ, Lucchesi BR: Myocardial ischemia: the potential therapeutic role of prostacyclin and its analogues; in Gryglewski RJ, Stock G (eds): Prostacyclin and Its Stable Analogue Iloprost. Berlin, Springer, 1987, pp 179–194.
6   Higenbottam T: The place of prostacyclin in the clinical management of primary pulmonary hypertension. Am Rev Respir Dis 1987;136:782–785.
7   Buchanan MR, Butt RW, Magas Z, et al: Endothelial cells produce a lipoxygenase-derived chemorepellent which influences platelet endothelial cell interactions – effect of aspirin and salicylate. Thromb Haemost 1985;53:306–311.
8   Buchanan MR, Haas TA, Lagarde M, et al: 13-Hydroxyoctadecadienoic acid is the vessel wall chemorepellent factor (LOX). J Biol Chem 1985;260:16056–16059.
9   Haas TA, Bastida E, Nakamura K, et al: Binding of 13-HODE and 5-, 12- and 15-HETE to endothelial cells and subsequent platelet, neutrophil and tumour adhesion. Eur J Pharmacol 1988;961:153–159.
10 Galton SA, Sneddon JM, Vane JR: The formation of 13-hydroxyoctadecadienoic acid (13-HODE) by endothelial cells (EC) from different species. Br J Pharmacol 1988;93:223P.

11 Furchgott RF: The role of endothelium in the responses of vascular smooth muscle to drugs. Ann Rev Pharmacol Toxicol 1984;24:175–197.

12 Pohl U, Holtz J, Busse R, et al: Crucial role of endothelium in the vasodilator response to increased flow in vivo. Hypertension 1986;8:37–44.

13 Lüscher TF, Deiderich D, Siebermann R, et al: Difference between endothelium-dependent relaxation in arterial and in venous coronary bypass grafts. N Engl J Med 1988; 319:462–467.

14 Sneddon JM, Vane JR: Endothelium-derived relaxing factor reduces platelet adhesion to bovine endothelial cells. Proc Natl Acad Sci USA 1988;85:2800–2804.

15 Moncada S, Radomski MW, Palmer RMJ: Endothelium-derived relaxing factor. Identification as nitric oxide and role in the control of vascular tone and platelet function. Biochem Pharmacol 1988;37:2495–2501.

16 Ignarro LJ, Byrns RE, Buga GM, et al: Endothelium-derived relaxing factor from pulmonary artery and vein possesses pharmacological and chemical properties identical to those of nitric oxide radical. Circ Res 1987;61:866–879.

17 Palmer RMJ, Ashton DS, Moncada S: Vascular endothelial cells synthesize nitric oxide from L-arginine. Nature 1988;333:664–666.

18 Thomas G, Ramwell PW: Peptidyl arginine deiminase and endothelium-dependent relaxation. Eur J Pharmacol 1988;153:147–148.

19 De Nucci G, Gryglewski RJ, Warner TD, et al: Receptor-mediated release of endothelium-derived relaxing factor and prostacyclin from bovine aortic endothelial cells is coupled. Proc Natl Acad Sci USA 1988;85:2334–2338.

20 De Nucci G, Thomas R, D'Orleans-Juste P, et al: The pressor effects of circulating endothelin are limited by its removal in the pulmonary circulation and by the release of prostacyclin and EDRF. Proc Natl Acad Sci USA 1988;85:9797–9800.

21 Rimele TJ, Sturm RJ, Adams LM, et al: Interaction of neutrophils with vascular smooth muscle: identification of a neutrophil-derived relaxing factor. J Pharmacol Exp Ther 1988;245:102–111.

22 Cynk E, Kondo K, Salvemini D, et al: Human monocytes and neutrophils inhibit platelet aggregation by releasing an EDRF-like factor. J Physiol (Lond) 1988;407:28P.

23 Salvemini D, Sneddon JM, Kondo K, et al: 'EDRF' released from human neutrophils and prostacyclin act synergistically to inhibit platelet aggregation. Br J Pharmacol 1988;95:728P.

Regina Botting, MD, The William Harvey Research Institute, St. Bartholomew's Hospital Medical College, Charterhouse Square, GB-London EC1 6BQ (UK)

Zor U, Naor Z, Danon A (eds): Leukotrienes and Prostanoids in Health and Disease.
New Trends Lipid Mediators Res. Basel, Karger, 1989, vol 3, pp 236–239

Endothelial Dependence of the Vasodilation to Arachidonic Acid in Rat Skeletal Muscle Arterioles

Akos Koller, Edward J. Messina, Michael S. Wolin, Gabor Kaley

Department of Physiology, New York Medical College, Valhalla, N.Y., USA

Our previous studies have been directed toward the characterization of the actions of arachidonic acid and its metabolites and their possible role in the regulation of vascular tone in the rat cremasteric microcirculation [1, 2]. In this preparation, topical application of arachidonic acid produces a concentration-dependent vasodilation that is eliminated by cyclooxygenase inhibition [3]. We have also shown that prostaglandin $(PG)E_2$ and PGI_2 are potent vasodilators [4, 5] and that $PGF_{2\alpha}$, the thromboxane A_2 mimic U46619, and leukotriene $(LT)C_4$ and LTD_4 are potent vasoconstrictors [6] of cremasteric arterioles. In addition, we have demonstrated that isolated microvessels obtained from the cremasteric musle synthesize PGE_2 and PGI_2 [7]. However, the cellular source of the vasodilator prostaglandins produced by topical application of arachidonic acid to this in vivo preparation remains to be established. In this study, we utilized local injury of endothelial cells with a light/dye technique, to determine if the vasodilator prostaglandins originate from this cell type upon arachidonic acid administration.

Methods

The left cremaster musle from 5- to 6-week-old male Wistar rats, anesthetized with sodium pentobarbital (30 mg/kg), were prepared for observation of third order arterioles (13–25 μm) by in vivo television microscopy, as previously described [4]. The internal diameter of arterioles was measured with an image-shearing monitor (IPM Model 907). The muscle was continuously suffused with bicarbonate-buffered (pH 7.35) Ringer-gelatin solution and maintained at 33.5 °C. All vasoactive agents were applied topically to the arteriole in 100 μl aliquots.

Light/dye injury of a 50–100 μm segment of the endothelium was produced by the exposure of the arteriole under study to filtered mercury light, (peak at 490 nm) for 100–140 s, in the presence of intravascular sodium fluorescein [8].

Results and Discussion

The vasodilator response of an arteriole to the application of 50 μM arachidonic acid, is shown in figure 1 (top). In these studies, adenosine was employed as a non endothelium-dependent control dilator agent since it is a direct relaxant of vascular smooth muscle [9]. Five minutes after light/dye injury, the arteriolar dilation to arachidonic acid is markedly attenuated (fig. 1, bottom). In contrast, the response to adenosine is unaffected. Normal reactivity of the same arteriole to arachidonic acid was maintained proximal and distal to the site of injury. It can also be seen in this figure that the basal diameter of the arteriole under study did not change markedly after light/dye injury.

The inhibition of vasodilation to topical application of arachidonic acid by the light/dye method, without altering the response to adenosine, suggests that the endothelium is the most likely source of the synthesis of vasodilator prostaglandins. As the vasodilation to adenosine was not affected by the light/dye intervention, it can be surmised that the reactivity of the arteriolar smooth muscle was not altered by light/dye injury. The

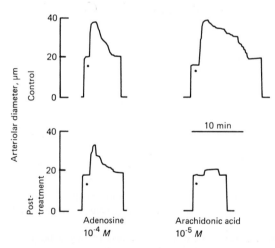

Fig. 1. A typical experiment demonstrating the effect of light/dye treatment on the dilator responses of a single arteriole (20 μm) to adenosine and arachidonic acid. In this record of the internal diameter of the blood vessel, the initial upswing of the electrogram measures the baseline diameter. At the dot symbol, after measurement of the control diameter, the agents were topically applied to the cremaster muscle. Dilation is measured as the increase in diameter above the baseline. As the response fades, the blood vessel diameter returns to control and then the electrogram is zeroed once more. The top two panels are those before endothelial injury by light/dye treatment, the bottom panels after. Note the significant inhibition of the response to arachidonic acid.

basal diameter of arterioles was slightly reduced after endothelial injury by light/dye treatment, suggesting that prostaglandins may play some role in the control of resting microvascular tone under these experimental conditions. We have also observed in this preparation that the arteriolar dilation to the calcium ionophore, A23187, a response which is mediated through prostaglandin generation [9] is inhibited by light/dye treatment. Thus, the endothelium appears to be involved in the mediation of prostaglandin-dependent vascular responses.

The light/dye method employed in the present study was originally used in the cerebral microcirculation by Rosenblum [8] to study vasodilation to acetylcholine. In the cerebral preparation, the responses to acetylcholine were inhibited by the light/dye injury suggesting that responses mediated by the cyclic GMP-associated endothelium-derived relaxing factor (EDRF) are also inhibited by this treatment.

These findings, as well as ours, indicate that responses of arterioles to various stimuli can be mediated through the formation by the endothelium of both vasodilator prostaglandins and EDRF. The present results further emphasize the importance of the role of microvascular endothelium in the mediation of the effects of various naturally-occurring vasocative substances as well as in the regulation of local blood flow in normal and pathological conditions.

References

1   Messina, E.J.; Weiner, R.; Kaley, G.: Inhibition of bradykinin vasodilation and potentiation of norepinephrine and angiotensin vasoconstriction by inhibitors of prostaglandin synthesis in rat skeletal muscle. Circ. Res. *37:* 430–437, 1975.
2   Messina, E.J.; Weiner, R.; Kaley, G.: Arteriolar reactive hyperemia: modification by inhibitors of prostaglandin synthesis. Am. J. Physiol. *6:* H571–H575 (1977).
3   Messina, E.J.; Rodenburg, J.; Slomiany, B.L.; Roberts, A.M.; Hintze, T.H.; Kaley, G.: Microcirculatory effects of arachidonic acid and a prostaglandin endoperoxide (PGH_2). Microvasc. Res. *19:* 288–296 (1980).
4   Messina, E.J.; Weiner, R.; Kaley, G.: Microvascular effects of prostaglandins E_1, E_2, and A_1 in the rat mesentery and cremaster muscle. Microvasc. Res. *8:* 77–89 (1974).
5   Messina, E.J.; Kaley, G.: Microcirculatory responses to PGI_2 and PGE_2 in the rat cremaster muscle. Adv. Prostaglandins Thromboxane Leukotriene Res. *7:* 719–722 (1980).
6   Messina, E.J.; Rodenburg, J.; Kaley, G.: Microcirculatory effects of leukotrienes, LTC_4, LTD_4, in rat cremaster muscle. Microcirc. Endothelium Lymphatics *5:* 355–376 (1988).
7   Myers, T.O.; Messina, E.J.; Rodrigues, A.M.; Gerritsen, M.E.: Altered aortic and cremaster muscle prostaglandin synthesis in diabetic rats. Am. J. Physiol. *249:* E374–E379 (1985).

8   Rosenblum, W.I.: Endothelial dependent relaxation demonstrated in vivo in cerebral arterioles. Stroke *17:* 494–497 (1986).

9   Kaley, G.; Rodenburg, J.M.; Messina, E.J.; Wolin, M.S.: Endothelium-associated vasodilators in rat skeletal muscle microcirculation. Am. J. Physiol. *256:* H720–H725 (1989).

Akos Koller, MD, Department of Physiology, New York Medical College, Valhalla, NY 10595 (USA)

Zor U, Naor Z, Danon A (eds): Leukotrienes and Prostanoids in Health and Disease.
New Trends Lipid Mediators Res. Basel, Karger, 1989, vol 3, pp 240–245

Endothelial Perturbation by Ionizing Irradiation: Effects on Prostaglandins, Lipid Chemoattractants and Mitogens

Amiram Eldor[a], Yaacov Matzner[a], Zvi Fuks[c], Larry D. Witte[d], Israel Vlodavsky[b]

Departments of [a]Hematology and [b]Oncology, Hadassah University Hospital, Jerusalem, Israel; [c]Department of Radiation Oncology, Memorial Sloan-Kettering Cancer Center, New York, N.Y., and [d]The Arteriosclerosis Research Center and Department of Medicine, Columbia University, College of Physicians and Surgeons, New York, N.Y., USA

Introduction

Early radiation injury is characterized by vascular damage and the initial site of damage appears to be the endothelial cell (EC) lining of the vessel wall [1]. Chronic irreversible tissue reactions to radiation include thrombotic occlusion of capillaries, enhanced atherosclerosis in larger blood vessels, inflammatory changes and late tissue fibrosis [2]. These processes may be mediated by EC products released as a result of cellular injury.

We have investigated the effects of ionizing irradiation on the capacity of cultured bovine aortic EC to (1) synthesize prostacyclin (PGI_2), a potent mediator involved in the defence against thrombosis and arteriosclerosis; (2) release of chemotactic factors which may be involved in postirradiation inflammatory reactions, and (3) release of mitogens which may be operative in tissue repair and in late radiation-induced fibrosis.

Materials and Methods

Cloned populations of EC were obtained from bovine aortic arch and cultured as previously described [3, 4]. Cells were irradiated after having formed a confluent and contact-inhibited cell monolayer. The medium was always replaced with a serum-free medium prior to irradiation. Cells were irradiated at approximately 1 Gy/min with a Philips 250 keV X-ray apparatus (HVL 0.5 μg Cu at SSD of 50 cm). During irradiation the cells were maintained at room temperature for a maximum of 10 min. Nonirradiated cells, which served as controls, were subjected to the same conditions as the irradiated cells. Following irradiation, the culture medium was removed, centrifuged, and assayed for the stable metabolite of PGI_2, 6-keto-$PGF_{1\alpha}$ (6KF) [3], neutrophil chemotactic activity [5] and mitogenic activity [6].

Results and Discussion

Release and Stimulated Production of PGI_2

Evaluation of the radiation effects on PGI_2 was performed in two ways: (a) measurement of the net accumulation of PGI_2 in the culture medium, and (b) measurement of the amount of PGI_2 produced by cells following a short stimulation (2 min) with arachidonic acid (20 μM). The net unstimulated accumulation of PGI_2 in the culture medium reflects the amount of PGI_2 produced by damaged and lysed cells and is observed with other types of cellular injury (e.g. freezing and thawing) [3]. The stimulated production of PGI_2 was assayed following removal of nonadherent cells from the cultures. The amounts of PGI_2 synthesized by the remaining adherent cells reflect the capacity of metabolically active and viable EC to produce PGI_2. As previously reported [3], radiation injury was associated with an increased accumulation of PGI_2 (measured as 6KF) in the culture medium, while the capacity of the viable EC to synthesize PGI_2, decreased (table 1). We have suggested that the decrease in PGI_2 production capacity is caused by damage to the enzymes involved in prostaglandin production [3].

Radiation damage is associated with oxidant stress and the production of free radicals. We therefore tested the ability of an oxygen radical scavenger, vitamin C, to protect the capacity of irradiated EC to produce PGI_2 [4]. Pretreatment of EC with low concentrations of vitamin C inhibited the radiation-induced release of PGI_2 into the culture medium (table 1). Vitamin C also enhanced the capacity of irradiated EC to produce PGI_2 upon stimulation with arachidonic acid (table 1). Treatment with this scavenger, however, did not protect the cells against the cytopathic effects of radiation [4]. Hence, vitamin C may protect the enzymes of the arachidonic acid cascade from the injurious effects of oxidative radicals.

Release of Chemotactic Activity

Irradiation of confluent EC cultures was associated with a rapid release into the culture medium of a chemotactic activity for human neutrophils, which was dependent both on the dose of radiation and the time between irradiation and sample collection (table 2). Significant chemotactic activity was detected 10 min after irradiation (10 Gy) and maximal activity was observed after 1 h. At this time interval, there were no

Table 1. Effect of vitamin C on PGI_2 production by irradiated EC: *(A)* PGI_2 accumulation in the culture medium, and *(B)* PGI_2 production capacity.

Vitamin C μg/ml		6KF, ng/ml		
		0 Gy	5 Gy	10 Gy
A	0	12.5 ± 3.5	14.6 ± 1.4	38.7 ± 8.8
	10	5.0 ± 0.9	6.3 ± 0.8	13.4 ± 2.1
	100	4.4 ± 0.5	6.4 ± 0.5	9.8 ± 0.6
		6KF, ng/10^5 cells		
		0 Gy	5 Gy	10 Gy
B	0	2.18 ± 0.6	0.69 ± 0.1	0.28 ± 0.01
	10	3.32 ± 1.0	1.0 ± 0.3	0.20 ± 0.03
	100	3.25 ± 0.9	1.6 ± 0.2	0.34 ± 0.09

A EC were pretreated with vitamin C and exposed to different radiation doses. The culture medium was collected 48 h later and assayed for 6KF. Five dishes were used for each data point and results are presented as mean $\pm$ SD.

B Cells were pretreated with vitamin C (30 min, 37 °C) and exposed to irradiation. 48 h later, the culture medium was removed and the cultures were washed twice with fresh culture medium to remove all nonadherent cells. Adherent cells which were viable (as shown by trypan blue exclusion) were stimulated with arachidonic acid (20 μM, 2 min, 37 °C) and assayed for 6KF. The cells were then dissociated and counted. Five dishes were used for each point and results are presented as mean $\pm$ SD.

Table 2. Generation of neutrophil chemotactic activity by EC monolayers

Radiation dose, Gy	Chemotaxis, μm		
	10 min	1 h	24 h
0	4	12	24
10	20	42	41
40	33	60	62
40 + ASA	36	62	64
40 + NDGA	12	24	33

EC cultures in serum-free medium were irradiated and incubated in the presence or abscence of either aspirin (ASA 100 μM) or NDGA (10 μg/ml) at 37 °C. The EC-conditioned medium was assayed in Boyden chambers for chemotactic activity as described [5]. The values for chemotaxis were corrected for random migration of 42 μm. The results are representative of 8 experiments.

apparent morphological changes detected in EC which remained viable and attached to the tissue culture plate. Hence, the rapid release of chemotactic activity from the irradiated EC was not due to cytolysis or loss of cell viability. This was also supported by the observation that there was no detectable chemotactic activity in supernatants obtained from EC that were lysed by three cycles of freezing and thawing. The chemotactic activity was characterized as a 5-lipoxygenase product [5]. It could not be destroyed by boiling or trypsin digestion and pretreatment of the irradiated EC with the lipoxygenase inhibitor NDGA resulted in a significant decrease of the released chemotactic activity. Pretreatment of the EC with the cyclooxygenase inhibitor, aspirin, had no effect. Further investigations indicated that this chemotactic activity is an arachidonic acid product which exhibits migration pattern on TLC and HPLC similar to that of leukotriene B_4 (LTB_4). However, LTB_4 could not be detected in the culture medium by a highly sensitive radioimmunoassay [5]. This rapidly released lipid chemoattractant, together with a chemotactic protein released 72 h after irradiation [7], may play a significant role in the acute and chronic inflammatory response resulting from the irradiation damage.

Release of Mitogenic Activity

Cultured bovine and human EC have the capacity to produce a variety of mitogens including platelet-derived growth factor (PDGF) and basic fibroblast growth factor (bFGF) [8, 9].

By using an assay of thymidine incorporation into the DNA of human fibroblasts [6], we found increased amounts of mitogenic activity in the conditioned medium of irradiated EC (table 3). The amounts of the released mitogenic activity were dependent on the dose of irradiation and the time that elapsed from the radiation injury. Only small quantities of

Table 3. Effect of irradiation on the release of mitogenic activity

Radiation dose, Gy	Mitogenic activity, units/dish
0	63
20	139
40	192
60	208
Lysed EC (1.1×10^6 cells/dish)	< 10

Confluent EC were subjected to ionizing irradiation. 48 h later the culture medium was collected and assayed for induction of ^{3}H-thymidine incorporation in 3T3 fibroblasts. One unit of mitogenic activity was defined as the amount of EC-conditioned medium required to produce 50% labelled nuclei in 3T3 cells [6].

mitogenic activity were detected following rapid lysis (freezing and thaw-ing) of nonirradiated EC. These results suggest that the enhanced mitogenic activity observed following irradiation of EC is due to de novo synthesis and release of growth-promoting factors from viable and metabolically active endothelial cells.

It has previously been shown that cultured EC secrete PDGF and an enhanced release of this mitogen was observed during activation of the coagulation process [10]. It is therefore conceivable that the release of this protein may be enhanced by radiation. Preliminary results from our laboratory indicate that conditioned medium of irradiated EC also stimu-lates the proliferation of bovine aortic EC, which are responsive to FGF but not to PDGF [9]. These and other results (i.e. sensitivity to heat and dithiothreitol) suggest that the mitogenic activity released by irradiated EC includes both PDGF and FGF-like mitogens.

The mitogens released by EC following irradiation may be involved in the pathogenesis of both early vascular damage and of the fibrosis which represents a prominent feature of the late radiation injury in normal tissues.

Acknowledgements

The authors are grateful to E. HyAm, R. Drexler and R. Atzmon for their excellent technical assistance. This work was supported by the Israel Atomic Energy Commission and the Council for Higher Education of Israel (A.E., I.V.), by NIH grant RO1-CA 30289 (I.V.), NIH grant HC 21006 (SCOR in Arteriosclerosis (L.D.W.), and the Kovshar Foundation (Y.M.).

References

1   Law MP: Radiation induced vascular injury and its relation to late effects of normal tissues. Adv Radiat Biol 1981;9:37–93.
2   Fajardo LF: Radiation induced coronary artery disease. Chest 1977;71:563–564.
3   Eldor A, Vlodavsky I, HyAm E, et al: The effect of radiation on prostacyclin (PGI₂) production by cultured endothelial cells. Prostaglandins 1983;25:263–279.
4   Eldor A, Vlodavsky I, Riklis E, et al: Recovery of prostacyclin production capacity of irradiated endothelial cells and the protective effect of vitamin C. Prostaglandins 1987;34:241–255.
5   Matzner Y, Cohn M, HyAm E, et al: Generation of lipid neutrophil chemoattractant by irradiated bovine aortic endothelial cells. J Immunol 1988;140:2681–2685.
6   Witte LD, Cornicelli JA, Miller RW, et al: Effects of platelet-derived and endothelial cell-derived growth factors on the low density lipoprotein receptor pathway in cultured human fibroblasts. J Biol Chem 1982;257:5392–5401.
7   Dunn MM, Drab EA, Rubin DB: Effects of irradiation on endothelial cell-polymor-phonuclear leukocyte interactions. J Appl Physiol 1986;60:1932–1941.

8   Fox PL, DiCorleto PE: Regulation of production of a platelet-derived growth factor-like protein by cultured bovine aortic endothelial cells. J Cell Physiol 1984;121:298–308.

9   Vlodavsky I, Fridman R, Sullivan R, et al: Aortic endothelial cells synthesize basic fibroblast growth factor which remains cell associated and platelet-derived growth factor which is secreted. J Cell Physiol 1987;131:402–408.

10   Harlan JM, Thompson PJ, Ross RR, et al: Thrombin induces release of platelet-derived growth factor-like molecules by cultured human endothelial cells. J Cell Biol 1986;103:1129–1133.

Amiram Eldor, MD, Department of Hematology, Hadassah University Hospital, PO Box 12000, Jerusalem 91120 (Israel)

Zor U, Naor Z, Danon A (eds): Leukotrienes and Prostanoids in Health and Disease.
New Trends Lipid Mediators Res. Basel, Karger, 1989, vol 3, pp 246–251

Cytoprotection by Prostacyclins in Myocardial Ischemia

Karsten Schrör

Institut für Pharmakologie der Heinrich-Heine-Universität, Düsseldorf, FRG

Introduction: *Definition of Cytoprotection*

'Cytoprotection' was introduced by Robert et al. [10] to separate an unique property of several prostaglandins in protecting the rat stomach mucosa against noxious stimuli from their antisecretory, i.e. specific and receptor-mediated actions. Later work has provided evidence for protection of the cell architecture from different types of injury in other tissues as well. Using prostacyclin or iloprost, cytoprotective effcts have been observed in brain tissue [9], hepatic cells [2], adrenergic nerve terminals [12, 13] and cardiac myocytes [1].

In this overview, cytoprotective actions of prostacyclin (including iloprost) and PGE_1 will be discussed for the ischemic myocardium. For the purpose of this discussion, cytoprotection will be defined as follows: Selective improvement of morphological integrity and functional behavior of the ischemic myocardium in absence of specific effects on the non-ischemic myocardium.

This definition has two important implications: First, cytoprotective actions are not restricted to the myocyte but might also include other cellular constituents of the heart as an organ, i.e. adrenergic nerve terminals and the vascular endothelium. Consequently, prevention of ischemia-induced accumulation of inflammatory cells in ischemic areas might also contribute to a cytoprotective effect for the jeopardized tissue. Second, cytoprotection is restricted to the injured parts of the heart and there is no direct effect on cell function in undamaged areas. In consequence, this would suggest that cytoprotection is not a specific response, i.e. not receptor-mediated. To understand how cytoprotection operates, one has to ask for a generalized phenomenon occurring specifically in injured tissue which is independent of the type of noxious stimulus and can be selectively affected by prostacyclins.

Ischemia-Induced Phospholipid Breakdown in Myocardial Ischemia –
A Target for Cytoprotection

Release of free fatty acids, including arachidonic acid (AA), from
membrane phospholipids is a generally accepted sign for cell membrane
destruction. The amount liberated depends upon the intensity and duration
of the noxious stimulus [3]. In contrast to the physiological situation, this
release of AA is uncontrolled and the free acid is no longer quantitatively
converted into eicosanoids but may now accumulate and exert actions by
its own. In the more complex in vivo situation, the (secondary) local
accumulation of inflammatory cells and platelets as well as interactions
between them, including 'precursor exchange', allow for the biosynthesis of
AA metabolites which are not formed in the nonischemic heart and merely
consist of compounds that aggravate cell injury [11].

One important problem arising is: How can the jeopardized tissue
protect itself from irreversible destruction by an uncontrolled action of
these 'self-made' proinflammatory mediators? Clearly, this may not be
expected from any specific, i.e. receptor-mediated response since the membrane architecture including receptor sites might have already been destroyed. Furthermore, this activity should be inhibitory, i.e. prevent further
phospholipid breakdown with the consequence of improved membrane
resistance and less fatty acid release and subsequent mediator formation. It
is also logical to suggest that this action will be selective for the injured
tissue and should not affect uninjured cells in the vicinity.

Reduced Prostacyclin Formation Despite Enhanced Eicosanoid
Release during Ischemia

Subsequent to injury-induced membrane disruption and AA release,
there is an enhanced generation of eicosanoids in all tissues studied so far.
In the heart, all major classes of eicosanoids have been detected in the
coronary effluent during ischemia or hypoxia and most of them are well
known to exert stimulatory action on cell function (thromboxane A_2,
HETEs, leukotrienes) [11]. In contrast to this enhanced generation of
stimulatory mediators, there is only a transient increase or even a reduced
formation of mediators that inhibit cellular activity, such as PGI_2. Green et
al. [5] have recently reported that 6/20 patients suffering an acute myocardial infarction failed to increase their PGI_2 production. This finding and
data from animal studies [14] suggest that endogenous PGI_2 production
might not be sufficient to exert an optimum beneficial effect on the injured
myocardium. Since this PGI_2 release occurs secondary to tissue injury, it

might reflect the stimulation of undamaged endothelium by products formed in the injured tissue. Interestingly, neither the myocytes nor any inflammatory cell such as neutrophils, platelets or macrophages are capable of significant PGI_2 production. In addition to a reduced capacity of injured endothelial cells to synthesize PGI_2, metabolites such as reactive oxygen species or fatty acid peroxides are generated in the ischemic myocardium which rapidly and selectively inactivate the prostacyclin synthase [11]. It is, therefore, conceivable to assume that the injured tissue needs exogenous supply with this tissue-protective eicosanoid which cannot be formed in sufficient amounts to prevent tissue destruction and to antagonize the bulk of proinflammatory mediators.

Direct Cytoprotective Actions of Prostacyclins in the Ischemic Myocardium

In 1979 and 1980, Lefer's group originally demonstrated that PGI_2 significantly improved the function of ischemic cat hearts [see 11, 14]. This was associated with a considerable improvement of biochemical indices of myocardial dysfunction. Cardioprotective effects of PGI_2, iloprost and PGE_1 were reported in numerous later studies on acute experimental myocardial ischemia. In contrast, inhibition of cyclooxygenase activity by indomethacin-type compounds was found to aggravate ischemic injury [11–17]. PGI_2 or indomethacin, given at the same doses to nonischemic hearts, did not affect these parameters. This suggests that PGI_2, formed endogenously in the vessel wall might antagonize for some extent tissue injury. However, the local concentration is not sufficient to exert a full productive effect. Probably, the cytoprotective action of PGI_2 and its analogs is due to direct membrane stabilization, resulting in less phospholipid cleavage and free fatty acid accumulation. This has been demonstrated by Darius et al. [4] in the ischemic rat heart (table 1).

Cytoprotective Effects of Prostacyclins in the Ischemic Myocardium via Inhibition of White Cells and Platelets

In addition to direct effects on the jeopardized cardiac myocyte, resulting in reduced loss of intracellular large molecules (purines, CK, SOD) and improved resistance against oxygen toxicity, prostacyclins exhibit a number of inhibitory actions on inflammatory cells and platelets, eventually resulting in reduced mediator release. PGI_2 [16], iloprost [11, 14]

Table 1. Phospholipid concentration (nmol P_i/mg protein) in ischemic (MI) and nonischemic (NMI) areas of rat hearts subjected to 6 h of LAD occlusion and treatment with iloprost (ILO) or vehicle (VEH)

Group	n	Total PL		PC		PE	
		MI	NMI	MI	NMI	MI	NMI
OP – VEH	7	256 ± 13*	286 ± 12	99 ± 4*	113 ± 4	86 ± 4*	96 ± 4
OP – ILO	8	300 ± 7	289 ± 4	112 ± 2	116 ± 3	103 ± 2	104 ± 2

Treatment with iloprost (100 ng/kg · min, i.v.) was started 20 min after LAD occlusion [4].
* $p < 0.05$ (MI vs. NMI).

and PGE_1 [15, 16] have been shown to prevent ischemia-induced leukocytosis and to inhibit neutrophil accumulation in the borderzone of ischemic myocardial tissue. Neither of these compounds exhibited any direct neutrophil-inhibitory actions in vitro at 'therapeutic' plasma levels. This suggests that the prevention of leukocyte activation results from inhibition of chemotactic mediator release from the injured myocardial tissue subsequent to improved tissue preservation [15]. In this context it is important to note that this anti-inflammatory action of prostacyclins does not retard the healing process, i.e. there is no reduced scar formation [7].

A great number of studies are available demonstrating inhibition of ischemia-induced platelet activation by prostacyclins [11, 14]. In vitro data suggest that this inhibitory action of prostacyclin occurs at the level of Ca^{++}-dependent, phospholipase-induced AA release [8]. However, prostacyclins also protected platelet- (and leukocyte)-free perfused hearts, made globally ischemic in vitro [13]. In pigs, selective stimulation of endogenous PGI_2 production reduces the infarct size without modifying the platelet count [6]. Thus, prevention of ischemia-induced platelet and neutrophil activation, eventually caused by interference with the release of platelet-stimulating and chemotactic factors from the injured myocardium, might contribute to the beneficial effects of the compounds but appears not to be an essential component for myocardial tissue protection.

Cytoprotective Effects of Prostacyclins in the Ischemic Myocardium via Inhibition of Ischemia-Induced Catecholamine Redistribution

Cardiac noradrenaline is located in the perivascular adrenergic nerve fibers. During ischemia, adrenergic nerve terminals lose their noradrenaline-accumulating activity with the consequence of catecholamine redistribution from the adrenergic neurons into the extraneuronal space, i.e. the ischemic myocardium [14]. Prostacyclins prevent this ischemia-induced catecholamine redistribution [12, 13] resulting in a complete inhibition of ischemia-induced increase of myocardial tissue cAMP levels [12]. Again, these actions are selective for the ischemic myocardium. In the non-ischemic myocardium there is no change in nerve stimulation-induced catecholamine overflow [13] nor any alteration in myocardial tissue cAMP [12].

Thus, in addition to direct cytoprotective actions on myocellular membrane integrity, prostacyclins also exhibit a number of indirect inhibitory effects on inflammatory cells and protect adrenergic nerve terminals from ischemic injury. These actions on platelets and white cells appear to be secondary to cardiac membrane destructions and AA release.

Conclusions

PGI_2, iloprost and PGE_1 improve tissue preservation in myocardial ischemia. This is due to a membrane-stabilizing activity (reduced phospholipid breakdown) and results in a reduced loss of large intracellular molecules (purines, CK, SOD) from the injured myocytes. This protection of injured cells from ischemic and hypoxic damage is a unique property of prostacyclins and has also been demonstrated for other tissues (liver, brain, kidney, adrenergic nerves).

The cytoprotective action of prostacyclins involves effects on other cells and mediator systems. There is reduced accumulation of neutrophils in the borderzone of ischemic myocardium, inhibition of ischemia-induced platelet activation and of the ischemia-induced catecholamine redistribution in the ischemic parts of the myocardium but no improved collateral perfusion. These actions are restricted to the ischemic myocardium and its vicinity and, therefore, might be secondary to improved tissue preservation.

References

1   Araki H, Lefer AM: Role of prostacyclin in the preservation of ischemic myocardial tissue in the perfused cat heart. Circ Res 1980;47:757–763.
2   Araki H, Lefer AM: Cytoprotective actions of prostacyclin during hypoxia in the isolated perfused cat liver. Am J Physiol 1980;238:176–181.
3   Chien KR, Sen A, Reynolds R, et al: Release of arachidonate from membrane phospholipids in cultured neonatal rat myocardial cells during adenosine triphoshpate depletion. Correlation with the progression of cell injury. J Clin Invest 1985;75:1770–1780.

4 Darius H, Osborne JA, Reibel DK, et al: Protective actions of a stable prostacyclin analog in ischemia induced membrane damage in rat myocardium. J Mol Cell Cardiol 1987;19:243–250.

5 Green K, Vesterqvist O, Rassmanis G, et al: Deficient prostacyclin formation after acute myocardial infarction. Lancet 1987;i:1037–1038.

6 Hohlfeld T, Thiemermann C, Schrör K: Protection from myocardial ischemic injury in cats and pigs by defibrotide – different mode of action? Abstr 4th Int Symp on Prostaglandins, 1988, p 43.

7 Jugdutt BI: Delayed effects of early infarct-limiting therapies on healing after myocardial infarction. Circulation 1985;72:907–914.

8 Lapetina EG, Schmitges CJ, Chandrabose K, et al: Cyclic adenosine 3′5′-monosphosphate and prostacyclin inhibit membrane phospholipase activity in platelets. Biochim Biophys Res Commun 1977;76:828–835.

9 Renkawek K, Herbaczynska-Cedro K, Mossakowski MJ: The effect of prostacyclin on the morphological and enzymatic properties of CNS cultures exposed to anoxia. Acta Neurol Scand 1986;73:111–118.

10 Robert A, Nezamis JE, Lancaster C, et al: Cytoprotection by prostaglandins in rats. Gastroenterology 1979;77:433–443.

11 Schrör K: Lipid metabolism in the normoxic and ischaemic heart. Basic Res Cardiol 1987;82(suppl 1):235–243.

12 Schrör K, Addicks K, Darius H, et al: PGI$_2$ inhibits ischemia-induced platelet activation and prevents myocardial damage by inhibition of catecholamine release from adrenergic nerve terminals. Evidence for cAMP as common denominator. Thromb Res 1981;21:175–180.

13 Schrör K, Funke K: Prostaglandins and myocardial noradrenaline overflow after sympathetic nerve stimulation during ischemia and reperfusion. J Cardiovasc Pharmacol 1985;7(suppl 5):50–54.

14 Schrör K, Smith EF III, Lefer AM: The cat as an in vivo model of myocardial infarction. Prog Pharmacol 1988;6:31–91.

15 Schrör K, Thiemermann C, Ney P: Protection of the ischemic myocardium from reperfusion injury by prostaglandin E$_1$ – inhibition of ischemia-induced neutrophil activation. Naunyn-Schmiedebergs Arch Pharmacol 1988;338:268–274.

16 Simpson PJ, Mickelson J, Fantone JC, et al: Reduction of experimental canine cardiac infarct size with prostaglandin E$_1$ – inhibition of neutrophil migration and activation. J Pharmacol Exp Ther 1988;244:619–624.

17 Simpson PJ, Mitsos SE, Ventura A, et al: Prostacyclin protects ischemic reperfused myocardium in the dog by inhibition of neutrophil activation. Am Heart J 1987;113:129–137.

Karsten Schrör, MD, Institut für Pharmakologie der Heinrich-Heine-Universität Düsseldorf, Moorenstrasse 5, D–4000 Düsseldorf 1 (FRG)

Zor U, Naor Z, Danon A (eds): Leukotrienes and Prostanoids in Health and Disease.
New Trends Lipid Mediators Res. Basel, Karger, 1989, vol 3, pp 252–256

Interactions of Leukotrienes, Platelet-Activating Factor and Thromboxane A_2 in Cardiac Anaphylaxis

Priscilla J. Piper, Hassan B. Yaacob, Julie D. McLeod

Department of Pharmacology, Hunterian Institute, Royal College of Surgeons of
England, Lincoln's Inn Fields, London, UK

Introduction

Anaphylactic shock in guinea pig isolated hearts is characterized by profound changes in cardiac function which include reduction in coronary flow and developed tension, arrhythmias, sinus tachycardia and left ventricular failure; this is accompanied by release of histamine, leukotrienes (LTs), platelet-activating factor (PAF) and thromboxane A_2 (TxA_2) [Capurro and Levi, 1975; Levi et al., 1984; Yaacob and Piper, 1988; Aehringhaus et al., 1984].

LTC_4 is the most potent cysteinyl-containing LT in causing an increase in coronary perfusion pressure (CPP) or reduction in coronary flow (CF) and cardiac developed tension (CDT) of guinea pig or rat perfused hearts [Letts and Piper, 1982]. PAF also increases CPP and reduces CDT, part of which is accounted for by LTC_4 [Piper and Stewart, 1986, 1987]. TxA_2 is also a powerful vasoconstrictor in the coronary circulation.

In this study we have investigated the generation and interaction of the phospholipid derived mediators, LTs, PAF and TxA_2, in cardiac anaphylaxis (CA) in guinea pig isolated hearts.

Methods

Hearts from male Dunkin-Hartley guinea pigs (300–350 g) were perfused via retrograde cannulation of the aorta with oxygenated Tyrode solution at 37 °C, either at constant flow (8 ml/min) or constant pressure. In some experiments hearts from animals previously sensitized to ovalbumin were used. CPP (mm Hg), CDT (g) and CF (ml/min) (as appropriate) were measured and coronary vascular resistance (CVR) calculated. LTs released into the cardiac effluent were either quantitated by bioassay on guinea pig ileum smooth muscle against LTC_4 [Samhoun and Piper, 1984] or by radioimmunoassay following HPLC. In some experiments indomethacin, CGS 8515 or WEB 2086 were administered to hearts 15 min prior to and during antigen challenge or injection of PAF or LTs. The LTD_4 antagonist L-660,711 [Jones et al., 1988] was given to hearts 15 min before and during injection of LTC_4 and LTD_4.

Materials

CGS 8515 (Ciba-Geigy), WEB 2086 (Boehringer), L-600,711 (Merck, Sharp & Dohme), indomethacin (Sigma), PAF (Bachem), LTC_4 (Miles Laboratories), anti-serum to TxB_2.

Results

Effects of PAF and LTs and Their Antagonism

Both PAF (50 pmol) and LTs cause potent reduction in CF and increase in CVR [Piper and Stewart, 1986, 1987; Letts and Piper, 1982]. The relative potency of LTs is $LTC_4 > LTD_4 > LTE_4$.

The PAF antagonist WEB 2086 (0.01–1.0 μM antagonized PAF-induced increase in CPP and decrease in CDT in a dose-dependent manner. In addition, WEB 2086 prevented PAF-induced generation of both LTC_4 and TxB_2 by 36 and 100%, respectively [Yaacob and Piper, unpubl. data].

The selective LTD_4 antagonist L-600,711 (2–19 μM) dose dependently antagonized the effect of LTD_4 on CVR and complete antagonism was obtained at 19 μM but at this dose L-660,711 only reduced the action of LTC_4-induced increase in CVR by 40%. The threshold concentration of L-660,711 on LTD_4 was 2 μM ($EC_{50} = 7.1$ μM) whereas a threshold concentration of this drug for LTC_4 was three times higher.

Antigen Challenge

Injection of ovalbumin into the coronary circulation of hearts from sensitized guinea pigs caused a biphasic increase in CPP [Yaacob and Piper, 1988] which remained elevated for more than 10 min and was accompanied by a reduction in CDT. The most severe phase of this reaction occurred at 1–2 min following challenge. LTs B_4, C_4, D_4 and E_4 were released into the coronary effluent. LTC_4 and LTD_4 were the major LTs generated, with smaller quantities of LTB_4 and LTE_4 being formed. The peak release of LTC_4 occurred at 2 min although LTs could be detected as early as 1 min and for longer than 7 min post-challenge. TxA_2 (3.8 pmol/min) was also released into the cardiac effluent, where the peak release occurred 1 min after challenge.

PAF (50 pmol) mimicked many of the features of CA causing an irreversible increase in CPP and reduction in CDT. These were accompanied by a release of LTC_4 (1.4 ± 0.2 pmol min^{-1}) and TxA_2 (1.2 ± 0.2 pmol min^{-1}).

Cyclo-Oxygenase Inhibition

Treatment of hearts with indomethacin ($2.8 \ \mu M$) inhibited anaphylactic release of TxA_2 ($n = 4$–6, $p < 0.05$) and exacerbated CA; the onset of the antigen-induced increase in CPP was significantly delayed but the magnitude markedly enhanced and the severity of dysrhythmias and reduction in CDT were increased. Generation of LTs was markedly potentiated (from 5.60 ± 1.36 to 7.75 ± 0.63 pmol min^{-1}) and the duration of release extended.

5-Lipoxygenase Inhibition

The 5-lipoxygenase inhibitor CGS 8515 ($1.0 \ \mu M$) markedly inhibited the antigen-induced output of LTs from sensitized hearts and also reduced the elevated release of LTs during CA in hearts treated with indomethacin (data not shown). The frequency of arrhythmias and the late phase (2–3 min) increase in CPP were also attenuated but this compound did not lessen the reduction in CDT and had no effect on the generation of TxA_2, reflecting its selectivity on 5-lipoxygenase [Yaacob and Piper, 1988].

Antagonism of PAF

WEB 2086 ($1.0 \ \mu M$) which totally blocked effects of PAF inhibited the total LT release during CA by 35% but did not affect the release of TxA_2. The antigen-induced increase in CPP was partially reduced by this dose of WEB 2086 ($n = 6$).

Discussion

The effects of LTs, TxA_2 and PAF on blood vessels and the microcirculation have been well documented. Although it is short acting, TxA_2 is a potent vasoconstrictor, while LTs and PAF exhibit selective effects on the cardiovascular system which include coronary vasoconstriction and a reduction in cardiac contractility. In guinea pig isolated perfused hearts the actions of PAF on CF and CDT was antagonized by WEB 2086. L-660,711 antagonized the action of LTD_4 on CVR and CDT. The effect of LTC_4 and CDT was slower in onset than that of LTD_4 [Letts and Piper, 1982] but on a molar basis this was unlikely to have been due to conversion of LTC_4 to LTD_4. Unlike its effect on CVR, the LTC_4-induced reduction in CDT was fully antagonized by L-660,711.

LTs, PAF and TxA_2 are mediators of CA but their contribution to the anaphylactic dysfunction have not been thoroughly investigated. In this study the use of indomethacin, CGS 8515 and WEB 2086 has demonstrated that an interaction exists between phospholipid-derived mediators in CA. These mediators may act together or synergistically to interfere with cardiac blood flow leading to ischemia in anaphylactic shock.

Indomethacin inhibited release of TxA_2 but exacerbated the anaphylactic increase in CPP and CDT. Furthermore, the peak output and duration of LT release was increased.

The importance of LTs was demonstrated by the observation that even in the presence of indomethacin, CGS 8515 had a cardioprotective action in CA and selectively inhibited LT release. This compound also attenuated the late phase of PAF-induced increase in perfusion pressure which has consistently been shown to be mediated by LTC_4 [Piper and Stewart, 1986, 1987].

WEB 2086 antagonized changes in CPP, CDT and the release of LTC_4 and TxA_2 stimulated by PAF. This compound also significantly reduced both the anaphylactic increase in CPP and LT output but was less effective than CGS 8515. In contrast, generation of TxA_2 in anaphylaxis was unaffected by WEB 2086. Since both PAF and LTs are generated in CA the possibility arises that PAF might stimulate antigen-induced LT release. However, this seems unlikely since a dose of WEB 2086 that totally antagonized the effects of PAF, only partially blocked the anaphylactic release of LTs and was less effective than CGS 8515.

From the experiments described above, it appears that TxA_2 probably accounts for the early increase in CPP while PAF and LTs are involved in the later phase of CA where LTs play a major role. This is consistent with our finding that the PAF antagonist, WEB 2086, had no effect on TxA_2 release during CA, further suggesting that the release of TxA_2 in CA is not triggered by PAF.

These results suggest that the persistent increase in perfusion pressure seen following the anaphylactic reaction may have been caused by an outburst and sequential release of these potent vasoconstrictors which may overshadow the activation and release of modulatory mediators such as prostacyclin and EDRF [see Piper and Stewart, 1987; Gryglewski et al., 1986]. The activity of the vasoactive mediators predominate at different time points during the course of the anaphylactic reaction. Different stages of CA were modified by the inhibitors and antagonists used. Indomethacin potentiated LT release and exacerbated CA, WEB 2086 only had a partial inhibitory effect, but CGS 8515 had a cardioprotective action. This investigation indicates that TxA_2 probably accounts for the early increase in CPP while PAF and LTs account for the later phase and the LTs play an important role in the cardiac dysfunction occurring in the anaphylactic reaction.

Acknowledgements

We thank the Faculty of Medicine, University of Malaya and the Malaysian Government for financial support (H.B.Y.) Dr. Sheldon Schaffer, Ciba-Geigy, Dr. Hubert Heuer, Boehringer Ingelheim, Dr. R. Zamboni, Merck, Sharp & Dohme for the gift of CGS 8515 and WEB 2086 and L-660,711, respectively.

References

Aehringhaus, U.; Peskar, B.A.; Wittenberg, H.R., et al.: Effect of inhibition of synthesis and receptor antagonism of SRS-A in cardiac anaphylaxis. Br. J. Pharmacol. *80:* 73–80 (1984).

Capurro, N.; Levi, R.: The heart as a target organ in systemic allergic reactions. Circ. Res. *36:* 520–528 (1975).

Gryglewski, R.J.; Moncada, S.; Palmer, R.M.J.: Bioassay of prostacyclin and endothelium-derived relaxing factor (EDRF) from porcine aortic endothelial cells. Br. J. Pharmacol. *87:* 685–694 (1986).

Jones, T.R., Guindon, Y.; Young, R.; Champion, E.; Charette, D.; Ethier, D.; Hamel, R.; Ford-Hutchinson, A.W.; Fortui, R.; Letts, G.; Masson, P.; McFarlane, C.; Piechuta, H.; Rokach, J.; Yoakun, C.; de Haven, R.N.; Maycock, A. Pong, S.S.: L-648,051, sodium 4-[3-(4-acetyl-3-hydroxy-2-propylphenoxy)-propyl-sulfonyl]-γ-oxo-benzene-butanoate: A leukotriene D_4 receptor antagonist. Can. J. Physiol. Pharmacol. *64:* 1532–1542 (1988).

Letts, L.G.; Piper, P.J.: The actions of leukotrienes C_4 and D_4 on guinea-pig isolated hearts. Br. J. Pharmacol. *76:* 169–176 (1982).

Levi, R.; Burke, J.A.; Hattori, Y., et al.: AGEPC, a putative mediator of cardiac anaphylaxis in the guinea-pig. Circ. Res. *54:* 117–124 (1984).

Piper, P.J.; Stewart, A.G.: Coronary vasoconstriction in the rat isolated perfused hearts induced by platelet-activating factor is mediated by leukotriene C_4. Br. J. Pharmacol. *88:* 595–605 (1986).

Piper, P.J.; Stewart, A.G.: Antagonism of vasoconstriction induced by platelet-activating factor in guinea-pig perfused hearts by selective platelet-activating factor receptor antagonists. Br. J. Pharmacol. *90:* 771–783 (1987).

Samhoun, M.N.; Piper, P.J.: Leukotriene F_4: Comparison of its pharmacological profile with that of the other cysteinyl-containing leukotrienes in guinea-pig ileum and lung parenchyma in vitro. Prostaglandins *28:* 623–628 (1984).

Yaacob, H.B.; Piper, P.J.: Inhibition of leukotriene release in anaphylactic heart by a lipoxygenase inhibitor CGS 8515. Br. J. Pharmacol. *95:* 1322–1328 (1988).

Priscilla J. Piper, MD, Department of Pharmacology, Hunterian Institute, Royal College of Surgeons of England, Lincoln's Inn Fields, GB-London WC2A 3PN (UK)

Localization of PLA$_2$: Regulation by Activated Protein Kinase C

Zor U, Naor Z, Danon A (eds): Leukotrienes and Prostanoids in Health and Disease.
New Trends Lipid Mediators Res. Basel, Karger, 1989, vol 3, pp 257–261

Immunogold Localization of Phospholipase A$_2$ in Rat Platelets

H. van den Bosch[a], *A.J. Aarsman*[a], *A.J. Verkleij*[b]

[a]Centre for Biomembranes and Lipid Enzymology, and [b]Department of Molecular Cell Biology, State University, Utrecht, The Netherlands

Introduction

Although arachidonate release from platelet phospholipids can proceed via different pathways, the action of phospholipase A$_2$ is thought to be a key event in this process [1]. Measurement of phospholipase A$_2$ activity in platelet lysates is strongly influenced by environmental factors, thus hampering a comparison of enzyme activities in reports from different laboratories. A comparison of platelet phospholipase A$_2$ activity was, therefore, made using a single assay system employing phosphatidylethanolamine as substrate.

Immuno cross-reactivity of phospholipase A$_2$ in various platelet extracts with monoclonal antibodies raised previously [2] against rat liver mitochondrial phospholipase A$_2$ was investigated to determine if immunogold electron microscopy could be attempted to study phospholipase A$_2$ localization in platelets. Previous reports in the literature on this subject have employed cell fractionation procedures in which conflicting results have been obtained. Several investigators reported from one third [3] to two thirds [4] of the enzyme activity to be recovered in the soluble fraction, with the remainder being distributed among mitochondrial, microsomal, granular and surface membrane fractions of human platelets. Others [5] have reported an exclusive membrane association of the phospholipase A$_2$, with intracellular membranes being 7-fold enriched in the enzyme compared to surface membranes. Studies with rat [6] and sheep platelets [7] again indicated a good deal of the enzyme to be present in soluble form. In addition, spontaneous aggregation of sheep platelets [8] and thrombin treatment of rat platelets [6] released the enzyme into the medium, suggesting a localization in α-granules. The availability of monoclonal antibodies recognizing rat platelet phospholipase A$_2$ allowed direct visualization of the enzyme by immunogold electron microscopic techniques.

Results and Discussion

Comparative experiments on platelet phospholipase A_2 activities using the same assay procedure revealed tremendous differences in the average measurable specific activities as indicated in table 1. Low activities were found in porcine, goat, bovine, human and sheep platelet lysates. Horse and rabbit contained moderate activities while by far the highest activity was detected in rat platelet lysates. A similar order was found when platelet phospholipase A_2 activities were expressed per milliliter blood. In those cases where reliable measurements could be made, i.e. with sheep, horse, rabbit and rat, over 90% of the lysate activity appeared to be solubilized by extraction with 1 M KCl. This strongly suggests that the enzyme is present either in soluble form or as a peripheral, rather than integral, membrane protein. It cannot be concluded definitively that this represents total platelet phospholipase A_2 activity. The assays were performed with 1-acyl-2-[1-^{14}C]linoleoylphosphatidylethanolamine and enzymes with an absolute specificity for phospholipids having other polar head groups, ether linkages at the sn-1 position or other fatty acids at the sn-2 position, if existing, are obviously not detected. On the other hand, phosphatidylethanolamine was

Table 1. Specific activities of phospholipase A_2 in platelet extracts and cross-reactivity with anti-rat liver mitochondrial phospholipase A_2 monoclonal antibodies

Species	Specific activity mU/mg	Cross-reactivity	
		dot blot	McAb-Sepharose
Porcine	<0.1	−	ND
Goat	<0.1	−	ND
Bovine	0.1	−	ND
Human	0.5	−	−
Sheep	1	±	±
Horse	6	−	ND
Rabbit	22	−	−
Rat	300	+++	+++

Platelet 1 M KCl extracts were assayed for phospholipase A_2 activity with 0.2 mM 1-acyl-2[1-^{14}C]linoleoylphosphatidylethanolamine (spec. act. 1,000 dpm/nmol) in the presence of 0.05% (w/v) Triton X-100 and 10 mM Ca^{2+} at pH 8.5. 1 mU is equivalent to 1 nmol linoleate released · min^{-1}. Cross-reactivity was determined in KCl extracts using dot blot and, when indicated, monoclonal antibody (McAb)-Sepharose binding experiments. ND = not done; − = no cross-reactivity; ± = weak cross-reactivity; +++ = strong cross-reactivity.

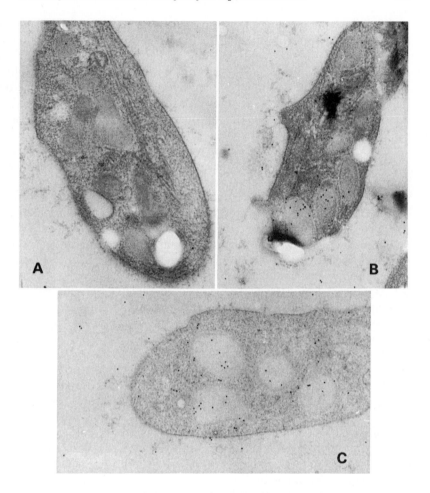

Fig. 1. Immunocytochemical localization of phospholipase A$_2$ in rat platelets. Thin sections of platelet pellets were obtained by a freeze-substitution procedure and incubated with (*B* and *C*) or without (*A*) specific monoclonal antibodies and then, consecutively, with swine anti-mouse immunoglobulins and protein A gold (particle diameter of 10 nm). ×65,000.

chosen as substrate because of its ability to get associated with the membranes in the presence of millimolar Ca^{2+}, thus enabling membrane-associated phospholipases A$_2$ to be detected with high sensitivity [9].

Immuno cross-reactivity assays by a dot blot procedure and, when performed, confirmed by monoclonal antibody-Sepharose binding experiments indicated that only rat and sheep platelets contained phospholipases A$_2$ that were recognized by monoclonal antibodies directed against rat liver mitochondrial phospholipase A_2 (table 1). Essentially all of the phospholipase A_2 activity present in the 1 M KCl extract of rat platelets was bound to monoclonal antibody-Sepharose and could be obtained in a homogeneous form, as judged by SDS-PAGE, in a single immunoaffinity chromatographic step after elution with glycine buffer at pH 2.5 (data not shown). The enzyme showed a molecular weight of 14 kd. Western blot experiments indicated that no other proteins in total rat platelet lysate were recognized by the antibodies.

The localization of this enzyme was subsequently studied by a freeze-substitution procedure in which the coupes were consecutively treated with monoclonal antibody, swine anti-mouse immunoglobulins and protein A gold. As can be seen in figure 1, the appearance of gold particles depended completely on the presence of the specific monoclonal antibodies. No gold particles were observed in the control experiments in which the coupes were only incubated with swine anti-mouse immunoglobulins and protein A gold (fig. 1A). Including the specific monoclonal antibodies in the incubation procedures revealed gold particles, a good deal of which were present over the α-granules (fig. 1B, C). This visualized directly that the enzyme is present as a soluble protein in the matrix of these granules. A striking feature of both figure 1B and C is that very few, if any, gold particles are associated with the surface membrane. Occasionally, gold particles are present over the cell body, but outside the α-granules. It cannot be concluded with certainty that these represent a membrane-associated form of the enzyme, because some gold particles are also seen outside the cell. These are believed to represent enzyme molecules that have been released from the granules after activation of platelets during the embedding procedure.

Obviously, it remains to be established whether the phospholipase A_2 in other platelet species shows a similar distribution. The presence of phospholipase A_2 in soluble form in α-granules in rat platelets, however, calls for a cautious interpretation of localization studies by cell fractionation techniques. It is quite feasible that some granules break during homogenization and that the released phospholipase A_2 gets associated in a nonspecific way with any adjacent membrane. The observation that the N-terminal sequence of a membrane-associated form of rat platelet phospholipase A_2 was identical to that of the enzyme secreted after thrombin treatment of cells seems to support this possibility [10].

References

1   Feinstein MB, Halenda SP: Arachidonic acid mobilization in platelets: The possible role of protein kinase C and G-proteins. Experientia 1988;44:101–104.

2   De Jong JGN, Amesz H, Aarsman AJ, et al: Monoclonal antibodies against an intracellular phospholipase A_2 from rat liver and their cross-reactivity with other phospholipases A_2. Eur J Biochem 1987;164:129–135.
3   Trugnan G, Bereziat G, Manier G, et al: Phospholipase activities in subcellular fractions of human platelets. Biochim Biophys Acta 1979;573:61–72.
4   Kramer RM, Checani GC, Deykin A, et al: Solubilization and properties of Ca^{2+} dependent human platelet phospholipase A_2. Biochim Biophys Acta 1986;878:394–403.
5   Lagarde M, Menashi S, Crawford N: Localization of phospholipase A_2 and diglyceride lipase activities in human platelet intracellular membranes. FEBS Lett 1981;124:23–26.
6   Horigome K, Hayakawa M, Inoue K, et al: Selective release of phospholipase A_2 and lysophosphatidylserine-specific lysophospholipase from rat platelets. J Biochem 1987;101:53–61.
7   Loeb LA, Gross RW: Identification and purification of sheep platelet phospholipase A_2 isoforms. J Biol Chem 1986;261:10467–10470.
8   Vadas P, Hay JB: The release of phospholipase A_2 from aggregated platelets and stimulated macrophages of sheep. Life Sci 1980;26:1721–1729.
9   Lenting HBM, Neys FW, Van den Bosch H: Hydrolysis of exogenous substrates by mitochondrial phospholipase A_2. Biochim Biophys Acta 1987;917:178–185.
10 Hayakawa M, Kudo M, Tomita M, et al: Purification and characterization of membrane-bound phospholipase A_2 from rat platelets. J Biochem 1988;103:263–266.

Prof. H. van den Bosch, Centre for Biomembranes and Lipid Enzymology, Transitorium 3, Padualaan 8, NL–3584 CH Utrecht (The Netherlands)

Zor U, Naor Z, Danon A (eds): Leukotrienes and Prostanoids in Health and Disease.
New Trends Lipid Mediators Res. Basel, Karger, 1989, vol 3, pp 262–267

Phospholipase A$_2$ and Lipocortin Effects?[1]

Edward A. Dennis, Florence F. Davidson, Mark D. Lister

Department of Chemistry, University of California, San Diego, La Jolla, Calif., USA

Phospholipase A$_2$ plays a key role in the release of free arachidonic from the *sn*-2 position of membrane lipids for the biosynthesis of the prostaglandins and leukotrienes, as summarized in figure 1 [for review, cf. 1]. How this phospholipase A$_2$ is regulated is a critical question on which more work is needed. The study of inhibitors of phospholipase A$_2$ is important for two reasons. First, inhibitors can provide mechanistic insight into the enzyme, and second, they are important for their therapeutic potential, the subject of this conference.

Inhibition of the cyclooxygenase by aspirin and other nonsteroidal anti-inflammatory drugs (NSAIDs) has been well studied by Vane and others [for review, cf. 2]. There is now also considerable interest in the inhibition of the 5′-lipoxygenase. However, if one could inhibit the phospholipase A$_2$ which acts prior to these, both pathways should be shut down. Indeed, it has been widely believed that the mode of action of the glucocorticoid steroids is to somehow inhibit the phospholipase A$_2$ step.

We and other investigators, looking at inhibitors of phospholipase A$_2$, have used almost exclusively the extracellular phospholipase A$_2$ either from snake venom or mammalian pancreas. These are pure enzymes that have been well characterized, thus allowing for mechanistic studies and the formulation of working hypotheses such as the 'dual phospholipid model' [3], which explains the enzyme's mode of action on the lipid-water interface as derived from kinetic results. Of course, we would all much rather have 'the relevant phospholipase A$_2$' from a cell actually involved in arachidonic acid release. For this reason, several years ago our laboratory began studies on the phospholipases in the macrophage [4]. We have initiated our studies

[1] Support was provided by NIH grant GM 20,501 and CA 09523.

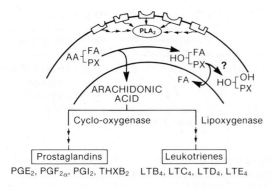

Fig. 1. The production of the prostaglandins, leukotrienes and related eicosanoids whether by the cyclooxygenase pathway or the lipoxygenase pathway depends on the availability of free arachidonic acid (AA). The control step for the production of free AA is believed to involve an as yet undefined membrane receptor event. Presumably, this would lead to the activation of a phospholipase either directly or through cellular mediators. It has been assumed that this phospholipase is localized in the plasma membrane of eicosanoid-producing cells, but other subcellular localizations have been suggested as well. The simplest and most logical candidate for the type of phospholipase would be phospholipase A$_2$ (PLA$_2$) because the bulk of the AA is found esterified in the sn-2 position of the membrane phospholipids. The other product of PLA$_2$ action on a diacyl phospholipid is a lyso phospholipid. This product is itself a biological detergent that is quite lytic to cells and must be either rapidly hydrolyzed by a lysophospholipase (LYSO PLA) or be reacylated to a diacyl phospholipid. When alkylether phosphatidyl-choline is the source of AA, the lyso product cannot be degraded further by LYSO PLA. Indeed, when it is acetylated, platelet-activating factor (PAF) is produced, itself a potent cellular activator. Reproduced with permission from reference 1.

with the P388D$_1$ macrophage-like cell line which is murine-derived. Phospholipases are ubiquitous in this cell, as they are in all cells studied. Of the many different phospholipases, we have thus far identified four major enzymes, as shown in table 1 [4]. Of these, we have purified to homogeneity and characterized the Ca^{2+}-independent, high pH optimal, soluble lysophospholipase [5] and we have solubilized in octylglucoside and purified the membrane-bound, Ca^{2+}-dependent, high pH optimum phospholipase A$_2$ [6]. The latter appeared to be the best candidate for the enzyme responsible for arachidonic acid release [7].

One potential inhibitor we have studied with this enzyme is lipocortin. The lipocortins are a family of proteins that have been defined as being inducible by steroids and as being inhibitors of the phospholipases. It has been suggested, and widely believed, that lipocortins are specific, noncompetitive inhibitors of phospholipase A$_2$. Parallel to, and recently converging on the study of the lipocortins has been the study of the calpactins. These

Table 1. Phospholipases in the P388D$_1$ murine macrophage-like cell line

Phospholipase	[Ca^{++}]	pH optimum	Localization
A$_1$	−	4	granules, lysosomes
A$_2$	−	7.5	cytoplasmic
A$_2$	+	7.5–9.5	membrane-bound
Lyso	−	7–9	soluble

are a group of abundant cytoskeletal proteins that bind Ca^{2+}, phospholipid and actin. They have received a great deal of attention because they are substrates of several tyrosine kinases, and are thereby linked to oncogenes. In the last few years, the calpactins have been shown to be the same proteins as the lipocortins. There are now more than six lipocortins that have been identified as part of a large family of related proteins. In particular, calpactin II, which is p35, the EGF receptor kinase substrate, is the same protein as lipocortin I, and calpactin I is the same protein as lipocortin II.

When we [8] initially looked for inhibition of phospholipase A$_2$ by the calpactins, we could not find any inhibition in any of our usual assays with either the pancreatic phospholipase A$_2$ or the cobra venom enzyme. However, when we turned to the use of autoclaved *Escherichia coli*, which was reported in the literature as the substrate used in most inhibition assays, one could in fact find inhibition by the calpactins with a 50% inhibition at about 1 μg/ml, which is what others reported. However, when we looked at the *E. coli* assay in more detail, we found that the substrate concentration was in the low micromolar range, which is an incredibly small amount of phospholipid compared to more standard phospholipase A$_2$ assays. We found that if we varied the phospholipid, the inhibition could be obliterated as shown in figure 2. The very low concentration of phospholipid in lipocortin assays is significant when one takes into consideration that the enzyme used is typically the pancreatic phospholipase A$_2$ (as shown here), an enzyme whose K$_m$ is usually in the millimolar range. As shown here, at 4 μM phospholipid, one gets practically 100% inhibition, and if the concentration is doubled to 8 μM, one gets no inhibition. This kind of dependence on substrate concentration is absolutely inconsistent with calpactin or lipocortin serving as a noncompetitive inhibitor. Rather, it is more consistent with a 'substrate depletion model' that we [8] proposed on the basis of kinetics and binding studies. Quite simply, the model states that the lipocortin binds to the phospholipid and depletes it, making it unavailable for the enzyme to act on until the concentration of phospholipid is high enough to saturate the lipocortin.

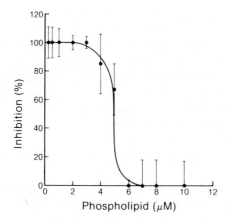

Fig. 2. Dependence of calpactin I (lipocortin II) inhibition on the concentration of extracted *E. coli* phospholipid vesicles. The concentrations of phospholipase A$_2$ (5.1 × 10^{-10} *M*) and calpactin I (1.1 × 10^{-7} *M*) were kept constant as the concentration of *E. coli* derived phospholipid vesicles was increased. Percent inhibition is shown as a function of the phospholipid substrate concentration. Reproduced with permission from reference 8.

This is shown in the following equation:

$$E + S \underset{}{\overset{K_s}{\rightleftharpoons}} ES \xrightarrow{k_p} E + P$$
$$+$$
$$I$$
$$K_0 \updownarrow$$
$$SI$$

Here we show the inhibitor binding the substrate to form a very tight substrate-inhibitor complex. In the case of calpactin, the inhibitor is Ca^{2+} dependent and prefers anionic lipids. Only when one increases the substrate concentration sufficiently, is the equilibrium driven to form the *ES* complex and on to product. It should be pointed out that while the equation is written as a stoichiometry of one to one, the calpactin may bind more than one phospholipid. It could do this by coating the surface of vesicles or membranes, or by intercalating and causing a phase change or precipitation, thereby affecting many more phospholipids. This model can explain all the kinetic data we obtained with various substrates and various enzymes, as well as, we believe, all the data in the literature for the apparent inhibition of the extracellular phospholipases.

Subsequent to suggesting this model last year [8], we carried out many additional experiments on different substrate systems and different enzymes. For example, we [9] looked at deoxycholate-phosphatidylcholine mixtures only to determine that it is a mixed micelle system which is totally unsuited to simple interpretation. Its micelles are very poorly defined and highly variable (cmc, micelle number, etc.), which has made it very difficult to carry out any meaningful kinetic studies with the pancreatic enzyme. The micelle structures can vary with concentration, and phase changes can occur due to calcium binding. Nevertheless, one can find conditions where inhibition by calpactin is observed and, as above, increasing the substrate concentration can overcome inhibition. For example, one can go from some inhibition at low phospholipid substrate, to no inhibition at higher substrate. In fact, at higher concentrations of substrate, one gets a slight activation which is probably due to the complicated kinetics of the deoxycholate micelle system.

Most recently, we [9] have looked at the macrophage phospholipase A_2. This phospholipase A_2 is similarly inhibited by calpactin I. If we look at the inhibition by the calpactin on various PC substrates for the macrophage phospholipase A_2, one sees similar dose dependency curves, but clearly less calpactin is necessary for the inhibition of enzyme activity when stearoyl,2-arachidonoyl PC is used in vesicles than for 1-stearoyl, 2-oleoyl PC, and much less than for dipalmitoyl PC. This dependency on the substrate is also suggestive that calpactin is actually binding to PC vesicles, rather than to the enzyme itself, and that the relevant affinities depend on the substrate form. If it were simply binding to the enzyme, one would expect similar dose responses with the various substrates since the enzyme itself works equally well on all of these substrates [7]. Finally, if one now looks at the phospholipid dependence with arachidonocyl PC as substrate, one can see that at low concentrations ($\sim 1~\mu M$ of substrate in vesicles), one can get a significant inhibition of the enzyme rate with calpactin, whereas at higher phospholipid concentrations, inhibition is reduced. With the macrophage enzyme and these amounts of calpactin, there is not a total inhibition as observed previously with the pancreatic enzyme. In this substrate depletion model, the extent of inhibition observed will depend on what the relative affinities are of the calpactin for phospholipid versus the enzyme for phospholipid. A similar binding affinity demonstrated by both proteins will yield a competition for substrate. A much tighter binding of substrate by the inhibitor protein relative to the enzyme results in total inhibition until the substrate concentration overcomes the inhibitor saturation level, as was probably the case with the pancreatic enzyme.

In conclusion, we believe that all of our data and all of the data in the literature on the inhibition of phospholipase A_2 by lipocortin can be explained by the substrate depletion model. It is clear that dexamethasone and other steroids do inhibit prostaglandin production in intact cells and animals, but there is no published data that unequivocably links this effect directly to the lipocortin family of proteins or shows a physiological role for the lipocortins in phospholipase A$_2$ inhibition. We, as much as anyone, would be delighted to have a protein produced in response to steroid that is a specific inhibitor of phospholipase A$_2$. However, before one postulates such a function for a given protein, it should ideally be tested against more stringent criteria than have heretofore been available. We would like to suggest that in vitro biochemical tests, such as we have conducted, are a necessity in that process.

References

1   Dennis, E.A.: The regulation of eicosanoid production: Role of phospholipases and inhibitors. Bio/Technology *5:* 1294–1900 (1987).

2   Vane, J.; Botting, R.: Inflammation and the mechanism of action of anti-inflammatory drugs. FASEB J. *1:* 89–96 (1987).

3   Roberts, M.F.; Deems, R.A.; Dennis, E.A.: Dual role of interfacial phospholipid in phospholipase A$_2$ catalysis. Proc. Natl. Acad. Sci. USA *74:* 1950–1954 (1977).

4   Ross, M.I.; Deems, R.A.; Jesaitis, A.J.; Dennis, E.A.; Ulevitch, R.J.: Phospholipase activities of the P388D$_1$ macrophage-like cell line. Arch. Biochem. Biophys. *238:* 247–258 (1985).

5   Zhang, Y.; Dennis, E.A.: Purification and characterization of lysophospholipase from P388D$_1$ macrophage-like cell line. J. Biol. Chem. *263:* 9965–9972 (1988).

6   Ulevitch, R.J.; Sano, M.; Watanabe, Y.; Lister, M.D.; Deems, R.A.; Dennis, E.A.: Solubilization and characterization of a membrane-bound phospholipase A$_2$ from the P388D$_1$ macrophage-like cell line. J. Biol. Chem. *263:* 3079–3085 (1988).

7   Lister, M.D.; Deems, R.A.; Watanabe, Y.; Ulevitch, R.J.; Dennis, E.A.: Kinetic analysis of the Ca^{2+} dependent, membrane-bound, macrophage phospholipase A$_2$ and the effects of arachidonic acid. J. Biol. Chem. *263:* 7506–7513 (1988).

8   Davidson, F.F.; Dennis, E.A.; Powell, M.; Glenney, J.: Inhibition of phospholipase A$_2$ by lipocortins: An effect of binding to phospholipids. J. Biol. Chem. *262:* 1698–1705 (1987).

9   Davidson, F.F.; Lister, M.D.; Dennis, E.A.: (in preparation, 1989).

Edward A. Dennis, PhD, Department of Chemistry, University of California, San Diego, La Jolla, CA 92093 (USA)

Zor U, Naor Z, Danon A (eds): Leukotrienes and Prostanoids in Health and Disease.
New Trends Lipid Mediators Res. Basel, Karger, 1989, vol 3, pp 268–271

Biosynthesis of PAF in Human Polymorphonuclear Leukocytes: The Role of Phorbol Esters and Protein Kinases

Mariano Sanchez Crespo, Maria Luisa Nieto

Instituto de Investigaciones Medicas de la Fundacion Jimenez Diaz, Centro
Asociado al CSIC, Madrid, Spain

The biosynthesis of PAF (1-O-hexadecyl/octadecyl-2-acetyl-*sn*-glycero-3-phosphocholine) in human polymorphonuclear leukocytes (PMN) has been extensively studied in the last years. This has allowed the delineation of two important features of this process that are now widely accepted. First, the generation of PAF only occurs after the activation of the cell by secretagogues in the presence of extracellular calcium, and, second, this generation may be modulated by interfering with the process of cell activation. A number of studies have emphasized the role of the enzyme 1-O-hexadecyl-2-lyso-*sn*-glycero-3-phosphocholine:acetyl-CoA acetyltransferase (EC 2.3.1.67) as the rate-controlling step of PAF biosynthesis [1–4]. A logical consequence of the above-mentioned concept should be the correlation of the events which occur during cell activation with the stimulation of both acetyltransferase activity and PAF biosynthesis. This has been carried out in preparations other than PMN by studying the role of calcium ions and a phosphorylation-dephosphorylation mechanism in rat macrophages and spleen microsomes [5, 6].

PAF can also be synthesized from alkyl-acetyl-*sn*-glycerol and CDP-choline through DTT-insensitive cholinephosphotransferase (EC 2.7.8.16). This biosynthetic pathway is analogous to the Kennedy or *de novo* pathway for the synthesis of choline glycerophospholipids, but no evidence has been as yet provided as to the involvement of this pathway in PAF biosynthesis in PMN. In this paper we summarize recent findings described in detail elsewhere [7, 8], which suggest that lyso-PAF:acetyl-CoA acetyltransferase from PMN is converted into the high activity form of the enzyme by phosphorylation linked to cyclic AMP-dependent protein kinase, in a similar manner to that already described in rat spleen microsomes. On the other hand, phospholipid-sensitive, calcium-dependent protein kinase (PrKC) causes a significant reduction of lyso-PAF:acetyl-CoA acetyltransferase activity in homogenates from stimulated PMN. Phorbol esters, a known family of PrKC activators, can influence PAF metabolism by enhancing synthesis through the *de novo* pathway and therefore involving DTT-insensitive cholinephosphotransferase. This is the first report on the involvement of this pathway in the assembly of PAF in PMN.

Results and Discussion

Homogenates of resting PMN incubated in the presence of the catalytic subunit of cyclic AMP-dependent protein kinase showed an up to 3-fold increase of acetyltransferase activity, in a dose- and time-dependent manner. This effect was prevented in the presence of the protein kinase inhibitor from porcine heart. In homogenates from PMN previously stimulated by complement-coated zymosan particles, the decay of acetyltransferase activity was partially prevented by the addition of soybean trypsin inhibitor, and almost completely inhibited when the medium also included inhibitors of alkaline phosphatase. In contrast, the addition of exogenous alkaline phosphatase caused a diminution of lyso-PAF:acetyl-CoA acetyltransferase activity. Addition of partially purified phospholipid-sensitive PrKC from rat spleen, to homogenates from ionophore A23187-stimulated PMN, reduced acetyltransferase activity by 63%, whereas only a 16% inhibition was observed on homogenates from resting PMN (table 1). No effect could be seen with heat-inactivated PrKC, which is an argument against inhibition from materials accumulated during the purification procedure. Omission of 1,2-diolein from the assay did not significantly modify the inhibitory effect of PrKC, whereas omission of phosphatidylserine abolished most of the effect. This could be due either to the presence of sufficient cofactor in the PMN preparation or to the features of spleen PrKC, whose activity seems to be more dependent on phosphatidylserine than on diacylglycerols [9]. Preincubation of PMN with 12-O-tetradecanoylphorbol-13-acetate (TPA) previous or simultaneously to the addition of ionophore A23187, reduced the increase in acetyltransferase produced by ionophore A23187, whereas the generation of superoxide ions was enhanced.

TPA was an active stimulator of PAF biosynthesis, but showed a time course more protracted than that observed in response to complement-coated zymosan particles and ionophore A23187, since PAF could be detected only after periods of incubation above 30 min after the addition of TPA. About 85% PAF generated in response to TPA remained cell associated, i.e., similar to the figures observed with other secretagogues. Incubation of PMN with TPA was not found to activate acetyltransferase

Table 1. Effect of PrKC on acetyltransferase activity in PMN homogenates

Addition	Enzyme activity, pmol/min/5 · 10^6 PMN)	
	Resting	A23187-treated
None	146 ± 11 (6)	492 ± 118 (6)
Phosphorylation mixture	123 ± 19 (6)	501 ± 110 (6)
PrKC	140 ± 13 (6)	446 ± 63 (6)
Phosphorylation mixture + boiled PrKC	138 ± 12 (3)	497 ± 58 (3)
EGTA + PrKC	133 ± 15 (3)	448 ± 43 (3)
PtdSer	136 ± 32 (3)	494 ± 56 (3)
(Phosphorylation mixture-PtdSer) + PrKC	125 ± 29 (3)	498 ± 43 (3)
Phosphorylation mixture + PrKC	111 ± 15 (6)	183 ± 21 (6)
(Phosphorylation mixture-diolein) + PrKC	121 ± 3 (6)	162 ± 29 (6)

PMN were incubated with medium or with 5 μM ionophore A23187 prior to homogenization. After 5 min of incubation in the presence of the additions indicated, acetyltransferase reaction was triggered by the addition of the substrates. The reaction was stopped after 10 min at 37 °C and the product formed separated and quantitated. PtdSer indicates 20 μg of phosphatidyl-serine/ml. (Phosphorylation mixture-PtdSer) indicates the absence of phosphatidylserine in the medium. (Phosphorylation mixture-diolein) indicates the absence of diolein in medium. EGTA + PrKC indicates the phosphorylation mixture omitting 0.5 mM CaCl$_2$. PrKC was used at a concentration of 43 U/ml. Numbers in parentheses indicate the number of experiments with duplicate samples. Data represent mean $\pm$ SD.

activity, whereas the reverse was true of the other secretagogues. TPA was found to enhance the incorporation of [^{3}H]methylcholine and both alkyl-labeled and acetyl-labeled 1-O-hexadecyl-2-acetyl-*sn*-glycerol into a lipid fraction comigrating with PAF. This incorporation showed a time course parallel to that of the generation of PAF in response to TPA. The present data indicate that lyso-PAF:acetyl-CoA acetyltransferase activity in PMN is modulated by a phosphorylation-dephosphorylation mechanism linked to cyclic AMP-dependent protein kinase. Phospholipid-sensitive PrKC does not seem to be involved in the mechanism of activation, but, most probably, in the generation of negative activation signals. TPA is the only agonist so far described that could initiate the biosynthesis of PAF through the DTT-insensitive cholinephosphotransferase pathway.

References

1   Wykle, R.L.; Malone, B.; Snyder, F.: Enzymatic synthesis of 1-alkyl-2-acetyl-*sn*-glycero-3-phosphocholine, a hypotensive and platelet-activating phospholipid, J. Biol. Chem. *255:* 10256–10258 (1980).

2   Alonso, F.; Garcia-Gil, M.; Sanchez Crespo, M.; Mato J.M.: Activation of 1-O-alkyl-2-lyso-glycero-3-phosphocholine:acetyl-CoA transferase during phagocytosis in human polymorphonuclear leukocytes. J. Biol. Chem. *257:* 3376–3378 (1982).

3   Albert, D.H.; Snyder, F.: Biosynthesis of 1-O-alkyl-2-acetyl-*sn*-glycero-3-phosphocholine (platelet-activating factor) from 1-alkyl-2-acyl-*sn*-glycero-3-phosphocholine by rat alveolar macrophages. J. Biol. Chem. *258:* 97–102 (1982).

4   Ninio, E.; Mencia-Huerta, J.M.; Heymans, J.M.; Benveniste, J.: Biosynthesis of platelet-activating factor (PAF-acether). V. Enhancement of acetyltransferase activity in murine peritoneal cells by calcium ionophore A23187. Biochim. Biophys. Acta *710:* 23–31 (1982).

5   Lenihan, D.J.; Lee, T.-C.: Regulation of platelet-activating factor synthesis: Modulation of 1-alkyl-2-lyso-*sn*-glycero-3-phosphocholine:acetyl-CoA acetyltransferase by phosphorylation and dephosphorylation in rat splenic microsomes. Biochem. Biophys. Res. Commun. *120:* 834–839 (1984).

6   Gomez-Cambronero, J.; Mato, J.M.; Vivanco, F.; Sanchez Crespo, M.: Phosphorylation of partially purified 1-O-alkyl-2-lyso-*sn*-glycero-3-phosphocholine:acetyl-CoA acetyltransferase from rat spleen. Biochem. J. *245:* 893–898 (1987).

7   Nieto, M.L.; Velasco, S.; Sanchez Crespo, M.: Modulation of acetyl-CoA:1-alkyl-2-lyso-*sn*-glycero-3-phosphocholine (lyso-PAF) acetyltransferase in human polymorphonuclears. The role of cyclic AMP-dependent and phospholipid-sensitive, calcium-dependent protein kinases. J. Biol. Chem. *263:* 4607–4611 (1988).

8   Nieto, M.L.; Velasco, S.; Sanchez Crespo, M.: Biosynthesis of platelet-activating factor in human polymorphonuclear leukocytes. Involvement of the cholinephosphotransferase pathway in response to the phorbol esters. J. Biol. Chem. *263:* 2217–2222 (1988).

9   Schartzman, R.C.; Raynor, R.L.; Fritz, R.B.; Kuo, J.F.: Purification to homogeneity, characterization and monoclonal antibodies of phospholipid-sensitive Ca-dependent protein kinase from spleen. Biochem. J. *209:* 435–443 (1983).

Mariano Sanchez Crespo, MD, Instituto de Investigationes Medicas de la Fundacion Jimenez Diaz, Centro Asociado al CSIC, Av. Reyes Catolicos 2, E-28040 Madrid (Spain)

Zor U, Naor Z, Danon A (eds): Leukotrienes and Prostanoids in Health and Disease.
New Trends Lipid Mediators Res. Basel, Karger, 1989, vol 3, pp 272–278

Novel Mechanisms of Glucocorticoid Actions in Inhibition of Phospholipase A_2 Activity: Suppression of Elevated $[Ca^{2+}]_i$, PI-PLC and PKC Activity Induced by Ca^{2+} Ionophore and Antigen

U. Zor[a], *T. Harell*[a], *J. Hermon*[a], *E. Her*[a], *E. Ferber*[b]

[a]Department of Hormone Research, The Weizmann Institute of Science,
Rehovot, Israel, and [b]Max-Planck-Institut für Immunbiologie, Freiburg, FRG

Cellular free arachidonic acid (AA) is an essential precursor for the generation of physiologically important eicosanoids [for review, see 1]. Under basal conditions (without activation), the level of cellular AA is quite low, and consequently serves as a rate-limiting factor in eicosanoid formation [2]. Upon stimulation by various agonists, the production and release of free AA markedly increases mainly due to activation of phospholipase A_2 (PLA_2). In some cells, under basal conditions, nonactivated PLA_2 is located exclusively in the cytosol [3, 4]. Upon cellular stimulation by Ca^{2+} ionophore, ras oncogene or diacylglycerol (DG), PLA_2 is translocated to the plasma membrane and activated [3, 4]. PLA_2 activity is absolutely dependent on high intracellural free Ca^{2+} concentration $[Ca^{2+}]_i$ [5]. In addition, recent evidence (albeit indirect) indicated that protein phosphorylation induced by protein kinase C (PKC) is possibly important for expression of full enzyme activity of PLA_2 [6–11].

5-Lipoxygenase (5-LOX) is a key enzyme in the formation of leukotrienes (LT) such as LTC_4. Like PLA_2 it is located in the cytosol under basal conditions, and translocated to and activated at the plasma membrane upon cellular stimulation by Ca^{2+} ionophore [12, 13]. Cellular 5-LOX activity is also absolutely dependent on high $[Ca^{2+}]_i$ [for review, see 1, 14]. In contrast to PLA_2, there is no indication that 5-LOX activity is regulated by protein phosphorylation induced by PKC or any other protein kinase.

Glucocorticoids (GC) are well-known inhibitors of PLA_2 [2] and 5-LOX [1, 14] activity. Their mechanisms of action, however, have not been fully elucidated [1, 4, 15]. Recently, we showed for the first time that GC prevented the elevation of $[Ca^{2+}]_i$ induced either by antigen or by ionomycin [1, 14]. The marked suppression by GC of antigen-induced LTC$_4$ formation is largely dependent on the reduction in [Ca^{2+}]$_i$ [1, 14].

GC induces various proteins, among them lipocortin, which binds to acidic phospholipids (such as PS and PI) and masks them from interacting with the appropriate enzymes [16, 17]. Recently, GC was also found to be a potent suppressor of PI-PLC activity [15], an inhibition which may lead to reduction in DG formation. These cofactors are essential for PKC activity and consequently possibly for PLA$_2$ activation. In this study we analyzed the mechanisms regulating PLA$_2$ activation, affected by the inhibitory action of GC on PI-PLC and PKC activity.

Results

Suppression by GC of AA Release Induces by Ca^{2+} Ionophore

Pretreatment with GC (HC and DEX) suppressed by about 50% the stimulation of ^{3}H-AA release induced by submaximal (0.1–0.3 μM) and maximal (3.0 μM) concentrations of ionomycin (fig. 1). In the absence of pretreatment, 1.0 μM ionomycin elevated [Ca^{2+}]$_i$ 12-fold above basal level (190 μM; data not presented). Following pretreatment with HC the elevation was reduced by 33%, but the [Ca^{2+}]$_i$ was still 8 times more than the resting concentration (fig. 1).

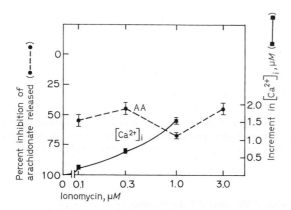

Fig. 1. Inhibition by GC of the stimulation of ^{3}H-AA release induced by ionomycin. HC or DEX (1 μM each) were added 16 h before challenge with various concentrations of ionomycin added for 2–3 min for measurement of [Ca^{2+}]$_i$ or for 30 min for ^{3}H-AA released to the medium. [Ca^{2+}]$_i$ was measured as previously described [1, 14]. ^{3}H-AA (0.2 μM/ 10^6 cells) was incorporated into phospholipids during 16 h incubation, after which the cells were washed to remove the remaining ^{3}H-AA (10%) from the medium.

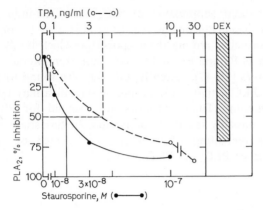

Fig. 2. Inhibition by staurosporine or TPA of Ca^{2+} ionophore-induced stimulation of AA release. TPA was added for 16 h. Staurosporine added for 10 min followed by washing. Both groups were treated for 30 min with Ca^{2+} ionophore A23187 (3 μM).

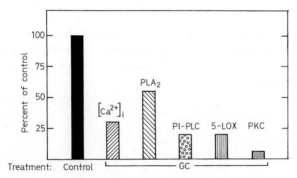

Fig. 3. Inhibitory effect of HC or DEX on $[Ca^{2+}]_i$, PLA_2, PI-PLC, 5-LOX and PKC. $[Ca^{2+}]_i$ and 5-LOX were determined in RBL cells as previously described. PLA_2 activity determined as in figure 1. PI-PLC activity determined by incorporation of ^{3}H-inositol to PI. Hydrolysis of PI by PI-PLC was attained by stimulation of RBL cells by IgE-DNP for 30 min. Translocation of PKC to plasma membrane was determined as described previously [4]. Mouse bone marrow macrophages were pretreated with DEX for 16 h. The cells incubated with DG analog for 4 h followed by measurement of PKC activity in plasma membrane and cytosol.

Relationship between PKC and PLA_2 Activity

Prolonged incubation (16 h) of RBL cells with relatively low concentrations (1–10 ng/ml) of TPA induces down-regulation of PKC (data not shown). This leads to marked inhibition (80%) of AA released, induced by Ca^{2+} ionophore A23178 (3 μM). Similarly, staurosporine – an inhibitor of PKC activity at concentrations of 10^{-8}–10^{-7} M – suppressed by 80% AA formation induced by Ca^{2+} ionophore. Both drugs were active to the same extent as DEX (fig. 2). GC markedly suppressed the translocation of PKC to the plasma membrane induced by DG and Ca^{2+} ionophore. This effect of the steroid was seen in macrophages (fig. 3) and in RBL cells (data not shown). GC markedly suppressed (70%) antigen-induced PI-PLC activity as measured by the formation of ^{3}H-IPs (fig. 3).

Discussion

Mode of Regulation of PLA$_2$ versus 5-LOX

Both PLA$_2$ and 5-LOX are activated by a high [Ca^{2+}]$_i$ [1, 5, 12]. However, while high Ca^{2+} is essential and sufficient for full expression of 5-LOX activity [1, 14], it is not sufficient for full expression of PLA$_2$ activity. Figure 1 clearly demonstrates that [Ca^{2+}]$_i$ was 8 times greater than the basal level in Ca^{2+} ionophore-stimulated RBL cells pretreated with GC, even though PLA$_2$ activity was inhibited by about 50% while 5-LOX activation was not impaired [1, 14]. These results suggest that GC modulate other messengers (besides [Ca^{2+}]$_i$) that are essential for PLA$_2$ activity. For reasons described below, the most likely enzyme to be involved in the regulation of PLA$_2$ activity is PKC [6–11].

Interaction between PKC and PLA$_2$

Indirect evidence (mostly related to the action of various inhibitors of PKC) suggests that protein phosphorylation induced by PKC may regulate PLA$_2$ activity [6–11]. The following criteria must be met in order to justify the involvement of PKC: (a) 'depletion' (down-regulation) of PKC by prolonged treatment with TPA should prevent Ca^{2+} ionophore or antigen-induced stimulation of PLA$_2$. We did indeed find this to be so (fig. 2), but this treatment was selective. Potentiation by TPA of bradykinin-induced PGE$_2$ formation was not affected [18]. (b) Inhibition of PKC should suppress PLA$_2$ activation. Staurosporine, which is a relatively selective potent inhibitor of PKC, markedly suppressed PLA$_2$ activation induced by antigen or by Ca^{2+} ionophore (fig. 2). Other less specific and less potent inhibitors of PKC, such as sphingosine and H-7, acted similarly (data not shown) though in some cell types, treatment with H-7 was only partially effective [18]. (c) Activation of PKC should result in translocation of PLA$_2$ to the plasma membrane followed by its activation. Such translocation and activation was induced by DG derivatives [4]. In addition to meeting these criteria, we have also demonstrated that Ca^{2+} ionophore induced translocation and activation of PKC, and that GC prevented the translocation of PKC to the plasma membrane induced by DG and Ca^{2+} ionophore [present study]. All these data strongly suggest an interaction between PKC and PLA$_2$.

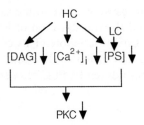

Fig. 4. HC action on factors which are essential for activation of PKC. HC reduced $[Ca^{2+}]_i$ [1, 14], possibly suppressed DG (see fig. 3; PI-PLC) and masked PS via the induced protein, lipocortin (LC). Reduction in the concentrations or availability of the above factors lead to prevention of PKC translocation to the membrane (see fig. 3) and consequently less protein phosphorylation, elements essential for PLA_2 activity.

The protein phosphorylated by PKC is not yet known. It is possible that it is PLA_2 itself, and thus PKC activates the enzyme directly. More likely, an inhibitor of PLA_2 such as lipocortin is the target for PKC action. Phosphorylation inactivates it or renders it susceptible to proteolytic enzymes [19]. It should also be borne in mind that, as stated above, there is evidence to suggest a direct effect of DG and/or TPA on PLA_2 without the involvement of PKC activity.

How does GC Prevent PKC Translocation and Activation?

Recently, GC were found to be potent inhibitors of PI-PLC (fig. 3) [15]. The mechanism of this inhibition is not yet clear, but it may involve masking of the substrate PI by lipocortin. It is possible that reduction in PI-PLC activity could result in suppression of DG production which is essential for PKC activity. Other enzymes could, of course, supply DG. We are studying whether GC actually inhibits DG formation.

GC induces several proteins, some of which bind to phosphatidylserine (PS) and may prevent the interaction of PS with PKC. PS located in the inner plasma membrane serves as an anchor for PKC. Masking of PS by lipocortin (LC) (fig. 4) could result in the suppression of PKC translocation and activation. Under certain conditions (antigen stimulation) the reduction in $[Ca^{2+}]_i$ by GC [1, 14] may also serve as a rate-limiting factor in PKC activation. Thus, shortage of $[Ca^{2+}]_i$, DG and masking of PS induced by GC may together be responsible for low PKC activity (fig. 4) and consequently inhibition of PLA_2 activity. We suggest that the dominant factor in the inhibitory action of GC on PKC activity is a lipocortin-type protein which prevents the association of PKC with PS.

Conclusion

GC have multiple mechanisms of action which culminate in quenching of cellular activity in general, and reducing $[Ca^{2+}]_i$, suppressing PLA$_2$, PI-PLC, PKC and 5-LOX activity in particular. Much more study has to be done to clarify these mechanisms. At this stage we are dealing only with the descriptive phase of a very interesting phenomenon.

Acknowledgements

This study is supported by the Institute Henri Beaufour, France, and Merck Frosst, Canada. We are grateful to Mrs. Malka Kopelowitz for typing the typescript, and to Dr. Sandra Moshonov for valuable editing. U. Zor is an incumbent of the W.B. Graham Professorial Chair in Pharmacology.

References

1   Zor U, Her E, Talmon J, et al: Hydrocortisone inhibits antigen-induced rise in intracellular free calcium concentration and abolishes leukotriene C$_4$ production in leukemic basophils. Prostaglandins 1987;34:29–41.

2   Irvine RF: How is the level of free arachidonic acid controlled in mammalian cells? Biochem J 1982;204:3–16.

3   Bar-Sagi D, Suhan JP, McCormick F, et al: Localization of phospholipase A$_2$ in normal and ras-transformed cells. J Cell Biol 1988;106:1649–1658.

4   Schonhardt T, Ferber E: Translocation of phospholipase A$_2$ from cytosol to membranes induced by 1-oleoyl-2-acetyl-glycerol in serum-free cultured macrophages. Biochem Biophys Res Commun 1987;149:769–775.

5   Van den Bosch H: Intracellular phospholipases A. Biochim Biophys Acta 1980;604:191–246.

6   Halenda SP, Zavoico GB, Feinstein MB: Phorbol esters and oleoyl acetoyl glycerol enhance release of arachidonic acid in platelets stimulated by Ca^{2+} ionophore A23187. J Biol Chem 1985;260:12484–12491.

7   Pfannkuche H-J, Kaever V, Resch F: A possible role of protein kinase C in regulating prostaglandin synthesis of mouse peritoneal macrophages. Biochem Biophys Res Commun 1986;139:604–611.

8   Liles WC, Meier KE, Henderson WR: Phorbol myristate acetate and the calcium ionophore A23187 synergistically induce release of LTB$_4$ by human neutrophils: Involvement of protein kinase C activation in regulation of the 5-lipoxygenase pathway. J Immunol 1987;138:3396–3402.

9   McIntyre TM, Reinhold SL, Prescott SM, et al: Protein kinase C activity appears to be required for the synthesis of platelet-activating factor and leukotriene B$_4$ in human neutrophils. J Biol Chem 1987;262:15370–15376.

10 Parker J, Daniel LW, Waite M: Evidence of protein kinase C involvement in phorbol diester-stimulated arachidonic acid release and prostaglandin synthesis. J Biol Chem 1987;262:5385–5393.

11   Ho AK, Klein DC: Activation of α_1-adrenoceptors, protein kinase C, or treatment with intracellular free Ca^{2+} elevating agents increases pineal phospholipase A_2 activity. Evidence that protein kinase C may participate in Ca^{2+}-dependent α_1-adrenergic stimulation of pineal phospholipase A_2 activity. J Biol Chem 1987;262:11764–11770.

12   Rouzer CA, Samuelsson B: Reversible, calcium-dependent membrane association of human leukocyte 5-lipoxygenase. Proc Natl Acad Sci USA 1987;84:7393–7397.

13   Wong A, Hwang SM, Cook MN, et al: Interactions of 5-lipoxygenase with membranes: Studies on the association of soluble enzyme with membranes and alterations in enzyme activity. Biochemistry 1988;27:6763–6769.

14   Her E, Weissman BA, Zor U: A novel mechanism of glucocorticosteroid action in inhibition of leukotriene C_4 production by basophilic leukemia cells: Suppression of the elevation of cytosolic free Ca^{2+} induced by antigen or ionomycin (submitted).

15   Zor U, Her E, Ostfeld I, et al: Glucocorticoid inhibition of Ca^{2+} and phospholipid-dependent enzymes regulating leukotriene C_4 formation and action in allergic and inflammatory responses; in Advances in PG, THA_2 and LT Research. New York, Raven Press, 1989.

16   Schlaepfer DD, Haigler HT: Characterization of Ca^{2+}-dependent phospholipid binding and phosphorylation of lipocortin I. J Biol Chem 1987;262:6931–6937.

17   Davidson FF, Dennis EA, Powell M, et al: Inhibition of phospholipase A_2 by 'lipocortins' and calpactins. An effect of binding to substrate phospholipids. J Biol Chem 1987;262:1698–1705.

18   Burch RM, Ma AL, Axelrod J: Phorbol esters and diacylglycerols amplify bradykinin-stimulated prostaglandin synthesis in Swiss 3T3 fibroblasts. Possible independence from protein kinase C. J Biol Chem 1988;263:4764–4767.

19   Haigler HT, Schlaepfer DD, Burgess WH: Characterization of lipocortin I and an immunologically unrelated 33-kDa protein as epidermal growth factor receptor/kinase substrates and phospholipase A_2 inhibitors. J Biol Chem 1987;262:6921–6930.

Prof. U. Zor, Department of Hormone Research, The Weizmann Institute of Science, Rehovot 76100 (Israel)

Zor U, Naor Z, Danon A (eds): Leukotrienes and Prostanoids in Health and Disease.
New Trends Lipid Mediators Res. Basel, Karger, 1989, vol 3, pp 279–282

Mechanism for Arachidonic Acid Liberation in PAF-Stimulated Human Polymorphonuclear Neutrophils

*Yoshinori Nozawa, Shigeru Nakashima, Akiyoshi Suganuma,
Masao Sato, Toyohiko Tohmatsu*

Department of Biochemistry, Gifu University School of Medicine, Gifu, Japan

Upon stimulation of neutrophils with a variety of stimuli, arachidonic acid (AA) is released and then converted to biologically active prostaglandins and leukotrienes. The release of AA from membrane phospholipids is the initial step in the synthesis of eicosanoids. Two pathways have been proposed for AA liberation in platelets; direct deacylation of phospholipids by phospholipase A_2 and the sequential action of phosphoinositide-specific phospholipase C followed by diacylglycerol (DG) lipase. The former is thought to play a major role for AA release in stimulated neutrophils [1–5]. Most investigators have demonstrated that increase in intracellular free Ca^{2+} concentration ($[Ca^{2+}]i$) is responsible for the activation of phospholipase A_2. However, the amount of Ca^{2+} needed for the in vitro assay of phospholipase A_2 activity is far above the level actually attained in stimulated neutrophils. On the other hand, several studies have shown recently that pertussis toxin pretreatment inhibits AA release in various types of cells including neutrophils [3–5], suggesting a GTP-binding regulatory protein (G protein)-mediated stimulation of AA release.

Platelet-activating factor (PAF) is a potent stimulant for neutrophils. PAF binds to its receptors and induces phosphoinositide breakdown [6]. However, little information is available regarding AA release in PAF-treated neutrophils. The present study was designed to investigate the effect of PAF on AA liberation and also to clarify the mechanism of phospholipase A_2 activation in human polymorphonuclear neutrophils.

Human neutrophils were isolated from normal human donors. Briefly, cells were obtained after 0.1% methylcellulose sedimentation for 60 min and layered over Ficoll-Paque for removal of mononuclear cells. To obtain [³H]AA-labeled cells, neutrophils were incubated for 2 h with [³H]AA in the presence of BSA. Where indicated, the medium was supplemented with pertussis toxin to obtain toxin-treated neutrophils.

PAF induced a marked increase in the amount of free [³H]AA in human neutrophils prelabeled with [³H]AA. The stimulation of [³H]AA release was observed in a dose-dependent fashion with a threshold concentration of about 0.8 nM. The PAF antagonist ONO-6240 (1-O-hexadecyl-2RS-O-ethyl-3-O-(7-thiazolioheptyl)glyceryl methanesulfonate) dose dependently inhibited PAF-induced [³H]AA liberation, indicating a receptor-mediated AA liberation (fig. 1). The amount of released free [³H]AA increased linearly with time in the first 2 min and then reached the plateau (fig. 2). After 5 min incubation with 400 nM PAF, the amount of released radioactivity corresponded to 5–7% of the total radioactivity incorporated into the cells. Concomitantly, decreases in radioactivity were observed in phosphatidylcholine (PC) and phosphatidylinositol (PI) fractions. In contrast, there were no significant decreases in other lipid fractions, such as phosphatidylethanolamine, phosphatidylserine and triacylglycerol. The inhibitors of phospholipase A_2, mepacrine (150 μM) and ONO-RS-082 (2-(p-amylcinnamoyl)amino-4-chlorobenzoic acid) (4 μM), effectively suppressed the liberation of [³H]AA without markedly affecting the level of [³H]DG (table 1), indicating that the liberation of AA is mainly catalyzed by phospholipase A_2. This is consistent with the previous finding [1] that deacylation of PI and PC by phospholipase A_2 occurred in Ca^{2+} ionophore A23187-treated human neutrophils.

Pretreatment of neutrophils with 150 μM TMB-8 (8-(N,N-diethyl-amino)octyl-3,4,5-trimethoxybenzoate), known as an intracellular Ca^{2+} antagonist, resulted in the marked inhibition of PAF-induced [³H]AA release. A high dose of quin 2, due to its high Ca^{2+} buffering capacity, is often used as a potent intracellular Ca^{2+} chelator. Neutrophils loaded with 40 μM quin 2 acetoxy methylester showed significant decrease of [³H]AA release in response to PAF, suggesting that intracellular Ca^{2+} is required for AA liberation.

When neutrophils were incubated for 2 h with pertussis toxin, which ADP ribosylates certain G proteins (Gi, Go, transducin) and reduces their ability to couple receptors to effector proteins [3, 4], the ability of PAF to stimulate [³H]AA liberation was inhibited. The inhibitory effect of pertussis toxin on PAF-induced AA liberation was dependent on the concentration of the toxin and the complete inhibition was observed at the dose of 200 ng/ml. As previously described in rabbit [3] and guinea pig [4] neutrophils, fMLP (formyl-methionyl-leucyl-phenylalanine)-stimulated [³H]AA release was also inhibited by the toxin.

PAF was found to elicit an increase in $[Ca^{2+}]i$ as monitored by fura 2. The fura 2 response to PAF was observed at concentrations as low as 8 pM and reached a maximal level at around 8 nM. When neutrophils were pretreated with pertussis toxin (200 ng/ml) for 2 h, fMLP-induced $[Ca^{2+}]i$

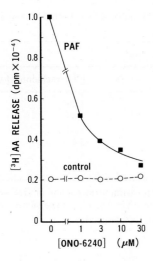

Fig. 1. Effect of PAF receptor antagonist, ONO-6240 on PAF-induced AA liberation in human neutrophils. [³H]AA-labeled human neutrophils were incubated with indicated concentrations of ONO-6240 for 5 min and then incubated for a further 5 min with 400 nM PAF. Each point is the mean of triplicate determinations. Two other experiments gave similar results.

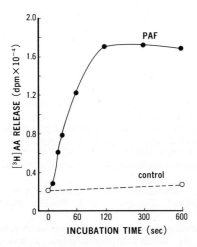

Fig. 2. Time course of AA release in response to PAF in human neutrophils. [³H]AA-labeled human neutrophils were incubated with 400 nM of PAF for the periods of time indicated. Each point is the mean of triplicate determinations. Two other experiments gave similar results.

Table 1. Effects of phospholipase A_2 inhibitors on PAF-induced AA liberation and DG formation in human neutrophils

	Radioactivity, dpm	
	[³H]AA	[³H]arachidonoyl-DG
Control	1,827	1,298
Mepacrine (200 μM)	1,613	1,499
ONO-RS-082 (4 μM)	1,909	1,304
PAF (400 nM)	12,662	3,418
Mepacrine + PAF	3,401	4,214
ONO-RS-082 + PAF	3,838	3,279

[³H]AA-labeled human neutrophils were incubated with or without mepacrine or ONO-RS-082 for 5 min and then incubated for a further 5 min with PAF. The results shown are mean values of triplicate determinations from one of two similar experiments.

rise was almost completely abolished. On the other hand, the fura 2 response to PAF was only marginally affected by the toxin treatment.

In summary, treatment of neutrophils with pertussis toxin resulted in inhibition of AA liberation in response to PAF. However, the fura 2 response to PAF was not significantly affected by the toxin treatment. These results indicate that the rise of $[Ca^{2+}]i$ is required but not sufficient to stimulate phospholipase A_2 and that a pertussis toxin-sensitive G protein may be involved in the activation of phospholipase A_2.

References

1   Walsh CE, DeChatelet LR, Chilton FH, et al: Mechanism of arachidonic acid release in human polymorphonuclear leukocytes. Biochim Biophys Acta 1983;750:32–40.
2   Balsinde J, Diez E, Schuller A, et al: Phospholipase A_2 activity in resting and activated human neutrophils. J Biol Chem 1988;263:1929–1936.
3   Okajima F, Ui M: ADP-ribosylation of the specific membrane protein by islet-activating protein, pertussis toxin, associated with inhibition of a chemotactic peptide-induced arachidonate release in neutrophils. J Biol Chem 1984;259:13863–13871.
4   Bokoch GM, Gilman AG: Inhibition of receptor-mediated release of arachidonic acid by pertussis toxin. Cell 1984;39:301–308.
5   Nakashima S, Nagata K, Ueeda K, et al: Stimulation of arachidonic acid release by guanine nucleotide in saponin-permeabilized neutrophils. Arch Biochem Biophys 1988;261:375–383.
6   Verghese MW, Charles L, Jakoi L, et al: Role of a guanine nucleotide regulatory protein in the activation of phospholipase C by different chemoattractants. J Immunol 1987;138:4374–4380.

Yoshinori Nozawa, MD, PhD, Department of Biochemistry, Gifu University
School of Medicine, Gifu 500 (Japan)

Reproductive Biology and Endocrine System

Zor U, Naor Z, Danon A (eds): Leukotrienes and Prostanoids in Health and Disease
New Trends Lipid Mediators Res. Basel, Karger, 1989, vol 3, pp 283–286

Immunocytochemical Localization of 5- and 12-Lipoxygenases and Cyclooxygenase in Nonpregnant Human Uteri

Ch.V. Rao, N. Chegini, Z.M. Lei

Department of Obstetrics and Gynecology, University of Louisville School of Medicine, Louisville, Ky., USA

Introduction

Human uterus consists of two distinct morphological compartments, i.e. endometrium and myometrium [1, 2]. Endometrium contains glandular and luminal epithelial cells and differentiated fibroblasts called stromal cells. Myometrium contains inner circular and outer elongated smooth muscle differing in several biophysical and biochemical properties [3]. In addition, endometrium and myometrium contain vascular smooth muscle, endothelium, connective tissue elements, tissue macrophages and various types of blood cells in the arterioles lumen [1, 2]. Many of these cells change during the cycle and pregnancy [1, 2].

Human uterus contains both lipoxygenase and cyclooxygenase pathways of arachidonic acid metabolism [4–8]. Whether the enzymes in these pathways are present in all or some of the cells in uterus is unknown. A majority of the previous studies on uterine eicosanoids production have used whole tissue fragments, minces or homogenates [4–8]; thus, they could not provide any information on cellular sources of eicosanoids. There are a few studies on eicosanoids production by separated glandular epithelium, stromal cells [9] and myometrium [10]. They also fail to reveal potential contribution of other cells to the uterine eicosanoids production. Immunocytochemical studies on whole uterine tissue using highly specific antibodies to eicosanoid-synthesizing enzymes is perhaps one of the best approaches to investigate the above question.

Materials and Methods

A monoclonal antibody to porcine leukocyte 5-lipoxygenase (5-LOX-1), polyclonal antibody to porcine leukocyte 12-lipoxygenase and highly purified 12-lipoxygenase were kindly provided by Prof. Shozo Yamamoto, Tokushima University School of Medicine, Japan. A monoclonal antibody to ovine seminal vesicle cyclooxygenase (Cayman Chemicals Co.) and a highly purified cyclooxygenase (Oxford Biomedical Research, Inc.) were purchased from commercial sources.

Six human uteri (three from proliferative and three from secretory phase) removed from premenopausal women for medical reasons were obtained from the operating room and processed immediately for light microscopy. Briefly, the tissues were fixed overnight at 22 °C in Bouin's solution and embedded in paraffin. Hydrogen peroxide (H_2O_2)-treated 5-μm thick sections were washed for 30 min with phosphate-buffered physiological saline (PBS), pH 7.3 and sequentially exposed to: (1) 1:50 dilution of normal horse (5-lipoxygenase and cyclooxygenase) or goat (12-lipoxygenase) serum in PBS for 30 min to saturate all the nonspecific IgG-binding sites; (2) 1:200 dilution of antibody to 5-lipoxygenase, 1:500 dilution of antibody to 12-lipoxygenase and 1:100 dilution of antibody to cyclooxygenase in PBS containing 0.5% BSA for 2 h; (3) 1:250 dilution of biotinylated horse anti-mouse IgG (5-lipoxygenase and cyclooxygenase) or biotinylated goat anti-rabbit IgG (12-lipoxygenase) in PBS for 1 h; (4) 1:100 dilution of avidin-biotinylated horseradish peroxide complex in PBS for 1 h, and (5) 0.1% 3,3′-diaminobenzidine (w/v) and 0.02% H_2O_2 (v/v) in PBS for 5 min.

After each step, the sections were rinsed 3 times for 5 min each with PBS. All the steps were carried out in a humidified chamber at 22 °C. Controls included: (a) the use of 5-lipoxygenase antibody preabsorbed with crude 5-lipoxygenase prepared from porcine leukocytes [11], 12-lipoxygenase and cyclooxygenase antibodies preabsorbed with corresponding highly purified enzymes, and (b) omission of unabsorbed primary antibody during the immunostaining procedure. The cellular differences and cycle-dependent changes of immunostaining were visually assessed independently by three different investigators.

Results

The specific immunostaining for 5- (fig. 1a–c) and 12-lipoxygenases (fig. 1d–f) and cyclooxygenase (fig. 2a–c) was seen in all the cells in endometrium and myometrium. Small cells interspersed between the stromal cells contained the most immunostaining for all three enzymes (fig. 1a, d, 2a). Although the true identity of these cells is not known, these could possibly be macrophages. Luminal and glandular epithelial cells contained more immunostaining for all three enzymes than stromal cells (fig. 1a, d, 2a). The distribution of 5-lipoxygenase and cyclooxygenase was similar in luminal and glandular epithelium (fig. 1a, 2a), whereas 12-lipoxygenase was higher in luminal compared to glandular epithelium (fig. 1d).

Myometrium contained less 5-lipoxygenase (fig. 1b) and cyclooxygenase (fig. 2b) and more (with the exception of macrophages) 12-lipoxygenase (fig. 1e) than endometrium. Arterioles smooth muscle and endothelium contained less 5- (fig. 1b) and 12-lipoxygenase (fig. 1e) and more cyclooxygenase (fig. 2b) than myometrial smooth muscle.

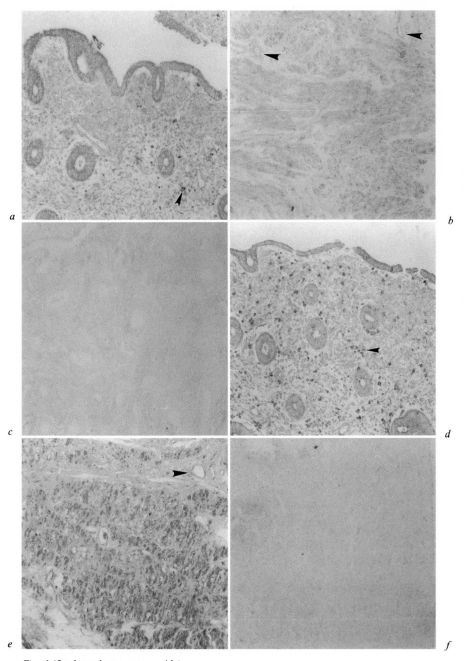

Fig. 1 (for legends see reverse side).

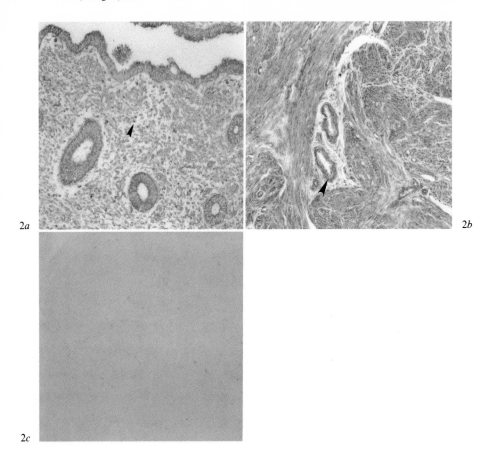

2a

2b

2c

Fig. 1. Light microscope immunocytochemical localization of 5-lipoxygenase (a–c) and 12-lipoxygenase (d–f) in nonpregnant human uterus. a, b Immunostaining of macrophages (arrowhead), luminal and glandular epithelial cells and stromal cells in endometrium. b, e Immunostaining of circular and elongated myometrial and arteriole smooth muscle and endothelial cells (arrowhead) in myometrium. c, f The complete disappearance of immunostaining from all the endometrial cells following preabsorption of the primary antibodies with the corresponding enzymes. × 140.

Fig. 2. Light microscope immunocytochemical localization of cyclooxygenase in nonpregnant human uterus. a Immunostaining in macrophages (arrowhead), luminal and glandular epithelial cells and stromal cells in endometrium. b Immunostaining in circular and elongated myometrial and arteriole smooth muscle and endothelial cells (arrowhead) in myometrium. c The complete absence of immunostaining in all the endometrial cells following preabsorption of the primary antibody with purified cyclooxygenase. × 140.

The immunostaining in all the cells in endometrium (fig. 1c, f, 2c) and myometrium (data not shown) disappeared following preabsorption of primary antibodies with the corresponding enzymes.

The immunostaining for 5-lipoxygenase increased in glandular epithelium with no change in other cells from proliferative to secretory phase. The 12-lipoxygenase decreased in all the uterine cells from proliferative to secretory phase. Immunostaining for cyclooxygenase increased in glandular epithelium with no consistent change in other cells from proliferative to secretory phase.

Discussion

Although human uterus is known to synthesize eicosanoids from lipoxygenase and cyclooxygenase pathways [4–8], the potential contribution of all different cells in uterus to this synthesis has never been established before. This study demonstrates for the first time that all the cells in uterus can potentially synthesize eicosanoids from 5- and 12-lipoxygenase as well as cyclooxygenase pathways. All three enzymes were found to a different degree in all the cells in uterus. The presumed macrophages contained the most intense immunostaining for all three enzymes. After this, the order of cellular immunostaining and the changes from proliferative to secretory phase differed for each enzyme. These differences suggest that these enzymes and products derived therefrom have different roles in uterus and that these enzymes are differently regulated.

There were previous studies on immunocytochemical localization of cyclooxygenase in human [12, 13] and ovine [14] uterus. Studies on human uterus reported divergent results, i.e., one study reported immunostaining only in glandular epithelium, whereas the second study demonstrated immunostaining in myometrial smooth muscle but not in uterine vasculature. Our results demonstrating the presence of cyclooxygenase in all the cells in uterus differed from these human studies, but are in substantial agreement with the study on ovine uterus which showed immunostaining in all the cells.

Since all the cells in uterus contain their own 5- and 12-lipoxygenases and cyclooxygenases, one could speculate that different cells in uterus produce their own needed eicosanoids. In addition, it is possible that eicosanoids produced by one cell type may regulate the eicosanoid production by other cell types in a paracrine manner.

To summarize: the 5- and 12-lipoxygenases and cyclooxygenases were found in all the cells including vascular and other cellular elements present in nonpregnant human uteri. The relative cellular distribution and changes from proliferative to secretory phase differed for each enzyme. This suggests that the eicosanoids formed by each of these enzyme pathways have different roles in uterine functions and that these enzymes are regulated differently.

References

1   Maximow AA, Bloom W: A Text Book in Histology. Philadelphia, Saunders, 1943.
2   Noyes RW, Hertig AT, Rock J: Dating the endometrial biopsy. Fertil Steril 1950;1:3–25.
3   Tuross N, Mahtani M, Marshall JM: Comparison of effects of oxytocin and prostaglandin $F_{2\alpha}$ on circular and longitudinal myometrium from the pregnant rat. Biol Reprod 1987;37:348–355.
4   Saeed SA, Mitchell MD: Formation of arachidonate lipoxygenase metabolites by human fetal membranes, uterine decidua vera and placenta. Prostaglandins Leukotrienes Med 1982;8:635–640.
5   Demers LM, Rees MCP, Turnbull AC: Arachidonic acid metabolism by the non-pregnant human uterus. Prostaglandins Leukotrienes Med. 1984;14:175–180.
6   Flatman S, Hurst JS, McDonald-Gibson RG, et al: Biochemical studies on a 12-lipoxygenase in human uterine cervix. Biochim Biophys Acta 1986;883:7–14.
7   Rees MCP, DiMarzo V, Tippins JR, et al: Leukotriene release by endometrium and myometrium throughout the menstrual cycle in dysmenorrhoea and menorrhagia. J Endocrinol 1987;113:291–295.
8   Abel MH, Kelly RW: Differential production of prostaglandins within the human uterus. Prostaglandins 1979;18:821–828.
9   Lumsden MA, Brown A, Baird DT: Prostaglandin production from homogenates of separated glandular epithelium and stroma from human endometrium. Prostaglandins 1984;28:485–496.
10 Vijayakumar R, Walters WA: Myometrial prostaglandins during the human menstrual cycle. Am J Obstet Gynecol 1981;141:313–318.
11 Kaneko S, Ueda N, Tonai T, et al: Arachidonate 5-lipoxygenase of porcine leukocytes studied by enzyme immunoassay using monoclonal antibodies. J Biol Chem 1987;262:6741–6745.
12 Rees MCP, Parry DM, Anderson ABM, et al: Immunohistochemical localisation of cyclooxygenase in the human uterus. Prostaglandins 1982;23:207–214.
13 Moonen P, Klok G, Keirse MJNC: Immunohistochemical localization of prostaglandin endoperoxide synthase and prostacyclin synthase in pregnant human myometrium. Eur J Obstet Gynecol Reprod Biol 1985;19:151–158.
14 Huslig RL, Fogwell RL, Smith WL: The prostaglandin forming cyclooxygenase of ovine uterus: relationship to luteal function. Biol Reprod 1979;21:589–600.

Ch.V. Rao, PhD, Department of Obstetrics and Gynecology, 438 MDR Bldg, University of Louisville, School of Medicine, Louisville, KY 40292 (USA)

Zor U, Naor Z, Danon A (eds): Leukotrienes and Prostanoids in Health and Disease.
New Trends Lipid Mediators Res. Basel, Karger, 1989, vol 3, pp 287–291

Mechanism of Action of Prostaglandin $F_{2\alpha}$ in the Rat Corpus Luteum

Michal Lahav, Hanna Rennert

Faculty of Medicine, Technion–Israel Institute of Technology, Haifa, Israel

Introduction

It is well established that prostaglandin $F_{2\alpha}$ ($PGF_{2\alpha}$) triggers luteal regression in many species, including the rat [1]. We and others have shown that, in isolated, intact rat corpora lutea (CL) and in luteal cell suspensions, $PGF_{2\alpha}$ prevented, or rapidly reversed, the luteinizing hormone (LH)-induced cyclic AMP (cAMP) accumulation and progesterone secretion [2–7]. As in broken cell preparations a direct effect of $PGF_{2\alpha}$ on adenylate cyclase or phosphodiesterase could not be demonstrated [3, 5], Behrman and co-workers proposed that calcium ions mediate the action of $PGF_{2\alpha}$ in the rat corpus luteum. This suggestion was based on the ability of ionophores and ouabain to inhibit cAMP accumulation in intact luteal cells, and on the inhibitory action of calcium ions at $10~\mu M$ or more on adenylate cyclase in luteal membranes [8, 9]. Moreover, Leung et al. [10] showed that $PGF_{2\alpha}$ stimulated polyphosphoinositide metabolism in cultured rat luteal cells.

We examined Behrman's hypothesis in isolated rat CL of pseudopregnancy. Very young CL were compared to CL near the end of their functional life span, since young CL are known to be relatively resistant to the luteolytic effect of $PGF_{2\alpha}$ in vivo and in vitro [3]. Our results are not compatible with the suggestion that the $PGF_{2\alpha}$-induced inhibition of cAMP accumulation in the rat CL depends on the intracellular messengers associated with polyphosphoinositide hydrolysis.

In our studies, the formation of CL (20–35/rat) was induced by administration of 15 IU pregnant mare serum gonadotropin (PMSG) to sexually immature animals; these CL remained functional for 11 days [4, 6]. Two- or 10-day-old CL were isolated and incubated in defined, oxygenated medium. Groups of CL (9–14 CL/sample) were exposed to hormones and drugs, usually for 90 min, and tissue cAMP (extracted by boiling in acetate buffer, pH4) and sometimes also progesterone in the medium were determined [4, 6].

Role of Calcium Ions in the Action of $PGF_{2\alpha}$ in Mature CL

In our model of pseudopregnant rats, day 10 CL are probably similar to the physiological target of $PGF_{2\alpha}$ as a luteolysin: they are functional, but close to the end of their functional life span, and thus are very sensitive to $PGF_{2\alpha}$. When day 10 CL were incubated in a calcium-depleted medium, containing 0.5 mM EGTA (in which the concentration of free calcium ions is 30 nM), there was a massive (64%) reduction in total calcium content in the tissue [6]. Nevertheless, the stimulatory effect of LH on cAMP accumulation, as well as the suppression by $PGF_{2\alpha}$ of this effect of LH, were not impaired [4, 6, 11]. Furthermore, $PGF_{2\alpha}$ did not stimulate the uptake of radioactive calcium when added either simultaneously with the isotope or 150 min later [4, 12]. Attesting to the validity of the latter experiments are the findings that calcium uptake was inhibited by lanthanum ions, and stimulated by preincubation of CL in calcium-poor medium, or by addition of the calcium ionophore A23187 [12]. Verapamil did not affect calcium uptake, as well as the actions of LH and $PGF_{2\alpha}$ on cAMP; thus, day 10 CL may lack voltage-dependent calcium channels [4, 12].

8-(N,N-diethylamino)octyl-3,4,5-trimethoxybenzoate (TMB-8) interferes with processes requiring intracellular calcium [references in 6, 13]. In day 10 CL, TMB-8, alone or in combination with calcium-depleted medium, did not prevent the inhibitory effect of $PGF_{2\alpha}$ on LH-induced cAMP accumulation. Three calmodulin inhibitors, trifluoperazine, W-7 and pimozide, also did not affect the action of $PGF_{2\alpha}$ [4, 6]. Interestingly, the first two drugs were also reported to inhibit protein kinase C [references in 6, 14].

When day 10 CL were incubated with LH together with increasing concentrations of phorbol 12-myristate 13-acetate (PMA, 0.02–10 μM), this activator of protein kinase C significantly augmented the effect of LH on cAMP at 1 and 10 μM, and had no effect at the lower concentrations [7].

Effect of $PGF_{2\alpha}$ on Polyphosphoinositide Metabolism

In a collaborative study with John S. Davis, in his laboratory at the University of South Florida in Tampa, we examined the effect of $PGF_{2\alpha}$ on phospholipid metabolism in intact CL isolated from pseudopregnant rats

[15]. In these experiments, formation of CL was induced by 8 IU, rather than 15 IU, of PMSG; since the life span of such CL is only 9 days [3, 15], 2- and 7-day-old CL were compared. We found that, in the young CL, $PGF_{2\alpha}$ stimulated $^{32}PO_4$ incorporation into phosphatidic acid and phosphatidylinositol, as well as formation of inositol phosphates. In contrast, in day 7 CL, $PGF_{2\alpha}$ had no effect; however, basal polyphosphoinositide turnover was significantly higher in the mature CL. LH did not stimulate polyphosphoinositide hydrolysis in CL of either age.

Thus, no correlation was found between the effectiveness of $PGF_{2\alpha}$ in suppressing luteal cAMP on the one hand, and in stimulating phospholipase C on the other hand; these results suggest that, in the rat CL, polyphosphoinositide hydrolysis is not the only mechanism of action of $PGF_{2\alpha}$.

Role of Calcium Ions in the Action of $PGF_{2\alpha}$ in Young CL

In day 2 CL from rats pretreated with 15 IU PMSG, maximal suppression of cAMP accumulation required a 100-fold higher concentration of $PGF_{2\alpha}$ compared to the amount required in day 10 CL. As in the mature CL, this inhibition was not prevented by interference with calcium availability. The effect of $PGF_{2\alpha}$ persisted in calcium-depleted medium, as well as in the presence of TMB-8 or trifluoperazine at concentrations which abolished the LH-induced progesterone secretion. PMA added simultaneously with LH augmented the effect of LH on cAMP at 1 μM, and had no effect at other concentrations [7].

Effect of Phosphodiesterase Inhibitors in Isolated CL and in Luteal Cell Suspensions

It was previously shown that, in isolated CL, the inhibitory effect of $PGF_{2\alpha}$ was not prevented by 3-isobutyl-1-methylxanthine (IBMX) at 0.1 mM [3]. However, this concentration is suboptimal in the rat CL [5, 7, and Lahav and Rafaeloff, unpubl.]. We found that, at 0.5 mM, IBMX abolished the effect of $PGF_{2\alpha}$ on cAMP in day 2 as well as in day 10 CL. The same was found for another phosphodiesterase inhibitor, Ro 20-1724, at 0.1 and 0.5 mM [7]. Dorflinger et al. [5] argued that, even at 0.5 mM, IBMX did not abolish the effect of $PGF_{2\alpha}$ on cAMP in luteal cell suspensions. We prepared luteal cells [16] from ovaries of rats pretreated with 50 IU PMSG and 25 IU hCG. We found that Ro 20-1724 at 0.01 and 0.05 mM, concentrations which optimally augmented LH-induced cAMP accumulation, significantly attenuated the inhibitory effect of $PGF_{2\alpha}$ from 70–85% to 50–55%; still, the remaining inhibition was highly significant [Rennert and Lahav, unpubl.].

These results are at present hard to interpret. The discrepancy between the effects of phosphodiesterase inhibitors in isolated CL and in luteal cell suspensions illustrate the need for caution in drawing conclusions from a single experimental system with regard to a physiological mechanism.

Acknowledgements

These studies were supported by the Israel Academy of Sciences and Humanities, The Dario and Mathilde Beraha Fund for Hormones and Cancer, and by the Chief Scientist of the Israel Ministry of Health.

References

1   Rothchild I: The regulation of the mammalian corpus luteum. Recent Prog Horm Res. Academic Press, 1981, vol 37, pp 183–298.
2   Lahav M, Freud A, Lindner HR: Abrogation by prostaglandin $F_{2\alpha}$ of LH-stimulated cyclic AMP accumulation in isolated rat corpora lutea of pregnancy. Biochem Biophys Res Commun 1976;68:1294-1300.
3   Khan MI, Rosberg S, Lahav M, et al: Studies on the mechanism of action of the inhibitory effect of prostaglandin $F_{2\alpha}$ on cyclic AMP accumulation in rat corpora lutea of various ages. Biol Reprod 1979;21:1175–1183.
4   Lahav M, Weiss E, Rafaeloff R, et al: The role of calcium ion in luteal function in the rat. J Steroid Biochem 1983;19:805–810.
5   Dorflinger LJ, Luborsky JL, Gore SD, et al: Inhibitory characteristics of prostaglandin $F_{2\alpha}$ in the rat luteal cell. Mol Cell Endocrinol 1983;33:225–241.
6   Lahav M, Rennert H, Sabag K, et al: Calmodulin inhibitors and 8-(N,N-diethyl-amino)octyl-3,4,5-trimethoxybenzoate (TMB-8) do not prevent the inhibitory effect of prostaglandin $F_{2\alpha}$ on cyclic AMP production in rat corpora lutea. J Endocrinol 1987;113:205–212.
7   Lahav M, Davis JS, Rennert H: Mechanism of the luteolytic action of prostaglandin $F_{2\alpha}$ in the rat. J Reprod Fertil 1989;Suppl 37:233–240.
8   Dorflinger LJ, Albert PJ, Williams AT, et al: Calcium is an inhibitor of luteinizing hormone-sensitive adenylate cyclase in the luteal cell. Endocrinology 1984;114:1208–1215.
9   Gore SD, Behrman HR: Alteration of transmembrane sodium and potassium gradients inhibits the action of luteinizing hormone in the luteal cell. Endocrinology 1984;114:2020–2031.
10 Leung PCK, Minegishi T, Ma F, et al: Induction of polyphosphoinositide breakdown in rat corpus luteum by prostaglandin $F_{2\alpha}$. Endocrinology 1986;119:12–18.
11 Lahav M, Rennert H, Barzilai D: Inhibition by vanadate of cyclic AMP production in rat corpora lutea incubated in vitro. Life Sci 1986;39:2557–2564.

12 Lahav M, Shariki-Sabag K, Rennert H: Lack of effect of prostaglandin $F_{2\alpha}$ and verapamil on calcium uptake by isolated corpora lutea from pseudopregnant rats. Biochem Pharmacol 1989;38:546–548.

13 Donowitz M, Cusolito S, Sharp GWG: Effects of calcium antagonist TMB-8 on active Na and Cl transport in rabbit ileum. Am J Physiol 1986;250:G691–G697.

14 Hidaka H, Hagiwara M: Pharmacology of the isoquinoline sulfonamide protein kinase C inhibitors. Trends Pharmacol Sci 1987;8:162–164.

15 Lahav M, West LA, Davis JS: Effects of prostaglandin $F_{2\alpha}$ and gonadotropin-releasing hormine agonist on inositol phospholipid metabolism in isolated rat corpora lutea of various ages. Endocrinology 1988;123:1044–1052.

16 Nelson S, Khan MI: Estradiol biosynthesis in the corpus luteum of the pregnant rat involves two different luteal cell populations. Program of the 69th Endocrine Society Meeting, 1987, p 216.

Michal Lahav, MD, Faculty of Medicine, Technion-Israel Institute of Technology, IL-Haifa 31096 (Israel)

Zor U, Naor Z, Danon A (eds): Leukotrienes and Prostanoids in Health and Disease.
New Trends Lipid Mediators Res. Basel, Karger, 1989, vol 3, pp 292–302

Regulation of Prostaglandin $F_{2\alpha}$ Output in Human Endometrial Epithelial Cells: Implications in Implantation

Frederick Schatz[a], Ronald E. Gordon[b], Neri Laufer[c], Erlio Gurpide[a]

Departments of [a]Obstetrics, Gynecology and Reproductive Science and
[b]Pathology, Mount Sinai Medical Center, New York, NY., and [c]Department of
Obstetrics and Gynecology, Hadassah Hospital, Jerusalem, Israel

Several lines of evidence indicate that endometrial prostaglandins (PGs) are mediators of the earliest implantational events, increased endometrial vascular permeability (EVP) and decidualization of the endometrial stroma. In nonprimates, these changes occur in localized areas adjacent to the sites of implantation where concentrations of PGs are elevated, and apparently controlled by a chemical or physical signal from the blastocyst. In women, each menstrual cycle is characterized by increased EVP and decidualization of the stroma throughout the endometrium consistent with a more generalized, nonblastocyst-regulated increase in endometrial PG production [for review, see 1].

Analysis of PGs in jet washes of the uterine lumen [2], or in extracts of endometrial biopsies [3], indicated that between approximately days 16 and the time during which implantation normally occurs in women, levels of $PGF_{2\alpha}$ increased severalfold. In vitro studies suggested that estrogens acted on the epithelium to regulate $PGF_{2\alpha}$ production in human endometrium. Thus, exogenous E_2 enhanced output of $PGF_{2\alpha}$ by explants of human secretory endometrium [4–6], whereas no estrogen effects were noted in proliferative tissue [4, 5]. In cultures of the predominant endometrial cell types on plastic dishes, exogenous E_2 elevated $PGF_{2\alpha}$ outputs in the epithelial cells (ECs) [7, 8], but not in the stromal cells (SCs) [8, 9].

This report reviews studies on the regulation of this response to estrogens in human EC monolayers on plastic dishes. Our recent results in culturing the ECs under conditions that have been shown to establish differentiated structure and function in other EC systems are also presented, together with a consideration of the potential that such polarized cells offer in studying the role of eicosanoids in implantation.

Methods

Tissues. Endometrial specimens were obtained from patients undergoing dilatation and curettage or hysterectomy.

Isolation of Glands. A small portion of each specimen was removed for histological dating, and glands were isolated and purified to virtual homogeneity as described previously [10].

$PGF_{2\alpha}$ Output by Cultures of Glandular Epithelia. Figure 1 shows the scheme used to study $PGF_{2\alpha}$ production in primary cultures of ECs that were derived from glands placed in 6-cm polystyrene culture dishes (Falcon Plastics, Los Angeles, Calif.) with 2.5 ml/dish of standard medium (Ham's F-10 supplemented with 10 μg/ml insulin, 1% of the antibiotic-antimycotic mixture and 10% charcoal-stripped calf serum). The dishes were kept in a 37 °C incubator for 24 h to allow attachment of glands (plating period). The medium containing floating glands and cells was replaced with fresh medium and the dishes returned to the incubator to enable attached glands to form monolayers (monolayer formation period). The medium was exchanged for fresh standard medium containing either the test compounds described in the Results section, or 0.1% ETOH, the vehicle used to add these compounds. After 64 h (testing period) media were collected and centrifuged, and the supernatants frozen until analyzed for $PGF_{2\alpha}$ content. Cells were harvested after trypsin-EDTA treatment and assayed for protein content.

Measurement of $PGF_{2\alpha}$. Levels of $PGF_{2\alpha}$ were measured in culture medium by radioimmunoassay as described previously [7].

Culture of Glandular Epithelia under Polarizing Conditions. Freshly isolated glands were suspended in basal medium (a phenol red-free mixture of Dulbecco's minimum essential medium (Gibco) and Ham's F-12 (Flow Laboratories), 1:1 v/v, supplemented with 100 U/ml penicillin, 100 μg/ml streptomycin, 0.25 μg/ml fungizone and 10% stripped calf serum) and transferred to the dual chamber (DC)-reconstituted basement membrane (RBM) system. The system consists of a cylindrical polystyrene insert (Millicell® CM, Millipore) that sits on small projections in the well of a tissue culture plate, thereby providing access of culture medium to the underside of a permeable (Biopore) filter, which is sealed to and constitutes the floor of the insert, and serves as a cell support. In these experiments, 12-mm diameter filters were each coated with 0.1 ml of the commercially available RBM, Matrigel (Collaborative Research) diluted 1:8 with basal medium and dried under a laminar flow hood. At the time the glands were plated, each insert was placed in a well of a standard 24-well tissue culture plate (Falcon Plastics) containing 0.5 ml of basal medium, which was sufficient to wet the RBM-coated filter, thereby rendering it transparent for phase microscopy. A fresh 0.1 ml aliquot of the diluted Matrigel was added, immediately followed by 0.5 ml of a suspension of the glands in basal medium. The plate was placed in an incubator, and the medium replaced in both upper and lower chambers every 3 days.

Aliquots of the gland suspension were also cultured in parallel in 3.5-cm polystyrene culture dishes containing 1.0 ml of basal medium, replacing the medium every 3 days.

Fixation and Microscopy. Incubations were terminated by removing the medium and rinsing the cells twice with Hank's balanced salt solution. The cultures were fixed immediately for 1 h with 0.1 M sodium cacodylate buffer containing 2% gluteraldehyde and 1% OsO_4 at pH 7.4, then the fixative was removed and the cultures were washed with a 0.2 *M* cacodylate solution. The RBM-coated filter containing glandular ECs was punched out of the Millicell insert, and subjected, in parallel with the cells on the plastic dishes, to dehydration in a graded series of ETOH steps before embedding both cultures in Epon 812. One-micron sections cut from the Epon blocks were stained with methylene blue and Azure II, and examined by light microscopy. Ultrathin sections were cut at areas judged to be representative of the specimen. These were stained with uranyl acetate and lead citrate and observed under a TEM 100 CX transmission electron microscope.

Results

Effects of E_2, Arachidonic Acid (AA) and A23187 on $PGF_{2\alpha}$ Output by Epithelial Cells on Plastic

Both AA, the obligatory substrate for PG synthase leading to $PGF_{2\alpha}$ formation, and A23187, a calcium ionophore stimulator of phospholipase activity [references in 11], were reported to enhance $PGF_{2\alpha}$ output by human endometrial explants [12]. Following the protocol described in figure 1, these agents were used to investigate the possibility that E_2-elevated $PGF_{2\alpha}$ output by human endometrial ECs involved increased PG synthase activity.

Table 1 compares the effects of E_2, AA and $E_2 + AA$ on $PGF_{2\alpha}$ outputs during 64 h incubations of primary cultures of ECs in plastic dishes [11]. As expected, exogenous E_2 elevated $PGF_{2\alpha}$ output in the ECs. The addition of AA at 4 and 20 µg/ml produced dose-dependent increases in $PGF_{2\alpha}$ output by the ECs. However, incubations with E_2 and AA at either concentration resulted in escalations of $PGF_{2\alpha}$ output to levels that were about 2.5 times the sum of the separate increases evoked by E_2 and AA alone. Although E_2 could enhance AA-elevated $PGF_{2\alpha}$ levels by inhibiting $PGF_{2\alpha}$ metabolism and/or increasing AA uptake, neither mechanism appears operative in the endometrial ECs, since control and E_2-treated EC cultures metabolized ^3H-$PGF_{2\alpha}$ [5], and removed ^3H-AA from the medium

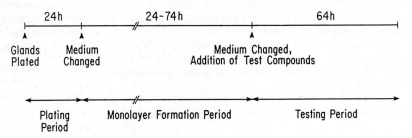

Fig. 1. Schema for testing the effects of E_2, AA, A23187, TAM and OHTAM on $PGF_{2\alpha}$ levels in cultures of glandular epithelium.

Table 1. Effects of E_2 and arachidonic on $PGF_{2\alpha}$ output by endometrial epithelial cells

| | Basal output, ng $PGF_{2\alpha}$/ mg protein $\times$ 64 h | Percent of controls | | | | |
| | | $+E_2$ $(10^{-8} M)$ | $+$arachidonic acid | | $+E_2$ $(10^{-8} M)$ $+$arachidonic acid | |
			4 μg/ml	20 μg/ml	4 μg/ml	20 μg/ml
Mean[1]	9.4	510	550	1,900	2,700	6,200
SE	± 3.0	± 130	± 100	± 180	± 830	$\pm 1,900$
n^2	5	5	5	3	5	3

[1] Averages from results of duplicate dishes.
[2] Cultures of epithelial cells derived from 1 specimen of early proliferative, 3 or midproliferative, 1 of late proliferative, and 1 of day 21 secretory endometrium.

[11] to a similar extent. Alternatively, the synergism between E_2 and AA is likely a reflection of E_2-increased PG synthase activity.

Table 2 serves to compare the effects of E_2 in standard culture medium, with those of E_2, A23187, and $E_2 + $A23187 in Ca^{++}-supplemented standard culture medium on $PGF_{2\alpha}$ output during 64 h incubations of primary cultures of ECs [11]. Estradiol evoked similar increases in $PGF_{2\alpha}$ output in both culture media. Carrying out incubations in the Ca^{++}-supplemented medium, A23187, which increased $PGF_{2\alpha}$ output in the ECs in an external Ca^{++} concentration-dependent fashion [11], together with E_2 elevated $PGF_{2\alpha}$ outputs to levels about twice the sum of that produced by each compound separately. The synergistic effects produced by incubations of E_2 and A23187 in the ECs, therefore supports conclusions drawn from similar effects obtained with E_2 and AA, that E_2 increases $PGF_{2\alpha}$ output by increasing PG synthase activity.

Effects of E_2, Tamoxifen (TAM), and 4-Hydroxytamoxifen (OHTAM) on $PGF_{2\alpha}$ Output by Epithelial Cells on Plastic

The incubation scheme shown in figure 1 was also followed to study OHTAM and TAM for possible estrogenic agonist and antagonist effects on $PGF_{2\alpha}$ output by the endometrial ECs [13]. Table 3, which summarizes the results of five experiments, indicates that OHTAM acted as an essentially pure antagonist of this response. Thus, $10^{-6} M$ OHTAM exerted almost no stimulatory effects, but counteracted virtually all of the marked elevation in $PGF_{2\alpha}$ output elicited by $10^{-8} M E_2$. As expected, OHTAM is a less effective antagonist at $10^{-7} M$. The antiestrogenicity of OHTAM was

Table 2. Effects of E_2 and A23187 acid on $PGF_{2\alpha}$ output by endometrial epithelial cells

	Control output, ng $PGF_{2\alpha}/$ mg protein × 64 h	Percent of controls			
		$+E_2$ ($10^{-8} M$)	$+E_2$ ($10^{-8} M$) $+Ca^{++}$ (1.5 mM)	$+A23187$ ($6 \times 10^{-7} M$) $+Ca^{++}$ (1.5 mM)	$+E_2$ ($10^{-8} M$) $+A23187$ ($6 \times 10^{-7} M$) $+Ca^{++}$ (1.5 mM)
Mean[1]	7.4	570	580	360	1,800
SE	±3.0	±150	±170	± 38	± 360
n[2]	5	4	3	5	5

[1] Averages from results of duplicate dishes.
[2] Cultures of epithelial cells were derived from 2 specimens of early proliferative, 3 from midproliferative, 1 of late proliferative, and 1 of day 21 secretory endometrium.

Table 3. Effects of E_2, OHTAM and TAM on $PGF_{2\alpha}$ output by endometrial epithelial cells

	Control output, ng $PGF_{2\alpha}/$ mg protein × 64 h	Percent of controls			
		$+E_2$ ($10^{-8} M$)	$+OHTAM$ ($10^{-6} M$)	$+E_2$ ($10^{-8} M$) $+OHTAM$ ($10^{-6} M$)	$+E_2$ ($10^{-8} M$) $+OHTAM$ ($10^{-7} M$)
Mean[1]	10.7	1,400	150	220	630
SE	± 3.8	± 350	± 15	± 63	±470
n[2]	6	6	5	6	2

[1] Averages from results of duplicate dishes.
[2] Cultures of epithelial cells were derived from 2 specimens of early proliferative, 2 of midproliferative, 1 of day 18 secretory and 1 of day 21 secretory endometrium.

compared with that of its parent compound TAM in two of the experiments. The results of these incubations (not shown), which were carried out with E_2 at $10^{-9} M$ and $10^{-8} M$, indicate that at least 1,000-fold molar excess of TAM was required to attain the antiestrogenicity of 100-fold molar excess of OHTAM. TAM was also found to be a virtually pure estrogen antagonist in this system.

The greater antiestrogenicity of OHTAM relative to TAM in these EC cultures corresponds to that reported in other estrogen target tissues where the different antagonistic activities correlate with binding affinities of specific estrogen receptors [references in 13]. These results suggest that specific receptors play a mediating role in E_2-enhanced PGF_2 output in the endometrial ECs. However, no correlation was found between specific estrogen receptor concentrations and the $PGF_{2\alpha}$ response to E_2 in explants of human secretory endometrium, or in fragments of endometrial carcinoma [14, 15].

Ultrastructure of Glandular ECs as a Function of Method of Culture

Figures 2 and 3 show morphological characteristics by TEM, of glands that were isolated from a specimen of proliferative endometrium and incubated for 2 weeks in parallel cultures in the RBM-DC system and on plastic dishes. In figure 2, a confluent, highly polarized monolayer of cuboidal ECs is evident on the surface of the RBM. The cells had a distinctive microvillous border, basal nuclei, and numerous organelles characteristic of a well-differentiated EC cytoplasm. The ECs were joined basally by interdigitating processes, and apically by tight junctions and desmosomes. Cultured on plastic (fig. 3), the glandular ECs bore little resemblance to either the characteristically highly polarized cells in situ, or to those in the RBM-DC system. Thus, they had only about one-sixth the depth of the cells in the RBM-DC system, their apical surfaces bore few microvilli, and few distinctive cytoplasmic organelles were seen. Cells overlapped in confluent areas, but no tight junctions or desmosomes were evident.

Discussion

Although in vivo studies in nonprimates have provided convincing evidence that endometrial-derived PGs are involved in implantation, neither the contribution by, nor hormonal regulation of, individual nonprimate uterine cells to their production are known [1]. In contrast, studies with primary cultures of the predominant cell types in human endometrium, the glandular epithelia and the stroma grown on plastic dishes, showed that E_2 elevates $PGF_{2\alpha}$ output only in the former. Synergistic increases in $PGF_{2\alpha}$ output resulted from incubating the ECs with E_2 together with AA, or A23187, a Ca^{++} ionophore liberator of AA from phospholipid stores. These findings suggest that the estrogen effect involves increased PG synthase activity, as appears to be the case in a number of cells and tissues in which various treatments increase prostanoid production [references in 11]. Moreover, estrogen-elevated $PGF_{2\alpha}$ output in the ECs may be mediated by specific estrogen receptors. Thus, 4-OH tamoxifen and

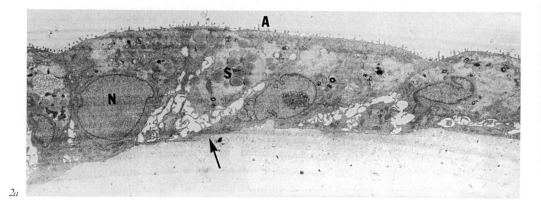

2a

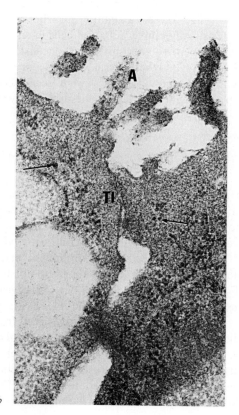

2b

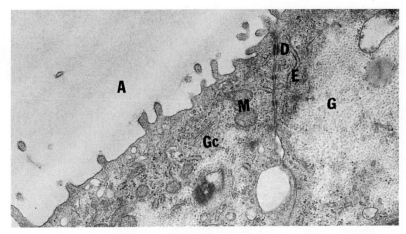

2c

Fig. 2. Electron micrographs of thin sections of gluteraldeyde-OsO$_4$-fixed glandular ECs after 14 days' culture in the RBM-DC system. *a* The glands have formed a monolayer of cuboidal to columnar ECs supported by a basement membrane at the basal aspects of the cells (arrow) and a microvillous border lining the apical (A) surface. *b, c* Tight junctions (TJ) and desmosomes (D) respectively, between adjacent cells. The nuclei (N) were located basally, and the well-differentiated cytoplasm contained membrane bound granules filled with lipid-like material, glycogen (G), mitochondria (M), Golgi complexes (Gc), granular endoplasmic reticulum and ribosomes (small arrows). The sections were stained with uranyl acetate and lead citrate. *a* ×3,850, *b* ×33,000, *c* ×20,000.

tamoxifen acted as virtually pure antagonists of E$_2$-elicited PGF$_{2\alpha}$ output in the ECs, exhibiting antiestrogenic potencies that corresponded to those reported for estrogen receptor-mediated processes in other tissues. Beyond PG production, a relevant in vitro model for implantation needs to deal with mechanisms whereby sequential ovarian steroid conditioning renders the apical endometrial EC surface receptive to attachment embryo [16]. However, the period of incubation on plastic dishes, in which E$_2$ enhances PGF$_{2\alpha}$ output in the endometrial ECs, is marked by a decline in differentiated function in vitro, namely basal PGF$_{2\alpha}$ outputs [7], activities of ornithine decarboxylase [17] and alkaline phosphatase [18]. This situation would appear to limit the usefulness of the cultured ECs in a model aimed at studying changing differentiated state.

Disappearance of differentiated function (functional polarity) occurs in many ECs during culture on impermeable glass or plastic surfaces along with loss of structural polarity (ECs flatten as structures associated in vivo with apical and basal surfaces disappear). However, retention of both structural and functional polarity have now been reported for ECs from diverse tissues that are cultured in contact with extracellular matrix-derived

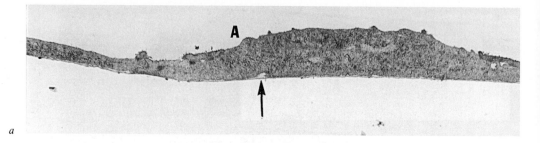

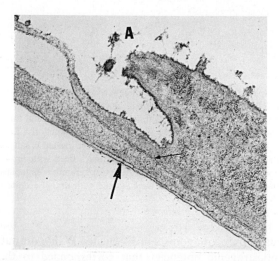

Fig. 3. Electron micrographs of thin sections of gluteraldehyde-OsO$_4$-fixed glandular ECs after 14 days' culture on tissue culture plastic. *a* The glands have formed a monolayer of attenuated cells on a basement membrane (arrow). Only a few microvilli are evident on the apical (A) border. *b* Neither tight junctions nor desmosomes are seen in areas of extensive cellular overlapping (small arrow), or elsewhere in the culture. The cytoplasm typifies that of immature cells, with microfiaments, ribosomes, and a few mitochondria as the predominant structures. The sections were stained with uranyl acetate and lead citrate. *a* ×9,000, *b* ×24,000.

proteins and/or on permeable supports that allow basolateral feeding [see 19, for review of EC polarity]. Accordingly, when human endometrial glands were cultured in the RBM-DC system in which these in situ conditions are approximated (see Method section) the resulting EC monolayer was strikingly more polarized than that in parallel cultures on plastic in terms of cell depth, distinctiveness of apical border, the presence of differentiated structures in the cytoplasm, and tight junctional complexes.

Structural polarity can be expected to affect AA metabolism by influencing access to epithelial cellular compartments in which lipoxygenase, PG synthase and esterifying enzymes are differentially located [20]. Therefore, to the extent that three-dimensional in situ cellular structure is better represented in the polarized than in nonpolarized EC monolayer, eicosanoid metabolism in the former should have more physiologic relevance.

The prevalence of junctional complexes in the polarized EC monolayer suggests a proclivity to form a 'tight' confluent monolayer in the dual-chambered system with which directional polarity can be studied, i.e., in which products secreted to (medium bathing) the apical side can be analyzed and manipulated separately from those secreted basolaterally. Directional polarity has now been demonstrated in a number of EC monolayers including a colonic epithelial cell line in which PGE_1 stimulates vectorial ion transport, but only when applied basolaterally [21]. Estradiol stimulates luminal secretion of endometrial PGs in nonpregnant pigs, whereas luminal PG secretion in pregnant pigs is thought to be regulated by blastocyst-derived E_2 [22]. In women, an elevation in $PGF_{2\alpha}$ levels during the secretory phase, in uterine luminal fluid [2], but not in uterine venous blood [20], suggests that its secretion occurs primarily in an apical direction by luminal and glandular endometrial ECs. Since $PGF_{2\alpha}$ accelerates hatching of blastocysts preceding implantation, whereas PGs are implicated as mediators of EVP and EVP-dependent decidualization of the stroma [reviewed in 1], preferential secretion of PGs by the endometrial ECs in an apical or basolateral direction has clear implications in implantation.

References

1 Kennedy TG: Interactions of eicosanoids and other factors in blastocyst implantation; in Hillier K (ed): Eicosanoids and Reproduction. Lancaster, MTP Press, 1987, pp 3–88.
2 Demers LM, Halbert DR, Jones DED, et al: Prostaglandin F levels in endometrial jet wash specimens during the normal menstrual cycle. Prostaglandins 1975;10:1057–1065.
3 Downie J, Poyser NL, Wunderlich M: Levels of prostaglandins in human endometrium during the normal menstrual cycle. J Physiol (Lond) 1974;236:465–472.
4 Abel MH, Baird DT: The effect of 17β-estradiol and progesterone on prostaglandin production by endometrium maintained in organ culture. Endocrinology 1980;106: 1599–1606.
5 Schatz F, Markiewicz L, Barg P, et al: In vitro effects of ovarian steroids on prostaglandin $F_{2\alpha}$ output by human endometrium and endometrial epithelial cells. J Clin Endocrinol Metabol 1985;61:361–367.
6 Tsang BK, Ooi TC: Prostaglandin secretion by human endometrium. Am J Obstet Gynecol 1982;142:626–633.

7   Schatz F, Gurpide E: Effects of estradiol on prostaglandin $F_{2\alpha}$ levels in primary cultures
    of epithelial cells from human proliferative endometrium. Endocrinology 1983;113:1274–
    1279.

8   Smith SK, Kelly RW: The effect of estradiol-17β and actinomycin D on the release of
    PGF and PGE from separated cells of human endometrium. Prostaglandins
    1987;34:553–561.

9   Schatz F, Markiewicz L, Gurpide E: Hormonal effects of PGF_2 output by cultures of
    epithelial and stromal cells in human endometrium. J Steroid Biochem 1986;24:297–
    301.

10 Satyaswaroop PG, Bressler RS, De la Pena MM, et al: Isolation and culture of human
    endometrial glands. J Clin Endocrinol Metabol 1979;48:639–641.

11 Schatz F, Markiewicz L, Gurpide E: Differential effect of estradiol, arachidonic acid,
    and A23187 on prostaglandin $F_{2\alpha}$ output by epithelial and stromal cells of human
    endometrium. Endocrinology 1987;120:1465–1471.

12 Leaver HA, Richmond DH: Effect of oxytocin, estrogen, calcium ionophore A23187 and
    hydrocortisone on prostaglandin $F_{2\alpha}$ and 6-oxy-prostaglandin $F_{1\alpha}$ production by cul-
    tured human endometrial and myometrial explants. Prostaglandins Leukotrienes Med
    1984;13:179–196.

13 Schatz F, Markiewicz L, Barg P, et al: In vitro inhibition with antiestrogens of estradiol
    effects on prostaglandin $F_{2\alpha}$ production by human endometrium and endometrial
    epithelial cells. Endocrinology 1986;118:408–412.

14 Tseng L, Gurpide E: Effects of progestins on estradiol receptor levels in human
    endometrium. J Clin Endocrinol Metabol 1975;41:402.

15 Markiewicz L, Gravanis A, Schatz F, et al: Prostaglandin production by human
    endometrial adenocarcinoma in vitro; in Baulieu EE, Iacobelli S, McGurie WL (eds):
    Endocrinology and Malignancy. Proc First Int Congr Cancer and Hormones. London,
    Parthenon Press, 1986, pp 420–427.

16 Psychoyos A: Endocrine control of egg implantation; in Greep RO, Astwood EB, Geiger
    SR (eds): Handbook of Physiology, Sect 7, vol II, part 2. Washington, American
    Physiology Society, 1973, pp 187–215.

17 Holinka CF, Gurpide E: Hormone-related enzymatic activities in normal and cancer
    cells of human endometrium. J Steroid Biochem 1981;13:183–192.

18 Holinka CF, Gurpide E: Ornithine decaboxylase activity in human endometrium and
    endometrial cancer cells. In Vitro 1985;21:697–706.

19 Simons K, Fuller SD: Cell surface polarity in epithelia. Ann Rev Cell Biol 1985;1:243–
    288.

20 Poyser NL: Prostaglandins in Reproduction. Chichester, Research Studies Press/Wiley,
    1980.

21 Weymer A, Huott P, Wilson L, et al: Chlorido-secretory mechanism induced by
    prostaglandin E_1 in a colonic epithelial cell line. J Clin Invest 1985;76:1828–1836.

22 Geisart RD, Thatcher WW, Roberte RM, et al: Establishment of pregnancy in the pig.
    III. Endometrial secretory response to estradiol valerate administered on day 11 of the
    estrus cycle. Biol Reprod 1982;27:957–966.

F. Schatz, PhD, Department of Obstetrics, Gynecology and Reproductive
Science, Mount Sinai School of Medicine, One Gustave Levy Place, New York,
NY 10029 (USA)

Zor U, Naor Z, Danon A (eds): Leukotrienes and Prostanoids in Health and Disease.
New Trends Lipid Mediators Res. Basel, Karger, 1989, vol 3, pp 303–309

Activation of Phospholipase D and Stimulation of Progesterone Production by Gonadotropin-Releasing Hormone in Ovarian Granulosa Cells

Mordechai Liscovitch, Abraham Amsterdam

Department of Hormone Research, The Weizmann Institute of Science, Rehovot, Israel

Ovarian granulosa cells differentiate into a mature, steroidogenic phenotype under the influence of multiple endocrine, paracrine, autocrine and neural signals. The influence of GnRH on granulosa cells is of particular interest, as this neuropeptide can exert both inhibitory and stimulatory control on the differentiation process, depending on the time of exposure to the hormone as well as on the maturational stage of the cells [reviewed in 1, 2]. Although superagonist analogs of GnRH have recently found considerable clinical application, the mechanism of GnRH action in the ovary remains poorly understood. The effects of GnRH in granulosa cells involve changes in phosphoinositide as well as in phosphatidylcholine turnover [3]. We have recently provided evidence that protein kinase C regulates phospholipase D phosphatidyltransferase activity in neural NG108-15 cells [4, 5]. Other studies showed that phospholipase D activity is stimulated by the chemotactic peptide formyl-Met-Leu-Phe in neutrophils and HL-60 granulocytes [6, 7] and by vasopressin in hepatocytes [8]. The signal-dependent formation of phosphatidic acid by phospholipase D-catalyzed hydrolysis of phosphatidylcholine may therefore represent a novel signal transduction pathway in mammalian cells. The present results indicate that activation of phospholipase D may participate in mediating the action of GnRH in rat granulosa cells.

Results and Discussion

In the presence of ethanol, mammalian phospholipase D catalyzes the formation of the unusual phospholipid phosphatidylethanol by its phosphatidyltransferase activity [9, 10]. This reaction appears to be unique to phospholipase D and is not catalyzed by the phospholipid base-

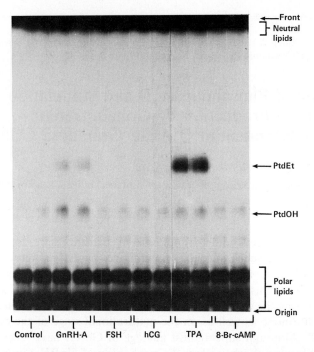

Fig. 1. [³H]phosphatidylethanol and [³H]phosphatidic acid production in granulosa cells. Preovulatory granulosa cells were prepared and cultured as previously described [17]. Cells were prelabeled with [³H]oleic acid (10 μCi/dish) for 4 h. The cells were then incubated with GnRH-A (250 ng/ml), FSH (250 ng/ml), hCG (10 U/ml), TPA (100 n*M*) and 8-Br-cAMP (1 m*M*) for 60 min in the presence of ethanol (0.5%). The incubations were terminated and the phospholipids were extracted and separated essentially as described [5]. TLC plate was prepared for fluorography by spraying with En³Hance and exposed for 24 h at −70 °C. PtdEt = phosphatidylethanol; PtdOH = phosphatidic acid.

exchange enzymes [9, 10]; it thus represents an unambiguous index for the signal-dependent activation of phospholipase D [7]. The gonadotropin-releasing hormone receptor agonist [*D*-Ala⁶,des-Gly¹⁰]-GnRH N-ethyl-amide (GnRH-A) stimulated accumulation of [³H]phosphatidylethanol and [³H]phosphatidic acid in preovulatory, cultured granulosa cells (fig. 1). FSH, hCG, and 8-Br-cAMP were ineffective. The protein kinase C activator 12-O-tetradecanoylphorbol-13-acetate (TPA), on the other hand, caused massive stimulation of [³H]phosphatidylethanol levels, but its effect on [³H]phosphatidic acid levels was less pronounced than that of GnRH-A (fig. 1). The response to GnRH-A was rapid in onset, reaching near maximal levels of phosphatidylethanol within 2 min; in contrast, accumulation

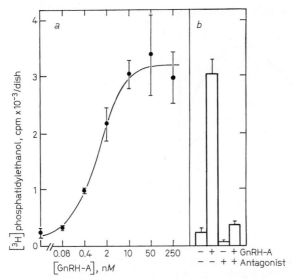

Fig. 2. Concentration dependence of GnRH-A-induced stimulation of [³H]phosphati-
dylethanol (a) and inhibition of GnRH-A-induced stimulation of [³H]phosphatidylethanol
production by a GnRH antagonist (b). Granulosa cells were prelabeled with [³H]oleic acid
(10 µCi/dish) for 3 h. The cells were then incubated with increasing concentrations of
GnRH-A (a) or with GnRH-A (10 nM) plus the GnRH antagonist [D-pGlu¹,D-Phe²,D-
Trp³,⁶]GnRH (200 nM) (b), as indicated, for 60 min. Ethanol (0.5%) was present in all the
incubations. Results represent the mean ± SD of triplicate determinations from a typical
experiment.

of [³H]phosphatidylethanol in the TPA-treated cells increased progressively
over a 60-min incubation period (not shown). The effect of GnRH-A on
[³H]phosphatidylethanol accumulation was concentration dependent, and
reached saturation at a concentration of 50 nM, with half a maximal effect
obtained at 1 nM (fig. 2a). This effect was completely inhibited by the
GnRH receptor antagonist [D-pGlu¹,D-Phe²,D-Trp³,⁶]GnRH (fig. 2b), indicating that the action of GnRH-A is specific. In contrast, stimulation of
phosphatidic acid formation by GnRH-A reached a maximum (188%
control) at 10 nM (fig. 3); this effect was inhibited in the presence of the
antagonist [D-pGlu¹,D-Phe²,D-Trp³,⁶]GnRH (not shown). At higher
GnRH-A concentrations the levels of [³H]phosphatidic acid in the stimulated cells declined, but remained above control value (135% control)
(fig. 3). To test the possibility that the biphasic effect of GnRH-A on
phosphatidic acid levels may have resulted from mobilization of phosphatidic acid phosphohydrolase, we measured the accumulation of phosphatidic acid in response to GnRH-A in the presence of an inhibitor of

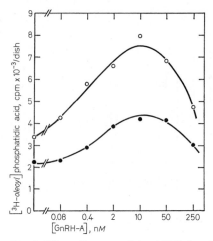

Fig. 3. Effect of propranolol on [³H]phosphatidic acid accumulation in response to GnRH-A. Granulosa cells were prelabeled with [³H]oleic acid (5 μCi/dish) for 24 h. The cells were then incubated with increasing concentrations of GnRH-A either in the presence (○) or absence (●) of propranolol (100 μM), for 20 min. Results represent the mean of duplicate determinations from a typical experiment.

phosphatidic acid phosphohydrolase, *DL*-propranolol [11]. Propranolol (100 μM) increased the basal levels of phosphatidic acid 1.5-fold and the net, GnRH-A-stimulated production of phosphatidic acid 2.3-fold. Propranolol could not, however, prevent the decrease in [³H]phosphatidic acid seen at high GnRH-A concentrations (fig. 3).

The pronounced effect of TPA on phosphatidylethanol production (fig. 1 and [5]) raised the possibility that the effect of GnRH-A on [³H]phosphatidylethanol levels is distal to activation of protein kinase C (following phospholipase C-catalyzed hydrolysis of phosphoinositides [cf. 3]. This possibility was ruled out in experiments showing that supramaximal concentrations of GnRH-A (50 n*M*) and TPA (1 μ*M*) caused additive stimulation of phosphatidylethanol accumulation, suggesting that the two agents do not act via the same mechanism; this conclusion is supported by the fact that 1-(5-isoquinolinesulfonyl)-2-methylpiperazine (H-7), a protein kinase C inhibitor, inhibited the effect of TPA 50%, but not that of GnRH-A (not shown). It may therefore be concluded that the effect of GnRH-A on phospholipase D activity is not mediated by protein kinase C. Recent studies suggested that regulation of plasma membrane phospholipase D activity involves a guanine nucleotide-binding (G-) protein [8]. Evidence that pituitary GnRH receptors can couple to a G-protein(s) was recently presented [12]. The possibility that ovarian GnRH

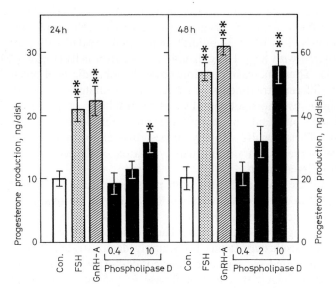

Fig. 4. Effect of treatment with exogenous phospholipase D on progesterone production by granulosa cells. Granulosa cells were cultured for 24 h. The medium was then replaced with McCoy's/BSA containing phospholipase D (from S. chromofuscus; 0.4, 2 or 10 IU/ml), GnRH-A (50 ng/ml) or FSH (250 ng/ml) for 4 h. The medium was replaced with normal growth medium and progesterone was determined in the medium 24 and 48 h later by RIA [18]. Results are expressed as the mean $\pm$ SEM of quadruplicate culture dishes. *Significantly different from untreated control ($p < 0.05$); **significantly different from untreated control ($p < 0.01$).

receptors are coupled to phospholipase D by a G-protein is currently under investigation in our laboratory.

The present results strongly support the hypothesis that GnRH activates a phospholipase D in preovulatory granulosa cells. Does this activation participate in mediating the stimulatory effect of GnRH on granulosa cell differentiation? We have found that the stimulatory action of GnRH on steroidogenesis could be mimicked by elevating endogenous phosphatidic acid levels in granulosa cells. Exogenous phospholipase D (at 10 IU/ml) caused a significant, 1.6-fold increase in progesterone production during 24 h of incubation (fig. 4; $p < 0.05$); lower concentrations of the enzyme were ineffective. After 48 h of incubation, phospholipase D stimulated progesterone production 2.7-fold ($p < 0.01$), while GnRH-A and FSH caused 3- and 2.6-fold stimulation, respectively (fig. 4). It is thus possible to mimic the effect of GnRH by treating granulosa cells with exogenous phospholipase D, a treatment which, as expected, causes elevated membrane levels of endogenous phosphatidic acid (not shown). In addition, propranolol, which increases endogenous phosphatidic acid levels (fig. 3), is itself a potent stimulator or steroidogenesis in granulosa cells; progesterone production amounted to 184 ± 34 ng/dish/48 h in cells treated with propranolol, as compared with 85 ± 14 ng/dish/48 h in control cells (mean $\pm$ SD, n = 3, p < 0.01). Thus, increasing endogenous phosphatidic acid levels by each of three different treatments (e.g., receptor-mediated activation by GnRH-A, exogenous phospholipase D and an inhibitor or phosphatidic acid degradation) all result in stimulating progesterone production by granulosa cells. These results strongly suggest that endogenous phosphatidic acid may serve as an intracellular mediator of GnRH action on granulosa cell differentiation.

Previous studies have implicated both phospholipase C [3] and phospholipase A_2 [13] in the mechanism of GnRH action in the ovary. In the present communication we provide evidence for a possible role of phospholipase D in mediating the GnRH signal. These different pathways are not, however, mutually exclusive. It has been amply demonstrated that phospholipase C is regulated by a guanine nucleotide-binding (G-) protein(s) [14]; recent studies suggest that phospholipase A_2 [15] and phospholipase D [8] can be similarly regulated. It is likely that a hormone-receptor complex could interact with more than one type of G-protein and that an activated G-protein could interact with more than one effector system; the stoichiometry of occupied receptors, G-proteins and effectors may thus be an important determinant of the signalling pathways which will be activated by a signal in a given cell [16].

Until recently, the possibility that phospholipase D may participate in signal transduction mechanisms has received relatively little attention [4–8]. The present results suggest that GnRH receptors can activate phospholipase D in ovarian granulosa cells, and that the resultant increase in endogenous phosphatidic acid levels participates in mediating the stimulation of progesterone production by these cells. It may be speculated that certain cellular constituents can be directly modulated by endogenous phosphatidic acid, produced by a signal-activated phospholipase D (or by other signal-activated mechanisms). The identity of these hypothetical constituents in granulosa cells and their role in controlling granulosa cell differentiation remain to be studied.

References

1   Hsueh AJW, Adashi EY, Jones PBC, et al: Hormonal regulation of the differentiation of cultured ovarian granulosa cells. Endocr Rev 1984;5;76–127.
2   Amsterdam A, Rotmensch S: Structure-function relationships during granulosa cell differentiation. Endocr Rev 1987;8:309–337.

3   Naor Z, Yavin E: Gonadotropin-releasing hormone stimulates phospholipid labeling in cultured granulosa cells. Endocrinology 1982;111:1615–1619.
4   Liscovitch M, Blusztajn JK, Freese A, et al: Stimulation of choline release from NG108–15 cells by 12-O-tetradecanoylphorbol-13-acetate. Biochem J 1987;241:81–86.
5   Liscovitch M: Phosphatidylethanol biosynthesis in ethanol-exposed NG108–15 neuroblastoma × glioma hybrid cells. Evidence for activation of a phospholipase D phosphatidyltransferase activity by protein kinase C. J Biol Chem 1989;264:1450–1456.
6   Cockcroft S: Ca^{2+}-dependent conversion of phosphatidylinositol to phosphatidate in neutrophils stimulated with fMet-Leu-Phe or ionophore A23187. Biochim Biophys Acta 1984;795:37–46.
7   Pai J-K, Siegel MI, Egan RW, et al: Phospholipase D catalyzes phospholipid metabolism in chemotactic peptide-stimulated HL-60 granulocytes. J Biol Chem 1988;263:12472–12477.
8   Bocckino SB, Blackmore PF, Wilson PB, et al: Phosphatidate accumulation in hormone-treated hepatocytes via a phospholipase D mechanism. J Biol Chem 1987;262:15309–15315.
9   Kobayashi M, Kanfer JN: Phosphatidylethanol formation via transphosphatidylation by rat brain synaptosomal phospholipase D. J Neurochem 1987;48:1597–1603.
10 Gustavsson L, Alling C: Formation of phosphatidylethanol in rat brain by phospholipase D Biochem Biophys Res Commun 1987;142:958–963.
11 Pappu AS, Hauser G: Propranolol-induced inhibition of rat brain cytoplasmic phosphatidate phosphohydrolase. Neurochem Res 1983;8:1565–1575.
12 Limor R, Schvartz I, Hazum E, et al: Effect of guanine nucleotides on phospholipase C activity in permeabilized pituitary cells: Possible involvement of an inhibitory GTP-binding protein. Biochem Biophys Res Commun 1989;159:209–215.
13 Wang J, Leung PCK: Role of arachidonic acid in luteinizing hormone-releasing hormone action: Stimulation of progesterone production in rat granulosa cells. Endocrinology 1988;122:906–911.
14 Litosch I, Fain JN: Regulation of phosphoinositide breakdown by guanine nucleotides. Life Sci 1986;39:187–194.
15 Jelsema CL: Light activation of phospholipase A_2 in rod outer segments of bovine retina and its modulation by GTP-binding proteins. J Biol Chem 1987;262:163–168.
16 Neer EJ, Clapham DE: Roles of G protein subunits in transmembrane signalling. Nature 1988;333:129–134.
17 Furman A, Rotmensch S, Kohen F, et al: Regulation of rat granulosa cell differentiation by extracellular matrix produced by bovine corneal endothelial cells. Endocrinology 1986;118:1878–1885.
18 Kohen F, Bauminger S, Lindner HR: Preparation of antigenic steroid-protein conjugates; in Cameron EHD, Hillier SG, Griffiths K (eds): Steroid Immunoassay. Cardiff, Alpha Omega, 1975, pp 11–23.

M. Liscovitch, PhD, Department of Hormone Research, The Weizmann Institute of Science, PO Box 26, Rehovot 76100 (Israel)

Zor U, Naor Z, Danon A (eds): Leukotrienes and Prostanoids in Health and Disease.
New Trends Lipid Mediators Res. Basel, Karger, 1989, vol 3, pp 310–313

The Role of Prostaglandins in Bone Formation

D. Sömjen[a], *E. Berger*[a], *Z. Shimshoni*[a], *A. Waisman*[b], *U. Zor*[b],
A.M. Kaye[b], *I. Binderman*[a]

[a]Hard Tissues Unit, Ichilov Hospital, Tel Aviv, and [b]Department of Hormone
Research, The Weizmann Institute of Science, Rehovot, Israel

Prostaglandins affect bone formation as well as bone resorption [1]. In
calvaria systems in vitro prostaglandin E_2 (PGE_2), a potent stimulator of
bone resorption, stimulates bone formation as well [2]. PGE_2 stimulates
cAMP production [3] and produces contractile shape changes that suggest
cell activation [4]. When injected locally in rats, PGE_2 produces both
resorption and increased bone formation in calvaria [5]. Given orally,
PGE_2 increases periosteal bone formation in femurs of dogs [6] and rats [7],
associated with increased numbers of osteoprogenitor cells including
osteoblasts [8].

In this study, we examined the action of PGE_2 on increased bone
formation in several types of skeletal derived cell cultures, by measuring
parameters correlated with cell proliferation such as DNA synthesis, ornithine decarboxylase (ODC) and creatine kinase BB (CKBB) activity, and
the accumulation of mRNA for CKBB and for the proto-oncogene c-fos
[9].

Stimulation of Cell Proliferation by Exogenous PGE_2

The formation of cAMP, a second messenger for PGE_2 action, is
increased transiently within 15 min by PGE_2 in cultured bone cells [10, 11]
as well as in other skeletal cells such as mandibular [12] and epiphyseal [13]
chondrocytes and the fibroblast-like ROS 25/1 cells [14].

In parallel, PGE_2 increases cell proliferation (table 1) in these and in
several other cell culture systems [12–14]. PGE_2 stimulates CK activity
in bone cells cultures significantly within 1 h [15] and maximally after 24 h,
in those cells in which DNA synthesis is increased (table 1). The increase in
CK-specific activity is due to the increase in the brain type (BB) isozyme of
CK [14, 15] which results from de novo synthesis of the protein [15]. In

Table 1. PGE$_2$ stimulation of ^{3}H-thymidine incorporation and the specific activity of CK

Cell type	DNA synthesis (PGE$_2$ treated/control)	CK-specific activity (PGE$_2$ treated/control)
Calvaria (including osteoprogenitor cells)	1.82 ± 0.07*	1.88 ± 0.08**
Osteoblast-enriched calvaria	1.02 ± 0.05	1.12 ± 0.12
Condylar cartilage	1.09 ± 0.09	–
Epiphyseal cartilage	1.84 ± 0.04*	–
ROS 25/1	1.43 ± 0.07*	1.47 ± 0.08*
ROS 17/2	1.05 ± 0.07	0.87 ± 0.08

After treatment with 500 ng/ml PGE$_2$ for 24 h, cells were either labeled with [^{3}H]thymidine and analyzed for incorporation into DNA [17] or CK was extracted and assayed [15]. Results are means ± SE of ratios for n ≥ 4. *p < 0.05, **p < 0.01, by Student's t test of experimental vs. control values.

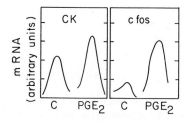

Fig. 1. Stimulation by PGE$_2$ of mRNA for CKB and for c-fos proto-oncogene. After 2 h exposure to PGE$_2$ (500 ng/ml), steady-state levels of mRNA for CK and c-fos in calvaria cells were measured by 'Northern blot' hybridization of total RNA with the cDNA plasmid pCKb (a gift from Dr. P. Benfield, Du Pont Laboratories, Wilmington, Del.) and with a c-fos probe (ATCC). Autoradiograms were quantified by laser densitometry.

calvaria cells, PGE$_2$ stimulates, within 2 h, the accumulation of mRNA for CK as well as mRNA for the proto-oncogene c-fos (fig. 1).

PGE$_2$ also stimulates ODC activity 9-fold in cultured bone cells within 4 h, followed by a rapid decline to basal activity after 6 h [16].

Stimulation of PGE$_2$ Formation by Mechanical Forces

Mechanical forces (MF) applied to cultured calvaria cells induce the production of cAMP via stimulation of PGE$_2$ formation [17]. Agents that inhibit phospholipase A$_2$ activity abolished the stimulation by mechanical force [18]. In contrast, in mandibular condyle where the formation of cAMP by MF is not mediated by PGE_2 [12], no inhibition of cAMP formation by inhibitors of phospholipase A_2 occurs. The results suggest that the perturbation of membrane phospholipid by MF activates phospholipase A_2, which leads to the release of arachidonic acid and increased synthesis of PGE_2 [18].

Conclusions

PGE_2 stimulates the formation of cAMP leading to bone resorption in a variety of skeletal derived cell types. In contrast, stimulation of bone formation (syntheses of DNA, the proto-oncogene c-fos and enzyme markers for proliferation) occurs mainly in fibroblast-like and osteoprogenitor cells of bone. Thus, PGE_2 mediates multiple facets of bone formation, either directly or by mediating hormonal and physical stimuli.

References

1   Raisz LG, Martin TJ: Prostaglandins in bone and mineral metabolism; in Peck WA (ed): Bone and Mineral Research. Amsterdam, Elsevier Science, 1984, vol 2, pp 286–310.
2   Nefussi JR, Baron R: PGE_2 stimulates both resorption and formation in bone in vitro: Differential responses of the periosteum and the endosteum in fetal rat long bone cultures. Anat Rec 1985;211:9–16.
3   Rodan GA, Rodan SB: Expression of the osteoblastic phenotype; in Peck WA (ed): Bone and Mineral Research. Amsterdam, Elsevier Science, 1984, vol 2, pp 244–285.
4   Shen V, Rifas L, Kohler G, et al: Prostaglandins change cell shape and increase intracellular gap junctions in osteoblasts cultured from rat fetal calvaria. J Bone Mineral Res 1986;1:243–249.
5   Baron R, Nefussi JR, Duflot-Vignery, A, et al: Failure of prostaglandin E to induce local bone resorption in vivo; in Horton JE, Tarplay TM, Davis WR (eds) Mechanisms of Localized Bone Loss (abstract). Calcif Tissue Int 1978;(suppl):433–434.
6   High WB: Effects of orally administered prostaglandin E_2 on cortical bone turnover in adult dogs: A histomorphometric study. Bone 1988;8:363–374.
7   Ueno K, Haba T, Woodburg D, et al: The effect of prostaglandin E_2 in rapidly growing rats: Depressed longitudinal and radial growth and increased metaphyseal hard tissue mass. Bone 1985;6:9–86.
8   Jee WSS, Ueno K, Kimmel DB, et al: The role of bone cells in increasing metaphyseal hard tissue in rapidly growing rats treated with prostaglandin E_2. Bone 1987;8:171–178.
9   Greenberg ME, Ziff EB: Stimulation of 3T3 cells induces transcription of c-fos proto-oncogene. Nature 1984;311:433–437.
10 Binderman I, Duksin D, Harell A, et al: Formation of bone tissue in culture from isolated bone cells. J Cell Biol 1974;61:427–439.

11   Binderman I, Sömjen D: Serum factors and low calcium modulate growth of osteoblast-like cells in culture; in Silberman M, Slavkin HC (eds): Current Advances in Skeletogenesis: Development, Biomineralization, Mediators and Metabolic Bone Diseases. Amsterdam, Excerpta Medica, 1982, pp 338–342.

12   Levy J, Shimshoni Z, Sömjen D, et al: Rat epiphyseal cells in culture: Responsiveness to bone-seeking hormones. In vitro 1988;24:620–625.

13   Shimshoni Z, Binderman I, Fine N, et al: Mechanical and hormonal stimulation of cell cultures derived from young rat mandibular condyle. Arch Oral Biol 1984;10:827–831.

14   Sömjen D, Yariv M, Kaye AM, et al: Ornithine decarboxylase activity in cultured bone cells is activated by bone-seeking hormones and physical stimulation; in Bachrach U, Kaye AM, Chayen R (eds) Advances in Polyamine Research. New York, Raven Press, 1983, vol 4, pp 713–718.

15   Sömjen D, Kaye AM, Binderman I: Stimulation of creatine kinase BB activity by parathyroid hormone and by PGE_2 in cultured bone cells. Biochem J 1985;225:591–596.

16   Sömjen D, Kaye AM, Rodan GA, et al: Regulation of creatine kinase activity in rat osteogenic sarcoma cell clones by parathyroid hormone, prostaglandin E_2 and vitamin D metabolites. Calcif Tissue Int 1985;37:635–638.

17   Sömjen D, Binderman I, Berger E: Bone remodelling induced by physical stress is prostaglandin mediated. Biochim Biophys Acta 1980;627:81–99.

18   Binderman I, Zor U, Kaye AM, et al: The transduction of mechanical force into biochemical events in bone cells may involve activation of phospholipase A_2. Calcif Tissue Int 1988;42:261–266.

D. Sömjen, MD, Hard Tissues Unit, Ichilov Hospital, Tel Aviv 64239 (Israel)

Zor U, Naor Z, Danon A (eds): Leukotrienes and Prostanoids in Health and Disease.
New Trends Lipid Mediators Res. Basel, Karger, 1989, vol 3, pp 314–318

Cyclopentenone Prostaglandins Induce Heat Shock Proteins in Association with G_1 Block of Cell Cycle Progression

Shuh Narumiya[a], *Kouji Ohno*[a], *Masanori Fukushima*[b],
Motohatsu Fujiwara[a]

[a]Department of Pharmacology, Kyoto University Faculty of Medicine, Yoshida,
Sakyo-ku, Kyoto, and [b]Department of Internal Medicine, Aichi Cancer Center,
Chikusa-ku, Nagoya, Japan

PGs of the A, D and E series cause inhibition of growth in cultured cells and in some cases induce differentiation. PGE_2 and D_2 are dehydrated enzymatically in culture medium to form cyclopentenone PGs, namely PGA_2 and Δ^{12}-PGJ_2 [1]. We have shown that the growth inhibitory activities of the PGs are actually exerted by these dehydration products, and PGs D and E serve only as precursors of the active compounds [2, 3]. Thus, the antiproliferative activity seems to correlate with the PGs with a cyclopentenone ring containing an α,β-unsaturated ketone. Although PGs usually express their actions by acting on a specific receptor on cell surface, we found that the cyclopentenone PGs are actively transported into cells by a specific carrier on cell membrane and accumulate in cell nuclei with binding to nuclear proteins [4, 5]. We further found that this uptake and accumulation are closely correlated with the expression of their growth inhibitory activities [4–6]. We have also examined the cellular mechanism of PG-induced growth inhibition by analyzing the PG effects on cell cycle progression in HeLa S3 cells of synchronized growth. We have shown that these PGs specifically act on cells in G_1 phase and block cell cycle progression at a point several hours prior to G_1/S boundary [7]. Thus, the cyclopentenone PGs taken up by cells appear to act on nuclei and induce the block of cell cycle progression. Then, what kind of events lie between the nuclear accumulation and the cell cycle block? The present study was undertaken to examine this issue.

Induction of Heat Shock Proteins by Cyclopentenone PGs

Since Santoro et al. [8]. already reported induction of specific proteins by cyclopentenone PGs, we analyzed if the PGs induce similar change in

(MW)   (MW)

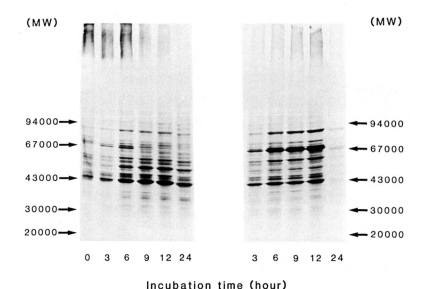

94000 ➤   ◄ 94000

67000 ➤   ◄ 67000

43000 ➤   ◄ 43000

30000 ➤   ◄ 30000

20000 ➤   ◄ 20000

       0   3   6   9 12 24   3   6   9 12 24

Incubation time (hour)

Fig. 1. SDS-polyacrylamide gel electrophoresis of proteins labeled with [^{35}S]methionine in control *(a)* and Δ^{12}-PGJ$_2$-treated cells *(b)* at various times of cell cycle. Reproduced from reference 10.

HeLa S3 cells and such change is related to the PG-induced G$_1$ block. HeLa S3 cells were enriched in G$_1$ phase and initiated to progress in cell cycle as described [7]. A PG was added to these cells at 0 h, and at 0, 3, 6, 9, 12, and 24 h cells were collected and labeled with [^{35}S]methionine. Figure 1 shows SDS-polyacrylamide gel electrophoresis of [^{35}S]-labeled proteins in the control (fig. 1a) and the PG-treated (fig. 1b) cells. Proteins of 68–70 kd were prominently labeled in control cells at 6 h. The synthesis of these proteins, however, decreased during further culture. Since the majority of control cells was in the G$_1$ phase at 6 h, this result suggests that they were proteins induced specifically and transiently in cells during G$_1$ phase of cell cycle. When these cells were treated with 12 μM Δ^{12}-PGJ$_2$, elevated synthesis of proteins of similar molecular weight was clearly observed (fig. 1b). When we compared the protein(s) induced in the G$_1$ phase of the control cells with the PG-induced protein(s) on two-dimensional (2-D) gel electrophoresis, they overlapped each other, indicating that they were identical proteins. The elevated synthesis of these proteins began at 3 h and was maintained at maximum from 6 to 12 h. While the elevated synthesis of these proteins continued, other labeled protein bands decreased in number, and total incorporation of the radioactivity markedly decreased at 24 h. The PG-induced elevation of the 68-kd protein synthesis was inhibited by

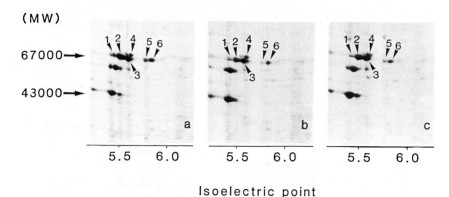

Isoelectric point

Fig. 2. Comparison of the Δ^{12}-PGJ$_2$-induced proteins and the 68-kd HSPs on 2-D polyacrylamide gel electrophoresis. *a* Labeled proteins of cells treated with Δ^{12}-PGJ$_2$. *b* Mixture of labeled proteins of cells treated with Δ^{12}-PGJ$_2$ and those of cells treated with heat shock. *c* Labeled proteins of cells treated with heat shock. Among the proteins numbered, proteins 1 and 4 are marker proteins found also in the control cells. Reproduced from reference 10.

the addition of actinomycin D to the cells, suggesting that the PGs act at the transcriptional level.

Because several proteins, termed heat shock proteins (HSPs), were induced in the cells at high temperature or by other stress, and such induction was associated with growth arrest in some systems [9], we have examined if the PG-induced 68-kd proteins belong to a HSP family. To this, we compared on 2-D gel electrophoresis the 68-kd proteins induced by the PGs with proteins induced in cells treated at high temperature (fig. 2). The PG treatment produced four new protein spots, proteins 2, 3, 5 and 6 in figure 2a. HSPs induced in cells by incubation at 43 °C for 90 min were also four 68-kd proteins, proteins 2, 3, 5 and 6 in figure 2c, and these HSPs 68 showed positions on the gel similar to those of the PG-induced proteins in figure 1a. In addition, when the labeled lysates from the PG- and heat-treated cells were mixed and analyzed on the same gel, the PG-induced 68-kd proteins and HSPs 68 exactly overlapped (fig. 2b). Thus, the proteins induced in cells by the PG treatment were HSPs 68. This conclusion was also confirmed by limited proteolysis of the PG-induced proteins and HSPs 68 [10].

Association of HSP Induction with PG-Induced G₁ Block

The above study has shown that the proteins induced by the cyclopentenone PGs are HSPs. We have further found that this induction has several properties in common with the PG-induced cell cycle block [10]. For example, it was already reported that Δ^{12}-PGJ$_2$ irreversibly blocks the cell cycle progression, whereas the block by PGA$_2$ is reversible [7]. Similarly, Δ^{12}-PGJ$_2$ caused marked and sustained induction of HSPs, whereas induction of HSPs by PGA$_2$ was weaker and reversible. The PGs induce G$_1$ block only in the G$_1$ phase cells [7]. Similarly, induction of HSPs by Δ^{12}-PGJ$_2$ was also observed only in the G$_1$ phase cells and not in the G$_2$/M phase cells. Thus, induction of HSPs 68 by cyclopentenone PGs in HeLa S3 cells appears to be associated with the G$_1$ block of cell cycle progression caused by the PGs. Our results further suggest that the PGs act on some target specifically present in G$_1$ phase cells to evoke the two phenomena. The identity of this PG target as well as the causative relationship between the HSPs and G$_1$ block will be clarified in future studies.

Acknowledgements

The authors thank Ono Pharmaceuticals for their generous supply of PGs. This work was supported in part by a Grant-in-Aid from the Ministry of Education, Science and Culture of Japan for Scientific Research (No. 62570105 to S.N.), a Grant-in-Aid from the Ministry of Health and Welfare of Japan for Comprehensive Ten Year Strategy for Cancer Control (to Ma.F.), and grants to S.N. from the Uehara Memorial Foundation, the Ciba-Geigy Foundation for Promotion of Science, and the Japanese Foundation on Metabolism and Diseases.

References

1   Kikawa Y, Narumiya S, Fukushima M, et al: 9-Deoxy-Δ^9,Δ^{12}-13,14-dihydro-prostaglandin D$_2$, a metabolite of prostaglandin D$_2$ formed in human plasma. Proc Natl Acad Sci USA 1984;81:1317–1321.
2   Narumiya S, Fukushima M: Δ^{12}-Prostaglandin J$_2$, an ultimate metabolite of prostaglandin D$_2$ exerting cell growth inhibition. Biochem Biophys Res Commun 1985;127:739–745.
3   Ohno K, Fujiwara M, Fukushima M, et al: Metabolic dehydration of prostaglandin E$_2$ and cellular uptake of the dehydration product: correlation with prostaglandin E$_2$-induced growth inhibition. Biochem Biophys Res Commun 1986;139:808–815.
4   Narumiya S, Fukushima M: Site and mechanism of growth inhibition by prostaglandins. I. Active transport and intracellular accumulation of cyclopentenone prostaglandins, a reaction leading to growth inhibition. J Pharmacol Exp Ther 1986;239:500–505.
5   Narumiya S, Ohno K, Fujiwara M, et al: Site and mechanism of growth inhibition by protaglandins. II. Temperature-dependent transfer of a cyclopentenone prostaglandin to nuclei. J Pharmacol Exp Ther 1986;239:506–511.
6   Narumiya S, Ohno K, Fukushima M, et al: Site and mechanism of growth inhibition by prostaglandins. III. Distribution and binding of prostaglandin A$_2$ and Δ^{12}-prostaglandin J$_2$ in nuclei. J Pharmacol Exp Ther 1987;242:306–311.

7   Ohno K, Sakai T, Fukushima M, et al: Site and mechanism of growth inhibition by prostaglandins. IV. Effects of cyclopentenone prostaglandins on cell cycle progression of G_1-enriched HeLa S3 cells. J Pharmacol Exp Ther 1988;245:294–298.

8   Santoro MG, Crisari A, Benabente A, et al: Modulation of growth of a human erythroleukemia cell line (K-562) by prostaglandins: Antiproliferative action of prostaglandin A. Cancer Res 1986;46:6073–6077.

9   Iida H, Yahara I: Durable synthesis of high molecular weight heat shock proteins in G_0 cells of the yeast and other eukaryotes. J Cell Biol 1984;99:199–207.

10   Ohno K, Fukushima M, Fujiwara M, et al: Induction of 68,000 dalton heat shock proteins by cyclopentenone prostaglandins; its association with prostaglandin-induced G_1 block in cell cycle progression. J Biol Chem 1988;263:19764–19770.

Shuh Narumiya, MD, Department of Pharmacology, Kyoto University
Faculty of Medicine, Yoshida, Sakyo-ku, Kyoto 606 (Japan)

Zor U, Naor Z, Danon A (eds): Leukotrienes and Prostanoids in Health and Disease.
New Trends Lipid Mediators Res. Basel, Karger, 1989, vol 3, pp 319–324

Activation of Arachidonate Metabolism in Rats after Inoculation of Tumour Cells: Preferential Formation of Thromboxane A$_2$ and Its Proliferative Effect on Tumour Cells

Santosh Nigam, Rüttger Averdunk

Prostaglandin Research, Department of Gynecological Endocrinology,
University Clinic Steglitz, Free University, Berlin, FRG

Introduction

During the last two decades it has become well established that growing tumours synthesize eicosanoids using both cyclooxygenase and lipoxygenase pathways [1, 2], and that endogenously synthesized eicosanoids are involved in the modulation of tumour cell replications [3]. We have reported previously that malignant tumours of the breast and gastrointestinal tract produce more prostacyclin and thromboxane A$_2$ (TXA$_2$) than benign tumours [4, 5], and that this elevation of prostanoid concentration in the plasma of patients remained unchanged even after surgical removal of the tumour. The mechanism for this elevation is still unknown. We conjectured that tumour cells might modulate the host cells to activate the prostanoid metabolism, thus facilitating tumour growth [6]. Since TXA$_2$ has been observed in several tumour cell lines as the main arachidonic acid (AA) metabolite, we investigated the tumour cell-induced TXA$_2$ synthesis in rats in relation to the tumour growth. We also investigated the effect of TXA$_2$ synthetase inhibitors on the tumour growth in rats.

Materials and Methods

Prostaglandins, thromboxane B$_2$, PGH$_2$ and TXA$_2$ analogue U 46619 were purchased from Cayman Chemicals, Ann Arbor, Mich., USA. Indomethacin and meclofenamic acid were supplied by Sigma, FRG and Warner Lambert, Mich., USA, respectively.

CAM cell line, an epithelial mammary cancer cell line of the rat, which produces a high concentration of eicosanoids in culture [6], was grown in RPMI 1640 medium with 10% fetal calf serum (FCS) as monolayer culture with a generation time of approximately 20 h. Following inoculation of 2–5×10^6 cells intraperitoneally into the rat, a solid or ascites tumour was observed after 4–6 weeks. Lewis or Wistar rats used for the experiments were kept on a normal diet.

Five milliliters of blood was withdrawn by puncturing the heart of anaesthetized animals into a polypropylene tube containing 0.2 ml of 0.1 mM EDTA. Indomethacin (1 μg/ml blood) and meclofenamic acid (5 μg/ml blood) were immediately added to inhibit the cyclooxygenase activity. For TXA$_2$ inhibition studies the rats were treated with indomethacin and BM 13177 (a gift from Boehringer Mannheim, FRG) as indicated in the Results section. Prostanoids were determined as described [4, 7]. Tumour cell proliferation was assessed by incorporation of ^{3}H-thymidine (1 μCi).

Results

Figure 1 shows the metabolic profile of AA in the plasma of the rat during tumour development. TXB$_2$ was the major metabolite during tumour growth, with TXB$_2 \gg$ PGE$_2 >$ FGF$_{2\alpha} \approx$ 6-keto-PGF$_{1\alpha}$. Although the qualitative profile of prostanoids did not change much during tumour growth, absolute values on day 28 were different from those found on day 1. Moreover, the supernatant of CAM cells cultured for 3 days showed that these cells preferentially produced TXB$_2$.

Tumour growth in the rat after the inoculation of CAM cells with a concomitant increase of TXB$_2$ levels in plasma led us to investigate the effect of the thromboxane synthetase inhibitor indomethacin on tumour growth. In order to check whether thromboxane has a proliferative effect on tumour cells, we treated the rat with thromboxane receptor antagonist BM 13177 and observed the AA metabolism for 6 weeks. Table 1 shows the tumour development in rats treated with indomethacin, BM 13177 or the vehicle alone. Tumour growth was almost unaffected despite the reduction of TXB$_2$ plasma levels with indomethacin treatment (202 $\pm$ 41 pg/ml). On the contrary, the treatment of rats with BM 13177 completely inhibited the tumour development. However, the TXB$_2$ level in plasma remained elevated.

To investigate the proliferative effect of thromboxane on CAM tumour cells, we measured the ^{3}H-thymidine uptake by CAM cells in culture with and without the addition of a stable TXA$_2$ analogue U 46619. Figure 2 shows that the thymidine uptake is dose dependently enhanced by U 46619, with a maximum of 140% at a concentration of 1 μM. A sharp decline at a concentration of 10 μM is caused by the U 46619-induced aggregation of CAM cells. This proliferative effect of U 46619 on CAM cells could be dose dependently inhibited by BM 13177 as illustrated in figure 3. Approximately 80% of the thymidine incorporation was inhibited by 1 μM BM 13177. Significant inhibition of thymidine uptake was observed at such a concentration as low as 10 nM.

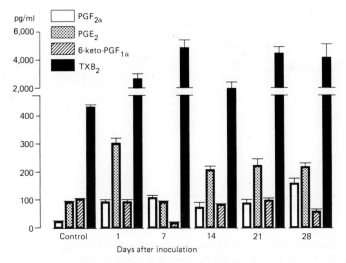

Fig. 1. Time course of changes in plasma concentrations of AA metabolites in the rat after the intraperitoneal inoculation of 5×10^7 CAM cells. Each column represents the mean ± SEM of 6 separate experiments. Control values are derived from prostanoid determinations in the supernatant of CAM cells cultured for 3 days (5×10^7 cells; n = 6).

Table 1. Tumour growth (size ≥ 2 cm) and concentration of TXB_2 in plasma of rats treated with either indomethacin or BM 13177 for 6 weeks after the inoculation of 5×10^7 CAM cells: the TXB_2 value represents the mean ± SEM of 12 separate experiments for vehicle and 18 separate experiments for indomethacin or BM 13177

	Tumours in rats		TXB_2, pg/ml
	detected	not detected	
Vehicle (n = 12)[1]	12	–	4,836 ± 664
Indomethacin (n = 18)[2]	16	2	202 ± 41
BM 13177 (n = 18)[3]	1	17	4,674 ± 585

[1] 2 ml Tris buffer were injected (i.p.) daily into the rat.
[2] 0.2 mg/kg body weight of indomethacin dissolved in 2 ml Tris buffer were injected (i.p.) daily into the rat.
[3] 50 mg/kg body weight of BM 13177 dissolved in 2 ml Tris buffer were injected (i.p.) daily into the rat.

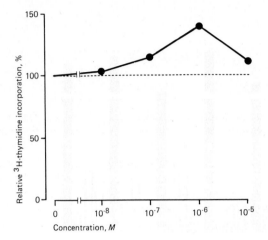

Fig. 2. Effect of U 46619 on the ³H-thymidine incorporation by CAM cells. Values are expressed relative to those of control cultures, and are averages of two cultures.

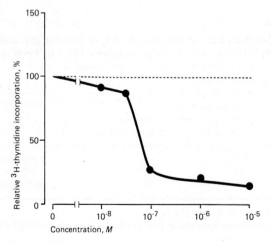

Fig. 3. Effect of thromboxane receptor antagonist BM 13177 on ³H-thymidine incorporation by CAM cells. Values are expressed relative to those of control cultures, and are averages of two separate cultures.

Discussion

This paper investigated the cyclooxygenase-mediated metabolic pattern of AA during tumour development in the rat. As already reported in a previous paper [6], TXB_2 was found to be the major metabolite preferentially formed during tumour growth. Three sources have been suggested by Chiabrando et al. [8] for the preferential increase of TXB_2: (a) re-distribution of endoperoxide metabolism through the most active enzymatic pathway; (b) preferential growth of selected cell subpopulations with different amounts of AA cascade enzymes, and (c) a selective AA cascade enzyme induction. However, we have shown very clearly in our previous report [6], that the intraperitoneal injection of the solubilized membrane fraction from CAM cell homogenate into the rat also activated the preferential synthesis of TXB_2. We also showed that infiltrating mononuclear phagocytes contributed predominantly to the TXB_2 synthesis.

Several authors have in the past reported the successful application of thromboxane synthetase inhibitors for the reduction of tumour growth and metastases [3, 9, 10]. Our results do not support these observations. Also, Chiabrando et al. [8] could not inhibit the tumour growth and metastatic spread with dazmegrel (UK-38,485) in M5076 (ovarian reticulosarcoma)-bearing mice. However, the treatment of rats with a TXA_2 receptor antagonist BM 13177 almost completely inhibited tumour growth. This demonstrates that TXA_2 might exert a receptor-mediated cell-proliferating effect on CAM cells. Furthermore, it does explain why treatment of the rats with thromboxane synthetase inhibitors did not stop tumour growth. The reason might be that no inhibitors of cyclooxygenase or thromboxane synthetase could ever achieve the complete suppression of TXA_2 synthesis. Apparently, a residual TXA_2 synthesis is sufficient to exert a proliferative effect on the tumour cells.

We have also shown that TXA_2 analogue U 46619 enhanced dose-dependently the thymidine incorporation into CAM cells, with the maximum enhancement of 140% at a dose of 1 μM. However, the inhibition of thymidine incorporation with BM 13177 was more effective than expected. 80% of the thymidine incorporation could be inhibited by 1 μM of BM 13177. This means that BM 13177 inhibited not only the effect of exogenous TXA_2 or PGH_2 by occupying the receptors, but also the endogenous synthesis of TXA_2 or PGH_2. This mechanism is under investigation.

Acknowledgements

This study was generously supported by grants from the Deutsche Forschungsgemeinschaft, FRG (Ni 242/2-1) and the Association for International Cancer Research, UK. The excellent technical help of Karin Hillbricht, Sabine Ziedrich and Gabriele Beyer is gratefully acknowledged. The authors also wish to thank Dr. Stegmeier, Boehringer Mannheim, for the kind gift of BM 13177 and Mrs. Herwig for revising the manuscript.

References

1   Levine L: Arachidonic acid transformation and tumour production. Adv Cancer Res 1981;35:49–79.
2   McLemore TL, Hubbard WC, Litterst CL, et al: Profiles of prostaglandin biosynthesis in normal lung and tumor tissue from lung cancer patients. Cancer Res 1988;48:3140–3147.
3   Honn KV: Prostacyclin/thromboxane ratios in tumour growth and metastasis; in Powles TJ, Bockmann RC, Ramwell PW (eds): Prostaglandins and Cancer. New York, Liss, 1982, pp 189–204.
4   Nigam S, Becker R, Rosendahl U, et al: The concentrations of 6-keto-PGF$_{1\alpha}$ and TXB$_2$ in plasma samples from patients with benign and malignant tumours of the breast. Prostaglandins 1985;29:513–528.
5   Nigam S, Rosendahl U, Benedetto C: Involvement of arachidonic acid metabolites in tumour growth and metastasis: role of prostacyclin and thromboxane in gynecological and gastrointestinal cancer of humans; in Muszbek L (ed): Hemostasis and Cancer. Florida, CRC Press, 1987, pp 231–242.
6   Nigam S, Averdunk R: Alteration of arachidonic acid metabolism in rats after inoculation of tumour cells and their subcellular fractions: role of mononuclear phagocytes as a major source of enhanced prostanoid synthesis; in Nigam S, McBrien DCH, Slater TF (eds): Eicosanoids, Lipid Peroxidation and Cancer. Heidelberg, Springer, 1989, pp 43–50.
7   Nigam S: Extraction of eicosanoids from biological samples; in Benedetto C, McDonald-Gibson RG, Nigam S, et al (eds): Prostaglandins and Related Substances: A Practical Approach. Oxford, IRL Press, 1987, pp 45–52.
8   Chiabrando C, Broggini M, Castelli MG, et al: Prostaglandins and thromboxane synthesis by M 5076 ovarian reticulosarcoma during growth: effects of a thromboxane synthetase inhibitor. Cancer Res 1987;47:988–991.
9   Drago JR, Al-Mondhiry HAB: The effect of prostaglandin modulators on prostate tumour growth and metastasis. Anticancer Res 1984;4:391–394.
10 Honn KV: Inhibition of tumour cell metastasis by modulation of the vascular prostacyclin-thromboxane A$_2$ system. Clin Exp Metastasis 1983;1:103–114.

Santosh Nigam, PhD, MB, BCh, Institut für Gynäkologische Endokrinologie,
Klinikum Steglitz, Freie Universität Berlin, Hindenburgdamm 30,
D–1000 Berlin 45 (FRG)

Zor U, Naor Z, Danon A (eds): Leukotrienes and Prostanoids in Health and Disease.
New Trends Lipid Mediators Res. Basel, Karger, 1989, vol 3, pp 325–330

Eicosanoid Modulation of Eicosanoid Release and of Cytostasis Are Interrelated Immunological Functions of Macrophages

I.L. Bonta[a], *G.R. Elliott*[a], *S. Ben-Efraim*[b], *J.A. Van Hilten*[a]

[a]Department of Pharmacology, Faculty of Medicine, Erasmus University, Rotterdam, The Netherlands, and [b]Department of Human Microbiology, Sackler School of Medicine, Tel Aviv University, Israel

Eicosanoids Modulate Macrophage Activation State

The release of eicosanoids is intimately related to the activation state of macrophages. The production of the cyclooxygenase metabolite PGE_2 is inversely correlated with the activation state. Increased levels of cyclic AMP inhibit macrophage functions and PGE_2, which activates the adenylate cyclase, is a deactivator of macrophages. Inhibitors of cyclooxygenase promote the activation of macrophages, as shown for example by enhanced release of lysosomal enzymes [1]. In contrast, the lipoxygenase pathway favors the activation of macrophages. Several immunological events are associated with increased biosynthesis of leukotrienes (LTs) [2]. Macrophages were shown to be responsive to exposure of either LTC_4 or LTD_4, both of them inducing the release of several products of the cyclooxygenase pathway. Using lysosomal enzyme secretion as a marker of cell activity, LTC_4 was also shown to trigger the activation of macrophages, thus enhancing the enzyme secretion, whereas PGE_2 inhibited this. Because lysosomal enzyme release was observed with a lower concentration of LTC_4 than necessary to induce the biosynthesis of PGE_2, it was proposed that the enzyme secretion is the primary event and that the subsequent release of PGE_2 serves to limit the activating function of the peptidoleukotriene. In that case full expression of LT-induced activation would only be observed in the absence of endogenous PGE_2. Indeed, inhibitors of cyclooxygenase promoted the LTC_4-induced release of a lysosomal enzyme [1]. The finding that LTs promote the production of PGE_2 indicates that the action of LTs is self-limiting and that eicosanoid formation is regulated by interactions between the different metabolites of arachidonic acid (AA).

The recent observation showing that the calcium flux-induced release of LTB_4 is counteracted by PGE_2 and augmented by inhibitors of cyclooxygenase [3] gives further support to the concept that the dynamic state of activation of macrophages is maintained by balanced interactions between endogenous PGE_2 and LTs. Macrophage cytotoxicity or cytostasis towards tumor cells is a characteristic expression of macrophage activation. Our studies, which were aimed to investigate the role of eicosanoids in the antitumor function of macrophages, represented a logical move.

Factors Mediating Antitumor Function of Macrophages

Tumor tissues transplanted between rodents, possessing identical histocompatibility antigens, can lead to immunologic rejection of the tumor. This has led to the proposal that tumors may result, at least in part, from failure of the immune response of the host to recognize and destroy cells bearing tumor antigens. This concept of immune surveillance directed the attention to the antitumor potential of macrophages, which are surveillance cells of the immune system. The antitumor potential of quiescent macrophages is negligible. But activated macrophages inhibit tumor cell growth during cocultures in vitro and destroy tumor targets by a non-phagocytic process. Even when their activity is not cytotoxic, or cytocidal, macrophages can inhibit tumor cell growth through cytostatic activity. The two events may be interrelated and several mechanisms have been proposed to be valid for both events. Some of these suggestions comprised that direct cell-to-cell contact is important. Others have indicated that soluble mediators, released by macrophages upon activation, are necessary. Such mediators could include lysosomal enzymes, cell damaging protein factors, i.e. a group of products referred to commonly as tumor necrosing factor (TNF). Recently, the involvement of interleukin-1 (IL-1) has been proposed [4]. O_2 metabolites – products of stimulated macrophages – have also deleterious effects on tumor cells.

In the original experiments to be discussed in this article, we used peritoneal resident macrophages from BALB/c mice. The target tumor cells included two murine cell lines: MOPC-315 plasmacytoma cells and P-815 mastocytoma cells.

Macrophage Cytostasis is Interrelated with Eicosanoid Release

Cyclooxygenase and lipoxygenase metabolites of AA can exert opposing effects on macrophage functions: for example, the lipoxygenase product

LTC_4 stimulated the secretion of macrophage beta-glucuronidase, a lysosomal enzyme, this effect was inhibited by PGE_2, a cyclooxygenase metabolite, and enhanced by indomethacin, a cyclooxygenase inhibitor [1]. Indomethacin has also been shown to enhance, and PGE_2 to inhibit A23187-induced LTB_4 synthesis in macrophages [3]. It appears therefore that indomethacin could stimulate macrophage functions by removing the inhibitory action of PGE_2 on LT formation, so increasing the effective concentration of these lipoxygenase metabolites. Mouse resident peritoneal macrophages could be activated by indomethacin, in vitro, to inhibit growth of MOPC-315 tumor cells and this cytostasis was enhanced by LTD_4 [5]. Thus in similarity with their effect on eicosanoid release, cyclooxygenase and lipoxygenase metabolites appeared to have opposing effects on expression of macrophage cytostatic function. We have further explored this phenomenon by examining the actions of PGE_2 and nordihydroguaiaretic acid (NDGA), a lipoxygenase inhibitor, on indomethacin-stimulated macrophage cytostasis. The results are now being published in extenso [6]. A brief account is given as follows.

Assessment of cytostasis of resident peritoneal macrophages from BALB/c mice towards MOPC-315 cells was carried out by a method published in detail elsewhere [5]. A coculture of macrophages and tumor cells was incubated with the test substances or the vehicles for 24 h. Thereafter, ^{3}H-thymidine (^{3}HTdR) was added for 16 h incubation, that was terminated by harvesting the cells onto glass fiber filter mats which were punched out to feed into a beta counter to measure the cellular incorporation of radioactivity. The effect of indomethacin and NDGA on macrophage LTB_4 release has been measured by RIA in the supernatant following 15 min incubation with the calcium ionophore A23187 ($10^{-6}\,M$).

Thymidine incorporation of MOPC-315 was decreased in the presence of macrophages. This cytostatic activity was further stimulated by indomethacin $10^{-5}\,M$, the effect of which was reversed by PGE_2. The stimulatory effect of indomethacin on macrophage cytostasis was also reversed by NDGA. LTB_4 synthesis in macrophages was also stimulated by indomethacin $10^{-7}\,M$ and inhibited by NDGA $10^{-5}\,M$. For full details, reference is made to the original paper [6].

The observations that the stimulatory effect of indomethacin on macrophage cytostasis required the concentration of $10^{-5}\,M$, whereas LTB_4 synthesis was stimulated by $10^{-7}\,M$, reinforce the earlier proposal [5] that the full effects of LTs are only observed in the absence of cyclooxygenase metabolites, in particular PGE_2. Thus, two interrelated processes could be important for indomethacin-stimulated macrophage cytostasis: first, inhibition of cyclooxygenase activity, and second, stimulation of LT formation. The stimulation by indomethacin of LT synthesis could have

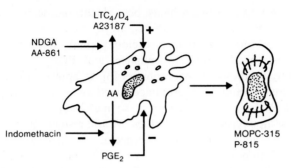

Fig. 1. Inhibitors and metabolites of eicosanoid pathways regulate macrophage cytostasis.

been due to substrate shunting or to removal of the inhibitory PGE_2. It is reasonable to postulate that NDGA reversed the indomethacin-stimulated cytostasis by inhibiting lipoxygenase activity. Thus, impairment of the lipoxygenase pathway leads to suppression of the cytostatic function of macrophages that have been activated by removal of endogenous PGE_2. This is complementary to the earlier finding which showed that LTD_4 additively enhanced the indomethacin-stimulated macrophage cytostasis towards MOPC-315. Congruent results have been obtained with A23187-activated macrophage cytostasis against P-815 cells using AA-861, which is more specific than NDGA in causing inhibition of lipoxygenase [7]. A schematic representation of eicosanoids involvement in macrophage cytostasis against MOPC-315 and P-815 cells is shown in figure 1.

Macrophages release a host of products on activation, including IL-1, TNF, active oxygen species and lysosomal enzymes, all of which could be involved in cytostasis. The release of some of these products is promoted by LTs and inhibited by PGE_2 [1, 2]. The MOPC-315 cell line is resistent to the effects of IL-1 and TNF (data not shown). This is likely to be valid for the P-815 cell line as well. Activated macrophages can exert cytostasis also by cell-to-cell contact. Provided this was involved in our experiments, neither the participation of active oxygen species, nor that of lysosomal enzymes is excluded.

Conclusions

The dynamic state of activation of macrophages is maintained by balanced interactions between endogenous PGE_2 and LTs. Factors which govern the balanced state of macrophage activity have parallel influences on, at least, two functions: eicosanoid release and in vitro cytostasis. Figure

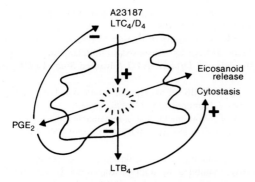

Fig. 2. Balanced regulation of eicosanoid-releasing and cytostatic function of the macrophage. Calcium ionophore and LTs stimulate macrophage functions. PGE_2 counteracts this stimulation, in part by inhibiting LT synthesis.

2 is an oversimplified view of this concept. It appears that the eicosanoid-secreting function of macrophages is positively related to the activity of the lipoxygenase pathway and negatively to the production of the cyclooxygenase metabolite PGE_2. LTs, either directly or indirectly (induced by calcium flux), enhance the release of eicosanoids, whereas PGE_2 suppresses this event. In similarity with this, the balance between lipoxygenase and cyclooxygenase metabolites is important in controlling macrophage cytostasis in vitro. Whereas an earlier finding by us showed that LTD_4 reinforces the effect of indomethacin on macrophage cytostasis, we now have evidence that indomethacin and/or calcium flux-induced stimulation of cytostasis is counteracted by PGE_2 and by inhibitors of lipoxygenase. This favors the view that lipoxygenase metabolites promote cytostasis by macrophages.

Acknowledgments

The original experiments have been sponsored by the Dutch Cancer Foundation (Koningin Wilhelmina Fonds). Some of the work was carried out during a sabbatical stay of Ivan L. Bonta as Elected Fellow of the Mortimer and Raymond Sackler Institute of Advanced Studies (Director: Prof. Yuval Ne'eman), Tel Aviv University, Israel. Shlomo Ben-Efraim is presently spending his sabbatical as Visiting Professor at the Faculty of Medicine, Erasmus University Rotterdam.

References

1   Schenkelaars EJ, Bonta IL: Cyclooxygenase inhibitors promote the leukotriene C_4 induced release of beta-glucuronidase from rat peritoneal macrophages: Prostaglandin E_2 suppresses. Int J Immunopharmacol 1986;8:305–311.

2   Rola-Pleszczynski M, LeMaire I: Leukotrienes augment interleukin-1 production by human monocytes. J Immunol 1985;135:3958–3961.
3   Bonta IL, Elliott GR: Eicosanoids as regulators of eicosanoid release in macrophages. Impact for exacerbation of tissue damage by nonsteroidal antiinflammatory drugs; in Folco G, Velo G (eds): Prostanoids and Drugs. NATO Advanced Study Course. New York, Plenum Press, 1989.
4   Onozaki K, Matsushima K, Aggarwal BB, et al: Human interleukin-1 is a cytocidal factor for several tumor cell lines. J Immunol 1985;135:3962–3968.
5   Ophir R, Ben-Efraim S, Bonta IL: Leukotriene D_4 and indomethacin enhance additively the macrophage cytostatic activity in vitro towards MOPC-315 tumor cells. Int J Tissue React 1987;9:189–194.
6   Elliott GR, Tak C, Pellens C, et al: Indomethacin stimulation of macrophage cytostasis against MOPC-315 tumor cells is inhibited by both prostaglandin E_2 and nordihydroguaiaretic acid, a lipoxygenase inhibitor. Cancer Immunol Immunother 1988;27:133–136.
7   Van Hilten JA, Elliott GR, Bonta IL: Specific lipoxygenase inhibition reverses macrophage cytostasis towards P815 tumor cells in vitro induced by the calcium ionophore A23187. Prostaglandins Leukotrienes Essent Fatty Acids 1988;34:187–192.

Prof. Dr. I.L. Bonta, Department of Pharmacology, Faculty of Medicine, Erasmus University Rotterdam, PO Box 1738, NL–3000 DR Rotterdam (The Netherlands)

Zor U, Naor Z, Danon A (eds): Leukotrienes and Prostanoids in Health and Disease.
New Trends Lipid Mediators Res. Basel, Karger, 1989, vol 3, pp 331–338

Biological Activities of Lipoxins

Sven-Erik Dahlén

Department of Physiology, and Institute of Environmental Medicine, Karolinska
Institutet, Stockholm, Sweden

Introduction

The lipoxins (LX) are formed in reactions involving interactions
between the 5- and 15-lipoxygenase, hence their trivial name (lipoxygenase
interaction products) [reviewed in 1]. This additional group of arachidonic
acid metabolites characteristically contain four conjugated double bonds
and three hydroxyl groups. The major products isolated from human
leukocytes were named LXA$_4$ (5,6,15-trihydroxyeicosatetraenoic acid) and
LXB$_4$ (5,14,15-trihydroxyeicosatetraenoic acid) [2] and their stereochemistry has been established (LXA$_4$: 5S,6R,15S-trihydroxy-7,9,13-*trans*-11-
cis-eicosatetraenoic acid; LXB$_4$: 5S,14R,15S-trihydroxy-6,10,12-*trans*-
8-*cis*-eicosatetraenoic acid) [3, 4] (fig. 1). More recently, the structures and
biological activities of other LX have been characterized [5]. Several
biosynthetic pathways may lead to formation of LX [3, 4, 6–8]. The
present chapter serves to give a brief overview of the biological activities so
far described for LXA$_4$ and LXB$_4$.

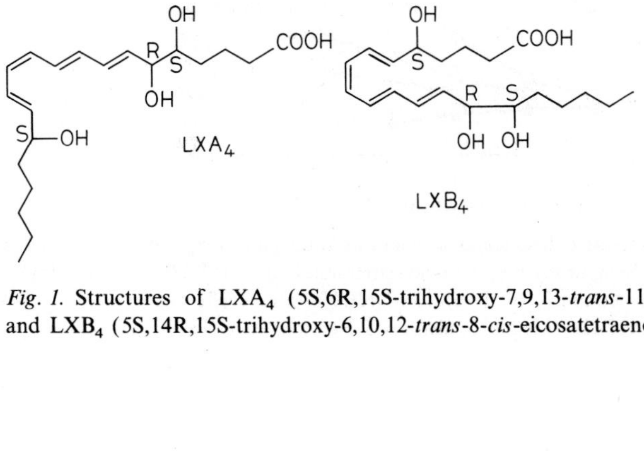

Fig. 1. Structures of LXA$_4$ (5S,6R,15S-trihydroxy-7,9,13-*trans*-11-*cis*-eicosatetraenoic
acid) and LXB$_4$ (5S,14R,15S-trihydroxy-6,10,12-*trans*-8-*cis*-eicosatetraenoic acid).

Vasodilation in vivo

In the hamster cheek pouch, prepared for intravital microscopy of the terminal vascular network, topical administration of LXA_4 (1 μM) induced a pronounced arteriolar dilation, but did not change venular diameters [9]. However, LXA_4 did not by itself affect FITC-dextran extravasation, used as a marker for microvascular permeability to plasma proteins, nor was leukocyte adherence to the endothelium of small venules stimulated by LXA_4 [9]. Therefore, the profile of activity of LXA_4 on the microvasculature differed from those known for lipoxygenase products such as leukotriene B_4 or the cysteinyl-containing leukotrienes (cysLTs: LTC_4, LTD_4 and LTE_4).

Smooth Muscle Contraction

LXA_4 (0.03–10 μM) induced dose-dependent and very long-lasting contractions of the guinea pig lung strip [9–11]. In contrast, LXA_4 had no direct effect on the ileum and trachea of the same species, again indicating that the activity of LXA_4 was different from those of the leukotrienes.

The contraction evoked by LXA_4 in the lung strip resembled the 'slow-reacting response' elicited with cysLTs. At the highest dose, the response to LXA_4 approached maximal tissue contractility, thus indicating that LXA_4 was a complete agonist (fig. 2). In this preparation, LXA_4 was

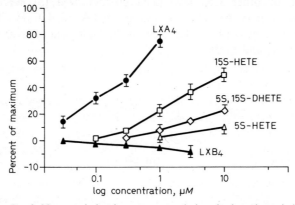

Fig. 2. Noncumulative dose-response relations in the guinea pig lung strip for LXA_4 and LXB_4, as well as 5(S)-hydroxy-6,8,11,14-eicosatetraenoic acid (5S-HETE), 5(S),15(S)-dihydroxy-6,8,11,13-eicosatetraenoic acid (5S,15S-DHETE) and 15(S)-hydroxy-5,8,11,13-eicosatetraenoic acid (15S-HETE). Means $\pm$ SE; n = 5–20.

approximately 10 times as potent as histamine (half maximal contractions for LXA_4 and histamine at 0.5 and 5 μM, respectively).

The contractile effect of LXA_4 in the lung strip was associated with release of thromboxane (TX) A_2 [10, 12]. The generation of TXA_2 was not a consequence of the contraction response per se [12]. Nevertheless, similar to what has been found for cysLTs under the present experimental conditions [13], indomethacin blocked release of TXA_2 from the lung strip without affecting the peak amplitude of the contraction response to LXA_4 [9, 10, 12].

In contrast, several structurally unrelated antagonists for cysLTs (FPL 55712, LY-171883, L-648,051, ICI 198,615) inhibited the contraction response to LXA_4 in a competitive manner [10, 11, and unpubl. data]. Since lipoxygenase inhibitors failed to alter the contraction induced by LXA_4 [10], it was concluded that LXA_4, rather than causing release of leukotrienes, directly activated the smooth muscle at a site with similar characteristics as the receptor(s) for cysLTs. This conclusion was substantiated by the finding that specific cross-desensitization could be induced between LXA_4 and the cysLTs in the guinea pig lung strip [10]. In this context, it is of interest that LXA_4 contracts human bronchi in a manner which is very similar to that observed in the guinea pig lung strip, including interactions with the receptors for cysLTs [10].

With the aid of stereochemically defined compounds [reviewed in 14], it has been possible to outline structure-activity relations (SAR) for LXA_4 in the guinea pig lung strip [3, 10, 13]. Evidently, the presence of alcohol groups at both carbon atoms 5 (C-5) and 6 (C-6) in LXA_4 are essential for spasmogenic activity. For example, 5(S)-HETE, 5(S),15(S)-DHETE, 14(R),15(S)-DHETE and 15(S)-HETE are considerably less active than LXA_4 (fig. 2). Likewise, LXB_4, which has its hydroxyls positioned at C-5, C-14 and C-15, causes but relaxation of the lung strip (fig. 2). In particular, the orientation of the two hydroxyls at C-5 and C-6 appears crucial, because 6S-LXA_4 is virtually inactive [3].

The structure-activity studies indicate that there are stereochemical similarities between LXA_4 and the cysLTs which explain the observed interactions. For example, both LXA_4 and the cysLTs have a system of conjugated double bonds at the same positions (7,9-*trans*-11-*cis*). More importantly, however, both LXA_4 and the cysLTs have polar groups at C-5 and C-6 with the relative orientation 5S,6R. It is known for the cysLTs that the 5S,6R orientation of the hydroxyl and cysteinyl substituent, respectively, is one very important determinant of both contractile activity [15] and binding to receptors [16]. Likewise, the 5S,6R hydroxyls were crucial for the spasmogenic activity of LXA_4 (see above).

Leukotriene Antagonism

As indicated above, LXA_4 failed to contract the guinea pig ileum [9, 17]. Therefore, although both the lung strip and the ileum are sensitive to cysLTs, the receptors in the two tissues apparently differ with respect to sensitivity for LXA_4. In fact, it was observed that LXA_4 causes a dose-dependent inhibition of the contraction response to LTC_4 in the ileum [10]. Thus, in the ileum, LXA_4 behaves as an antagonist, whereas it is a full agonist for contraction in the lung strip. The observations with LXA_4 enforce other indications that there are multiple and tissue-specific receptors for cysLTs [15]. In addition, in view of the antagonism exerted by LXA_4 in the ileum, it is of interest that one potent antagonist of cysLTs, SKF-104,353 [18], also has structural features (polar groups with the relative orientation 2S,3R) which suggests a specific interaction with the 5S,6R center required for optimal binding of cysLTs.

Relaxation of Isolated Arteries

In contrast to their differential activity in the lung strip, LXA_4 and LXB_4 both relaxed the guinea pig aorta and the human pulmonary artery (fig. 3) [19]. The responses were predominantly, if not exclusively, endothelium dependent. The potency of LXA_4 and LXB_4 was identical, and both compounds caused dose-dependent relaxations in concentrations which were similar to that of acetylcholine $(0.1-10\ \mu M)$. Pretreatment with

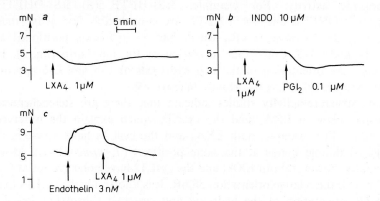

Fig. 3. LXA_4 causes relaxation of strips of human pulmonary arteries precontracted with either *(a, b)* prostaglandin $F_{2\alpha}$ ($PGF_{2\alpha}$ $0.3-0.5\ \mu M$) or *(c)* the novel vasoconstrictor peptide endothelin [26]. As indicated by the tracings *(b)*, indomethacin (10 μM, 30 min pretreatment) blocks the relaxant effect of LXA_4 but not that of prostaglandin I_2 (PGI_2).

indomethacin effectively annulled the relaxations induced by the LX (fig. 3), but left the response to acetylcholine unaffected. Furthermore, in the guinea pig aorta, the pharmacological characterization suggested that the relaxant cyclooxygenase product liberated by LXA_4 and LXB_4 was prostaglandin (PG) I_2.

Obviously, the SARs for the relaxant property of LX in isolated arteries ($LXA_4 = LXB_4$) differed from those required for the spasmogenic action in airway smooth muscle (LXA_4 but not LXB_4). Since 15-HETE also may evoke cyclooxygenase-dependent relaxation of the guinea pig aorta [20], it is suggested that the polar group at C-15 is important for the relaxation of vascular smooth muscle induced by LX. In line with this proposal, 15-HPETE caused vasodilation in the hamster cheek pouch with similar characteristics as that induced by LXA_4 [9].

Conclusions

The biological activities reported for the LX are summarized in table 1. Most information is available concerning the actions of LXA_4. LXA_4 thus constricts guinea pig lung parenchyma and human bronchi, but fails to contract ileum and trachea from guinea pigs. In the ileum, LXA_4 rather

Table 1. Biological activities of LX

LXA_4 but not LXB_4	
Contraction	GP lung strip (GPLS) human bronchus (HB) rat tail artery
Release of TXA_2	GP lung human leukocytes
Interactions with cys LTs (partial agonist/antagonist)	GPLS, GP ileum, and HB
Activation of protein kinase C (PKC)	human placental PKC
LXA_4 and LXB_4	
Relaxation (indomethacin-sensitive)	GP aorta human pulmonary artery
Arteriolar dilation	hamster cheek pouch rat kidney
Immunomodulation	inhibition of cytotoxicity (human natural killer cells)

acts as an antagonist of LTC_4. In addition, LXA_4 stimulates release of the spasmogen TXA_2 from the guinea pig lung, and this mechanism may to some extent be present also in human leukocytes [21]. Furthermore, in vivo, LXA_4 induces arteriolar dilation in the hamster cheek pouch (see above) and rat preglomerular circulation [22], whereas it constricts the rat tail artery [23]. Interestingly, LXA_4 proved to be a very potent activator of protein kinase C isolated from human sources [24].

In contrast, both LXA_4 and LXB_4 can relax the guinea pig aorta and the human pulmonary artery by an action presumably involving release of PGI_2 (see above). Finally, both LX inhibit cytotoxicity induced by natural killer cells [25].

It is therefore apparent that LX have several biological activities which are distinct from those of other arachidonic acid metabolites. In addition, interactions occur between LX and other eicosanoids, as indicated by release of cyclooxygenase products and antagonism of leukotrienes. Considered together, the LX have the potential to function as mediators or modulators of several important biological responses, for example inflammation, immunological reactions and intracellular signal transduction. Since the biochemical pathways required for the formation of LX are extensive in many cells [discussed in 1, 10], it is of interest to further explore the functions of LX and related lipoxygenase products. Finally, it is likely that these novel compounds will prove to have additional biological activities in systems yet to be tested.

Acknowledgements

Supported by grants from the Swedish Medical Research Council (project 14X-4342), the Swedish Association Against Chest and Heart Diseases, the Swedish Association Against Asthma and Allergy (RmA), the Institute of Environmental Medicine, the Swedish Environment Protection Board (5324067-7), and Karolinska Institutet.

References

1   Samuelsson B, Dahlén S-E, Lindgren JÅ, et al: Leukotrienes and lipoxins: Structures, biosynthesis, and biological effects. Science 1987;237:1171–1176.
2   Serhan CN, Hamberg M, Samuelsson B: Lipoxins: Novel series of biologically active compounds formed from arachidonic acid in human leukocytes. Proc Natl Acad Sci USA 1984;81:5335–5339.
3   Serhan CN, Nicolaou KC, Webber SE, et al: Lipoxin A: Stereochemistry and biosynthesis. J Biol Chem 1986;261:16340–16345.
4   Serhan CN, Hamberg M, Samuelsson B, et al: On the stereochemistry and biosynthesis of lipoxin B. Proc Natl Acad Sci USA 1986;83:1983–1987.

5   Nicolaou KC, Marron BE, Veale CA, et al: Identification of a novel 7-*cis*-11-*trans*-lipoxin A_4 generated by human neutrophils: Total synthesis, spasmogenic activities, and comparison with other geometric isomers of lipoxin A_4 and B_4. Biochim Biophys Acta, in press.

6   Puustinen T, Webber SE, Nicolaou KC, et al: Evidence for a 5(6)-epoxy tetraene intermediate in the biosynthesis of lipoxins in human leukocytes. FEBS Lett 1986;177:255–265.

7   Corey EJ, Mehrotra MM: A stereoselective and practical synthesis of 5,6(S,S)-epoxy-15(S)-hydroxy-7(E),9(E),11(Z),13(E)-eicosatetraenoic acid (4), possible precursor of lipoxins. Tetrahedron Lett 1986;27:5173–5175.

8   Kühn H, Wiesner R, Alder L, et al: Formation of lipoxin B by the pure reticulocyte lipoxygenase. FEBS Lett 1986;208:248–252.

9   Dahlén S-E, Raud J, Serhan CN, et al: Biological activities of lipoxin A include lung strip contraction and dilation of arteriols in vivo. Acta Physiol Scand 1987;130:643–648.

10 Dahlén S-E, Franzén L, Raud J, et al: Actions of lipoxin A_4 and related compounds in smooth muscle preparations and on the microcirculation in vivo; in Wong PY-K, Serhan CN (eds): Lipoxins: Biosynthesis, Chemistry and Biological Activities, New York, Plenum Press, 1988, pp 107–130.

11 Spur, BW, Jacques C, Crea AE, et al: Lipoxins of the 5-series derived from eicosapentaenoic acid; in Wong PY-K, Serhan CN (eds): Lipoxins: Biosynthesis, Chemistry and Biological Activities. New York, Plenum Press, 1988, pp 147–154.

12 Wikström E, Westlund P, Nicolaou KC, et al: Lipoxin A_4 causes generation of thromboxane A_2 in the guinea-pig lung. Agents Actions 1989;26:90–92.

13 Dahlén S-E, Hedqvist P, Westlund P, et al: Mechanisms for leukotriene-induced contractions of guinea pig airways: Leukotriene C_4 has a potent direct action whereas leukotriene B_4 acts indirectly. Acta Physiol Scand 1983;118:393–403.

14 Webber SE, Veale CA, Nicolaou KC: The total synthesis of lipoxins and related compounds; in Wong PY-K, Serhan CN (eds): Lipoxins: Biosynthesis, Chemistry and Biological Activities. New York, Plenum Press, 1988, pp 61–78.

15 Krell RD, Brown FJ, Willard AK, et al: Pharmacologic antagonism of the leukotrienes; in Chakrin LW, Bailey DM (eds): The Leukotrienes, Chemistry and Biology. Orlando, Academic Press, 1984, pp 271–299.

16 Aharony D, Falcone RC, Krell RD: Inhibition of ^{3}H-leukotriene D_4 binding to guinea pig lung receptors by the novel leukotriene antagonist ICI-198,615. J Pharmacol Exp Ther 1987;243:921–926.

17 Cristol JP, Sirois P: Comparative activity of leukotriene D_4, 5,6-dihydroxy-eicosatetraenoic acid and lipoxin A_4 on guinea pig lung parenchyma and ileum smooth muscle. Res Commun Chem Pathol Pharmacol 1988;59:423–426.

18 Hay DWP, Muccitelli RM, Tucker SS, et al: Pharmacologic profile of SK&F 104,353: A novel potent and selective peptidoleukotriene receptor antagonist in guinea pig and human airways. J Pharmacol Exp Ther 1987;243:474–481.

19 Matsuda H, Dahlén S-E, Haeggström J, et al: Lipoxins A_4 and B_4 relax isolated arteries from guinea pig and man. Eur J Pharmacol, in press.

20 Matsuda H, Dahlén S-E, Kumlin M, et al: Pharmacodynamics of 15-HPETE and 15-HETE in isolated arteries from guinea pig, rabbit, rat and man. Prostaglandins, in press.

21 Conti P, Reale M, Cancelli A, et al: Lipoxin A augments release of thromboxane from human polymorphonuclear leukocyte suspensions. FEBS Lett 1987;225:103–108.

22 Badr KF, Serhan CN, Nicolaou KC, et al: The action of lipoxin A on glomerular microcirculatory dynamics in the rat. Biochem Biophys Res Commun 1987;145:408–414.

23 Lam BK, Wong PY-K: Biosynthesis and biological activities of lipoxin A_5 and B_5 from eicosapentaenoic acid; in Wong PY-K, Serhan CN (eds): Lipoxins: Biosynthesis, Chemistry and Biological Activities, New York, Plenum Press, 1988, pp 51–60.

24 Hansson A, Serhan CN, Haeggström J, et al: Activation of protein kinase C by lipoxin A and other eicosanoids. Intracellular action of oxygenation products of arachidonic acid. Biochem Biophys Res Commun 1986;134:1215–1222.

25 Ramstedt U, Ng J, Wigzell H, et al: Action of novel eicosanodis lipoxin A and B on human natural killer cell cytotoxicity: Effects on intracellular cAMP and target cell binding. J Immunol 1985;135:3434–3438.

26 Yanagisawa M, Kurihara H, Kimura S, et al: Endothelin: A novel potent vasoconstrictor peptide produced by vascular endothelial cells. Nature 1988;332:411–415.

Sven-Erik Dahlén, MD, Department of Physiology, and Institute of
Environmental Medicine, Karolinska Institutet, S–104 01 Stockholm (Sweden)

Subject Index

A-64077 50–54
AA-861 328
4-Acetylaminoantipyrine 210–212
Acetylcholine 65, 78, 82, 232, 238, 334, 335
Acetyltransferase activity 269, 270
Actinomycin D 10, 11, 13–15 18, 19, 22
Adenosine 237
Adenylate cyclase 232
Aggregation, *see* Platelet aggregation
Air hypersensitivity 106–108
Alkaline phosphatase 269
Allergan 102
4-Aminoantipyrine 210–213
Analgesics 125–128
Anaphylactic shock, *see* Shock, anaphylactic
Anterior hypothalamic/preoptic region 185
Anterior pituitary 31
Antiallergic activity 56–61
Antibodies 3, 35, 36
 monoclonal 30, 38, 258–260
 polyclonal 31
Antigen challenge 79–81, 253
Antigens 78, 81
Antihistamine activity 57
Anti-inflammatory activity 52, 53
Antipyresis 128
Antitumor function 326
Arachidonic acid 1, 8, 10, 14, 25, 34, 35, 110, 111, 175, 177, 178, 247, 272–274, 294, 295

release 279–282
tumors 319–323
vasodilation 236–238
Arteries 334, 335
Arterioles 236–238
Arthus reaction 52, 53, 57
Aspirin 8–11, 13–15, 36, 91, 117, 243
Asthma 62, 78–81, 100–104, 135
Astrocytomas 167–169
Atopic eczema 144
ATP 11
Atropine 63, 64

B cells 138–140
Basic fibroblast growth factor, *see* bFGF
Benoxaprofen 144
bFGF 243, 244
Biotin 30
Blood-brain barrier 184, 185
BM 13177 320–323
BN 52021 130–132, 145, 228
BN 52063 145
BN 52111 132
Bone
 formation 310–312
 resorption 310
Borohydride reduction 36
Bovine 38
Brain 175–181
 injury 171–174
 microvessels 161–164
 pyrogens 184, 185
Bronchi 79